CLASSIFICATION AND CLUSTERING

Publication No. 37
of the Mathematics Research Center
The University of Wisconsin at Madison

ACADEMIC PRESS RAPID MANUSCRIPT REPRODUCTION

CLASSIFICATION AND CLUSTERING

Edited by J. Van Ryzin

Proceedings of an Advanced Seminar
Conducted by the Mathematics Research Center
The University of Wisconsin at Madison
May 3–5, 1976

Academic Press, Inc.
New York • San Francisco • London 1977
A Subsidiary of Harcourt Brace Jovanovich, Publishers

ACADEMIC PRESS, INC.
111 Fifth Avenue, New York, New York 10003

United Kingdom Edition published by
ACADEMIC PRESS, INC. (LONDON) LTD.
24/28 Oval Road, London NW1

Library of Congress Cataloging in Publication Data

Main entry under title:

Classification and clustering.

(Publication of the Mathematics Research Center, University of Wisconsin-Madison ; no. 37)
Includes index.
1. Discriminant analysis–Congresses. 2. Cluster analysis–Congresses. I. Van Ryzin, John. II. Wisconsin. University–Madison. Mathematics Research Center. III. Series: Wisconsin. University–Madison. Mathematics Research Center. Publication ; no. 37.
QA3.U45 no. 37 [QA278.65] 510'.8s [519.5'3]
77-7139

ISBN 0–12–714250–9

PRINTED IN THE UNITED STATES OF AMERICA

Contents

List of Contributors

Susan W. Ahmed, Department of Biostatistics, The University of North Carolina at Chapel Hill, Chapel Hill, North Carolina 27514

Frank B. Baker, Department of Statistics, The University of Wisconsin at Madison, Madison, Wisconsin 53706

Pi Yeong Chi, National Institute of Environmental Health Sciences, Research Triangle Park, North Carolina 27709

Jerome Cornfield, Department of Statistics, George Washington University, Washington, D.C. 20006

Rosalie A. Dunn, Biostatistics Section, The Veterans Administration Research Center, Washington, D.C. 20422

K. S. Fu, Department of Electrical Engineering, Purdue University, West Lafayette, Indiana 47907

Seymour Geisser, Director, School of Statistics, University of Minnesota, Minneapolis, Minnesota 55455

I. J. Good, Department of Statistics, Virginia Polytechnic Institute and State University, Blacksburg, Virginia 24061

J. A. Hartigan, Department of Statistics, Yale University, New Haven, Connecticut 06520

Lawrence J. Hubert, Department of Statistics, The University of Wisconsin at Madison, Madison, Wisconsin 53706

Raul Hudlet, IIMAS, Universidad Nacional Autonoma de Mexico, Apartado Postal 20-726, Mexico, 20 DF

Richard Johnson, Department of Statistics, The University of Wisconsin at Madison, Madison, Wisconsin 53706

Joseph Kruskal, Mathematics and Statistics Research Center, 600 Mountain Avenue, Bell Telephone Laboratories, Murray Hill, New Jersey 07974

Peter A. Lachenbruch, School of Public Health, Department of Biostatistics, The University of North Carolina at Chapel Hill, Chapel Hill, North Carolina 27514

David W. Matula, Department of Computer Science, Southern Methodist University, Institute of Technology, Dallas, Texas 75275

Hubert V. Pipberger, Department of Medicine, The Veterans Administration Research Center, Washington, D.C. 20422

C. Radhakrishna Rao, Indian Statistical Institute, 7, S.J.S. Sansanwal Marg, New Delhi 110029, India

Robert R. Sokal, Department of Ecology and Evolution, State University of New York, Stony Brook, New York 11790

Herbert Solomon, Department of Statistics, Stanford University, Stanford, California 94305

J. Van Ryzin, Department of Statistics, The University of Wisconsin at Madison, Madison, Wisconsin 53706

Grace Wahba, Department of Statistics, The University of Wisconsin at Madison, Madison, Wisconsin 53706

L. A. Zadeh, Department of Electrical Engineering, University of California at Berkeley, Berkeley, California 94720

Preface

This volume contains the Proceedings of the Advanced Seminar on Classification and Clustering held in Madison, Wisconsin, May 3–5, 1976, sponsored by the Mathematics Research Center, University of Wisconsin at Madison, with financial support from the National Science Foundation under grant No. MCS 75-21351 and the United States Army under Contract No. DAAG29-75-C-0024.

At the Advanced Seminar thirteen invited speakers presented papers, twelve of which appear in these proceedings. We very much regret that it has not been possible to include the manuscript of Professor Louis Guttman's excellent talk on "Categorical and Monotone Data Analysis." Nevertheless, the contribution of his talk and his lively discussion is gratefully acknowledged. In addition, this volume contains four papers by authors here at the University of Wisconsin — Madison who gave talks in an evening session consisting of twelve short presentations on research work in classification and clustering.

The six sessions at the Advanced Seminar were chaired by:

John Van Ness, University of Texas at Dallas, Dallas, Texas;
Robert V. Hogg, University of Iowa, Iowa City, Iowa;
P. R. Krishnaiah, Wright-Patterson Air Force Base, Dayton, Ohio;
Ingram Olkin, Stanford University, Stanford, California;
Douglas Tang, Walter Reed Army Institute of Research, Washington, D.C.;
Herman Chernoff, Massachusetts Institute of Technology, Cambridge, Massachusetts.

These people not only did an excellent job of chairing the sessions but all contributed much to the discussion.

The members of the program committee were Bernard Harris, Lawrence Hubert, Richard Johnson, and the editor as Program Chairman, all of the University of Wisconsin at Madison. Mrs. Gladys Moran, Program Secretary, handled all of the organizational details in a most efficient and pleasant way. Mrs. Dorothy Bowar accomplished with cheer and care the difficult task of typing the manuscripts and corresponding with the authors relative to corrections and like matters. Finally, thanks are extended to Ben Noble, Director of the Mathematics Research Center for his continual support of this effort, to J. Michael Yohe, Associate Director, for help in the financial planning, and to Dean E. David Cronon, College of Letters and Science, for his welcoming address on those sunny pleasant May days.

The advanced seminar's attendance numbered more than two-hundred and seventy-five — far exceeding expectations. This in itself was justification enough for the effort involved by all concerned. It also is a great testimony to the importance of the research problems being attacked by the scientists in the field of classificaiton and clustering, be they mathematicians, biological scientists, social scientists, computer scientists, statisticians, or engineers. All these disciplines to one degree or another were represented in the attendees and speakers. It is only hoped that in some small way this volume can contribute to the diversity and advancement of the twin fields of classification and clustering. We thank all those who have participated.

J. Van Ryzin

Clustering and Classification: Background and Current Directions

Robert R. Sokal

In this paper I will provide a general background into the nature and purpose of clustering and classification. Coming as it does in the middle of the symposium, some of what I have to say has already been alluded to by other speakers. Much of what I propose to say will be familiar to at least some of you who have been intimately concerned with classification problems for a considerable time. Yet I hope there is some value in my attempt to describe the field as I see it and to delineate the directions that I believe it should take.

It is frequently stated that classification is one of the fundamental processes in science. Facts and phenomena must be ordered before we can understand them and to develop unifying principles explaining their occurrence and apparent order. From this point of view, classification is a higher level intellectual activity necessary to our understanding of nature. But since classification is the ordering of objects by their similarities (Sneath and Sokal, 1973) and objects can be conceived of in the widest sense including processes and activities -- anything to which a vector of descriptors can be attached, we recognize that classification transcends human intellectual endeavor and is indeed a fundamental property of living organisms. Unless they are able to group stimuli into like kinds so as to establish classes to which favorable or avoidance reactions can be made, organisms would be ill-adapted for survival. Such reactions can be learned (among organisms capable of learning), or instinctive, programmed into the genome of the population through natural selection. So what we taxonomists do is a very natural activity and the procedures of cluster analysts for setting up classes of similar objects merely systematize and quantify one of the most fundamental processes of humans and other living organisms.

Attempts to develop techniques for automatic classification, necessitated the quantification of similarity. The ability to perceive any two objects as more similar to each other than either is to a third, must surely have been present in the ancestors of the human species. Many animal choice experiments involve the very same principle. When a bird, for example, is given a choice between two different models, either as mates or threat objects, the bird's nervous system must be making a judgement of relative similarity of either model to a learned pattern residing in its nervous system. One can quantify estimates of resemblances by humans in distinct ways. One can ask a single person to make an overall similarity judgement between objects and to express this as a number from 0 to 1 or in terms of inequalities for pairs or triplets of objects. Social scientists frequently follow this approach, but unless the perception of only one person is the object of study, a single table of resemblances as quantified by any one individual is sufficient. Customarily, a sample of subjects is asked to make the estimates of resemblance, and averages of these resemblances are used to obtain a classification of the objects.

In much classificatory work, it would be impractical to obtain estimates of taxonomic similarity in an assemblage of objects from a sample of subjects. Furthermore, scientific practice generally eschews judgement based on majority vote or popularity. So a second approach to the quantification of estimates of resemblance is through attempts to find the basis for similarity judgements. This is usually done by the detailed description of characteristics on the basis of which it is believed that similarities are expressed. This approach has led to a particularization and atomization of the descriptors of the objects to be classified. Long lists of descriptors, vectors of character states, are attached to each object and the classification is carried out on a <u>data</u>

matrix, composed of a collection of such vectors. Important theoretical issues turn on the nature of the fundamental unit characters but since these issues differ with the field of application, I shall not go into them here.

The consequence of such procedures is that the objects to be classified, -- operational taxonomic units (OTU's for short), are represented in a space whose dimensions are the characters. This attribute space (A-space) is formally of *n* dimensions (for *n* characters), but because of correlations of characters across the OTU's can usually be reduced to fewer dimensions with little loss of information. Such reduction of dimensionality is carried out by ordination methods discussed below.

The comparative study of objects in A-space across a variety of disciplines will rapidly convince the observer that only in unusual circumstances are objects aggregated in compact regions of the full attribute hyperspace. Although aggregated with respect to many of the dimensions of the space, OTU's from a given cluster may well be distributed along the entire range of some character dimensions. Different OTU's will deviate from the cluster along different character axes. What this means is that classes of objects can be defined without resorting to uniformity of position of the objects on all character axes or even on any given character axis. While any one OTU must resemble all others of its cluster in most of its characteristics, it need not resemble them in all characters. Class membership is thus defined by "majority vote" (the greatest number of shared character states) and no one characteristic becomes defining for membership in a given class or taxon. Such relationships have been found empirically by researchers investigating supposedly natural classifications in a variety of disciplines and it has been formalized as a principle of classification by Beckner (1959). Classes defined in this manner are called polythetic (Sneath, 1962) and most methods of cluster analysis and classification aim at obtaining polythetic classes. In the converse system, monothetic classes or taxa, membership is defined by common values for all or at least some characters. This makes monothetic classifications useful for constructing taxonomic keys, but the resulting arrangement of natural objects is frequently unsatisfactory.

Acceptance of polythetic classification has important consequences. Membership in a taxonomic class is a function of the distribution of the OTU's in A-space. It is not a two-valued function such as belonging or not belonging, as is the case in monothetic classifications, but belongingness is determined along a continuous scale. In polythetic classifications taxonomic structure becomes a statistical rather than a purely geometric problem.

It follows from what has been said so far that quantitative polythetic classifications require many characters for a correct grouping of the objects. Such classifications are frequently called natural; their members will be in some sense more similar to each other than they are to members of other taxa. Classifications based on a great variety of characters will likely be of general utility while those based on only a few characters should be less useful generally, except for the special purposes relevant to the chosen characters. For purposes of general classification and for determining the natural order among objects or phenomena, a natural classification based on all available information (characters) is clearly preferable. For special purposes, where only some properties of the objects are of interest to the investigator or where the classification is to serve the needs of some special practical application, a special classification is indicated. Using few rather than many characters in effect weights the characters employed by unity and those omitted by zero and brings up the general subject of weighting which has been a very controversial one in taxonomy. I shall not discuss this at length since it has been treated adequately elsewhere (Sneath and Sokal, 1973). Suffice it to state that while weighting of characters for identification purposes (allocation of known OTU's to predetermined classes) is an accepted part of taxonomic procedure, no consistent scheme for weighting characters before undertaking a classification has yet been proposed. When characters are few and weighted, only special classification can emerge.

Natural polythetic classifications permit two types of predictions concerning character states. These states should be homogeneous within taxa and heterogenous among them. Knowing that OTU j is a member of taxon A permits one to make a prediction about the value of a character state X_{ij} of character i for j with a given probability of success, based on the prior probability of the distribution of states for that character in the taxon. A second type of prediction is based on the correlation of characters across taxa. In a natural classification, it is expected that the distribution of character states that in fact were not studied until after the establishment of the classification would conform to the taxonomic structure already established. So in a natural classification, if a member j of taxon A has a given character state X_{hj} for a newly described character h, it might be predicted that the character state X_{hk} for OTU k, a recognized member of taxon A for which this new character has not been studied, would equal X_{hj}.

The most common mathematical basis for classifying objects has been the computation of pair functions between pairs of OTU's over the characters. This results in matrices of similarity S_{jk} or dissimilarity U_{jk} coefficients between all possible pairs jk of OTU's. These coefficients are fundamentally of three kinds. Distance coefficients, applied to interval and ordinal scale characters, are of the general form

$$d_r(j,k) = \left(\frac{1}{n} \sum_{i=1}^{n} |X_{ij} - X_{ik}|^r \right)^{\frac{1}{r}}$$

where j and k are OTU's, X_{ij} refers to the state of character i for OTU j, n is the number of characters, and r is a positive integer. Two cases are especially useful: the city block or Manhattan distance ($r = 1$) and the taxonomic distance ($r = 2$). Similarities between OTU's described by binary coded or nominal data are estimated by association coefficients. These furnish the ratio of observed identies in character states for the pair of OTU's to the total possible number of identities, generally the number of characters. A general form of the association coefficient is Gower's (1971) general similarity coefficient

$$\underline{S}_{jk} = \sum_{i=1}^{n} \underline{w}_{ijk} \underline{s}_{ijk} / \sum_{i=1}^{n} \underline{w}_{ijk}$$

where $0 \leq \underline{s}_{ijk} \leq 1$ is a similarity between the states of character $\underline{i}$ for OTU's j and k, $\underline{w}_{ijk}$ is a weight assigned to that character and $\underline{n}$ is the number of characters. Complements of the association coefficients are distances in the metric peculiar to the coefficient. Another common way of expressing the similarity between OTU's has been to compute correlation coefficients between OTU's over characters.

The pair functions discussed so far are suitable for data matrices where each vector represents an OTU. Most early applications in biology and psychology used single objects as OTU's and in many applications of cluster analysis this is still the appropriate model. However, applications originating in anthropology were based on samples studied differences between samples, so that each operational taxonomic unit (a population sample) was represented not only by a vector estimating its location but also by a variance-covariance matrix. A first attempt at dealing with these problems was the coefficient of racial likeness developed by Pearson (1926) which, while allowing for variances, did not allow for the covariances between characters as it estimated distances between population samples. The formula for this coefficient is

$$\text{C. R. L.} = \left[\frac{1}{n} \sum_{i=1}^{n} \left(\frac{(\bar{X}_{iJ} - \bar{X}_{iK})^2}{(s_{iJ}^2/t_J)+(s_{iK}^2/t_K)}\right)\right]^{\frac{1}{2}} - \frac{2}{n}$$

where $\bar{X}_{iJ}$ stands for the sample mean of the $\underline{i}$th character for sample J, $\underline{s}_{iJ}^2$ for the variance of the same, and $\underline{t}_J$ for the sample size of J. The currently widely employed and well-known multivariate techniques of Mahalanobis' generalized distances and canonical variate analysis (see Morrison, 1967) derive from problems posed originally in anthropological research and are applied in the classification of population samples. A matrix of generalized distances can be clustered in the same way as other distance matrices.

Despite the wide application of cluster analysis there is no consensus on the definition of clusters. There is an intuitive understanding among most workers that members of a cluster are closer to each other than to other individuals but the specifics of this relationship are not spelled out. Various parameters of clusters have been used to define them: the density of OTU's in the attribute hyperspace, the volume occupied by the cluster, connectedness among cluster members, and gaps between adjacent clusters as compared to cluster diameters.

Cluster analysis imposes relationships on objects based on the numerical values of pair functions between them. These relationships are designed to bring out underlying structure but frequently they impose structure according to the specifications of the clustering algorithm. Thus cluster analysis not only uncovers inherent order, regularities, or natural laws, but also fits the data to some preconceived model. In many studies, there is no clear evidence that the investigator distinguishes between these properties of the analysis. However, in skillful hands the heuristic alternation of descriptive and proscriptive clustering approaches can yield satisfying insights. The agglomerative and hierarchic clustering methods preferred by the majority of users have built-in structuralist assumptions which often cannot be met by the data. For example, perhaps the commonest underlying model of various cluster analyses is that of tight hyperspheroidal clusters of objects separated from other similar clusters by gaps that are substantial when compared to the cluster diameter. Since many natural objects have dispersions in attribute space quite different from this model, it is not surprising that hierarchic cluster analysis as displayed by dendrograms, would often represent the relationship rather poorly. Part of the difficulty is that relationships as shown by a dendritic hierarchy are basically one-dimensional and unable to express the often quite divergent similarities between individual members of different clusters.

Such considerations have led to increased emphasis on scaling approaches for representing taxonomic structure. That is, objects are projected in the attribute space and taxonomic structure is determined from the constellation of points in that space. Since this space is hyper-dimensional, one frequently employs methods for reducing its dimensionality. Familiar approaches are principal components analysis, principal coordinate analysis and nonmetric multidimensional scaling (Orloci, 1975; Sneath and Sokal, 1973). These techniques were developed largely in the social sciences. In biology, the first move away from clustering of discrete entities and towards scaling came in plant ecology where workers had become dissatisfied with the discrete, mutually exclusive classes of plant associations imposed on them by traditional concepts. The term ordination became established for this approach and has been adopted by biologists and others for such work. Ordination is now probably the preferred approach when taxonomic structure of objects is of primary interest, although the rapidity of most agglomerative, hierarchic clustering methods makes such an analysis and the resulting dendrogram a quick and easy first step in data analysis.

Not all clustering methods depend on pair functions of the OTU's. Some approaches partition the data matrix with the aim of minimizing interpart dispersion and maximizing interpart dispersion. The well-known k- means clustering algorithm by MacQueen (1967) is a case in point. Such a clustering technique is not hierarchic and except for the arbitrary decision on how many parts (clusters) to obtain does not impose structure on the data. Many applied problems such as redistricting, routing, and optimal allocation problems are most effectively solved by a non-hierarchic method of this type.

Although clustering and ordination methods will undoubtedly be improved in the future, we seem to have at our disposal an adequate array of techniques to serve many purposes. I believe that the major effort in classificatory work in the next few years should be devoted to comparisons of different approaches and to tests of significance of classifications. Work in these fields has so far been quite unsatisfactory. In fact, none of the three recent books on cluster analysis (Anderberg, 1973; Hartigan, 1975; Späth, 1975) even deals with these problems.

To carry out a significance test in cluster analysis, we have to formulate an appropriate null hypothesis first. Possible null hypotheses might include

(i) All similarity coefficients $\underline{S}_{jk}$ equal to zero.

(ii) All $\underline{S}_{jk} = 1$.

(iii) All $\underline{S}_{jk}$ are equal among themselves, but $\neq 0$ or 1.

Another approach might be to examine the distribution of OTU's in character hyperspace and to test the observed distribution against some expected one. Expectations could be uniform distributions, random distributions (for example the Poisson distribution), or multivariate normality of the points. Models of the generative processes for positions of OTU's in attribute hyperspace may lead to contagious distributions, similar to those observed in actual data. However, the nature of the data in many cluster analyses is such that tests of the null hypotheses against these expectations are of little interest since the data are known to possess structure. Thus when we group six species of butterflies or twenty Indo-European languages, we know pretty well beforehand that they are different from each other ($\underline{S}_{jk} \neq 1$) and moreover that they are unlikely to be equally similar to each other (i.e. equidistant in attribute space). Thus a more relevant test of significance is whether the taxonomic structure developed by the clustering algorithm is different from a prior preconceived classification. This should be an extremely useful test yet it is not often performed and there has been no special development of this beyond the matrix correlations originally proposed by Sokal and Rohlf (1962).

One approach to significance testing has been devising optimality criteria for classifications. This can be done in several ways. One can compare the matrix of original pair functions with another matrix that represents the relationships implied by the classification. For a hierarchic classification, the relations are represented by an ultrametric and the comparison is often made by the cophenetic correlation coefficient which is simply the product-moment correlation between the similarity values S_{jk} and the values implied by the classification C_{jk}. A second measure of goodness of fit is a normalized Minkowski metric and is given by $(\sum_{jk} |U_{jk} - C_{jk}|^{r})^{\frac{1}{r}} / (\sum_{jk} U^{r})^{\frac{1}{r}}$ where $\sum_{jk}$ is over all pairs of OTU's j, k $(j \neq k)$ and $0 \leq r \leq 1$. Another such measure of goodness of fit is the stress measure employed in nonmetric multidimensional scaling where the distances implied by the nonmetric scaling are compared to a ranked similarity matrix.

A second approach would be to test whether the OTU's in the study have been arranged in a nonarbitrary manner. Do the clustering levels in a particular dendritic arrangement correspond to differences in S_{jk} among the members or could such differences be obtained by random allocation of the OTU's to this specific partition? This is a superficially simple test, complicated, however, by the lack of independence of the S_{jk} values. Innovative approaches along these lines have recently been proposed by Ling (1973) and Lennington and Turner (1976). Alternatively to working with S_{jk} tests can be carried out by ranking the S_{jk}'s in a similarity matrix and comparing rank sums representing intrataxon and intertaxon relationships. It can easily be shown that good clustering algorithms in such cases produce partitions of the rank orders of the similarity matrix that would be significant were the S_{jk}'s on which the ranks are based independent. The method can be extended beyond a single level by partitioning the dendrogram by means of phenon lines (Sneath and Sokal, 1973) and examining the rank sums of similarity coefficients representing the phenon class along the similarity axis of the dendrogram. The difference between the rank sums can be used as a measure of optimality of the OTU allocation to the taxa but the problem becomes more difficult if one wishes to investigate whether the particular partition represented by the dendrogram is optimal. The number of possible partial orders is so large, that it is probably impractical to obtain a solution by enumeration. Questions to be asked are whether the difference in the criterion (e.g. rank sums) would be sharpened by (a) Maintaining the topological structure but altering membership of the taxa (i.e., by reallocating OTU's among taxa whose size is predetermined); or (b) By doing the above and in addition altering taxomonic structure (i.e., the size of taxa and topology of the dendrogram).

Another way to approach optimality has been by the optimization of inherent criteria. For example, Ward's (1963) clustering algorithm, another hierarchical method, permits those clusters to merge that yield the least increment in the criterion function $\sum_{k} \sum_{t_J} \sum_{n} (\underline{X}_{ij} - \bar{X}_{iJ})^2$ where $\underline{k}$ is the number of clusters, $\underline{t}_J$ is the number of OTU's in cluster $\underset{\sim}{J}$, $\underline{n}$ is the number of characters, $\underline{i}$ indexes characters, $\underline{j}$ indexes numbers of OTU's in cluster $\underset{\sim}{J}$, $\underline{X}_{ij}$ is the character state for character $\underline{i}$ and OTU $\underset{\sim}{j}$, while $\underline{\bar{X}}_{iJ}$ is the mean for character $\underline{i}$ and cluster $\underset{\sim}{J}$. Such methods will not necessarily give a globally optimal solution while optimizing the criterion for each clustering level.

The above methods either find how well a classification fits an original similarity matrix or how well the data are partitioned at any one clustering level. The investigation of significant structure might be pursued from a different perspective, the study of the distribution of the similarity coefficients represented by the similarity matrix. Distributions of similarity coefficients between and within taxa have been studied by several authors (Hutchinson, Johnstone and White, 1965, Tsukamura, 1967, Sneath, 1972), but there has not been a systematic study of the distributions of similarity coefficients in large taxonomic samples and of the implications of the different forms that these distributions can take. A promising metric for such investigations would be the distribution of distances among all points. Such distributions have been studied in the plane by Dacey (1963, 1969).

Yet another approach would test homogeneity within and among clusters in terms of the variables (characters) defining them. When each OTU is a population sample for which multivariate normality can be assumed, one can use the technique of multivariate simultaneous test procedures (Gabriel, 1968) to test whether there are significant differences among the parts of the partition and whether there is significant heterogeneity within the partition. When OTU's are single point estimates one might use the clusters to generate a variance and dispersion matrix. A similar approach has been used by Rohlf (1970).

There will be cases when such tests can be carried under further constraints. For example, when there are relationships which guide the connection of adjacent points but are not themselves part of the descriptive vectors of the OTU's. An example is the spatial or geographical location of points as examined in regional analysis or geographic variation studies. In such studies Sokal and Riska (1977) have connected only points which are geographically contiguous following some definition of geographic contiguity. The simultaneous test procedures are then carried out geographically contiguous clusters that were initially grouped together because their means were similar as well. One then tests the statistical homogeneity of the resulting clusters. Similar three-dimensional spatial aggregations should occur in geological and mining research and other examples involving time as one dimension can easily be constructed. Such cases would be useful in the study of phylogeny and in archeological and historical research. Obviously the space in which the guiding connections are made could be an abstract space defining relationships of revelance to the problem under study.

The approaches discussed so far do not explicitly examine the structure of the clusters in terms of the kinds of interrelationships within each structure which might best be visualized by graph theoretical means. In this connection, the graph theoretical work by Professor Matula discussed later in this volume (Matula p. 95) will be of especial interest. It does not now provide solutions to the significance problem, but does at least furnish suggestions of ways to approach it. One possible line of inquiry is to consider cluster configurations at different cutoff points. Consider the ratio of similarity values above a cutoff point to the total number of elements in a similarity matrix to be the probability of connecting two vertices in a random graph. One might then predict the distributions of types of subgraphs resulting from such conditions, and should be able to test the observed distribution of subgraphs against these expectations.

Successful quantification of taxonomic structure, and of optimality criteria will lead to progress in the development of adaptive clustering algorithms, another important and needed direction in which cluster analysis should be moving.

If statistical test for elucidating and validating taxonomic structure can be developed, I believe that cluster analysis rather than ordination would again be the method of choice for classificatory work.

References

Anderberg, M. R. 1973, Cluster Analysis for Applications, Academic Press, New York, 359 pp.

Beckner, M. 1959, The Biological Way of Thought, Columbia U. Press, New York, 200 pp.

Dacey, M. F. 1963, Two-dimensional random point patterns: A review and an interpretation, Reg. Sci. Assoc. Pap. 11: 41-55.

Dacey, M. F. 1969, Proportion of reflexive nth order neighbors in spatial distribution, Geogr. Analysis 1: 385-388.

Gabriel, K. R. 1968, Simultaneous test procedures in multivariate analysis of variance, Biometrika 55 : 489-504.

Gower, J. C. 1971, A general coefficient of similarity and some of its properties, Biometrics 27 : 857-871.

Hartigan, J. A. 1975, Clustering Algorithms, John Wiley and Sons, New York , 351 pp.

Hutchinson, M., K. I. Johnstone, and D. White 1965, The taxonomy of certain thiobacilli, J. Gen. Microbiol. 41 : 357-366.

Lennington, R. K. and R. H. Flake 1976, Statistical evaluation of a family of clustering methods, in Proceedings of the Eight International Conference on Numerical Taxonomy, Ed. G. Estabrook, W. H. Freeman, San Francisco, pp. 1-37.

Ling, R. F. 1973, A probability theory of cluster analysis, J. Amer. Stat. Assoc. 68 : 159-164.

MacQueen, J. 1967, Some methods for classification and analysis of multivariate observations, in Proceedings of the Fifth Berkeley Symposium on Mathematical Statistics and Probability, Eds. L. M. Le Cam and J. Neyman, Vol. 1, U. California Press, Berkeley, pp. 281-297.

Morrison, D. F. 1967, Multivariate Statistical Methods, McGraw-Hill, New York, 338 pp.

Orloci, L. 1975, Multivariate Analysis in Vegetation Research, Dr. W. Junk, the Hague 276 pp.

Pearson, K. 1926, On the coefficient of racial likeness, Biometrika 18 : 105-117.

Rohlf, F. J. 1970, Adaptive hierarchical clustering schemes, Systematic Zool. 19 : 58-82.

Sneath, P. H. A. 1962, The construction of taxonomic groups, in Microbial Classification, Eds. G. C. Ainsworth and P. H. A. Sneath, Cambridge U. Press, Cambridge, pp. 289-332.

Sneath, P. H. A. 1972, Computer taxonomy, in Methods in Microbiology, Eds. J. R. Norris and D. W. Ribbons, Vol. 7A, Academic Press, London, pp. 29-98.

Sneath, P. H. A. and R. R. Sokal 1962, Numerical taxonomy, Nature, 193 : 855-860.

Sneath, P. H. A. and R. R. Sokal 1973, Numerical Taxonomy, W. H. Freeman, San Francisco, 573 pp.

Sokal, R. R. and B. Riska 1977, Partitioning geographic variation patterns, MS in preparation.

Sokal, R. R. and F. J. Rohlf 1962, The comparison of dendrograms by objective methods, Taxon 11 : 33-40.

Späth, H. 1975, Cluster-Analyse-Algorithmen, R. Oldenbourg, Munich, 217 pp.

Tsukamura, M. 1967, A statistical approach to the definition of bacterial species, Jap. J. Microbiol. 11 : 213-220 .

Ward, J. H., Jr. 1963, Hierarchical grouping to optimize an objective function, J. Amer. Statist. Assoc. 58 : 236-244.

Contribution No. 183 from the Graduate Program in Ecology and Evolution at the State University of New York at Stony Brook. Preparation of this review was aided by Grant B035233 from the National Science Foundation.

Department of Ecology and Evolution
State University of New York
Stony Brook, New York 11794

The Relationship between Multidimensional Scaling and Clustering

Joseph Kruskal

Introduction.

Clustering and multidimensional scaling are both methods for analyzing data. To some extent they are in competition with one another. Much more, however, they stand in a strongly complementary relationship. They can be used together in several ways, and these joint uses are often desirable. The main theme of this paper is to describe the relationship between the two methods.

In this paper I will first make some comments about what clustering is and what it is for. After that, I will describe multidimensional scaling briefly for the sake of those not already familiar with it. Then I will go on to the central theme of my talk, the relationship between multidimensional scaling and clustering. Finally, I will describe some applications of clustering to astronomy which should be, but are not famous in the field of clustering. These bear on some of the remarks I make elsewhere in the paper.

Dimensions of Clustering Methods.

Elsewhere in this volume, I. J. Good has given a large number of dimensions which describe alternative approaches for clustering. I would like to give a small number of dimensions which seem most important.

Many of the basic concepts of clustering belong to the biological inheritance of humans and many other animals. It appears that the concept of "similarity" is built into the human nervous system. A human being growing up under primitive conditions, but with a reasonable set of life experiences, would doubtless form many clusters spontaneously: the cluster of people, the cluster of birds, the cluster of trees, and so forth. He or she would no doubt perceive a cat as more similar to a squirrel than a cat to an ant. What distinguishes clustering as discussed in this volume from the spontaneous human activity I've just described is that we derive the clustering systematically from data.

This brings us to our first dimension which is illustrated in Figure 1. There are three main types of data used in clustering. The first type I shall call multivariate data, the second type proximity data, and the third type clustering data. Multivariate data gives the values of several variables for several individuals. We shall describe such data by x_{ij} where i corresponds to the individual and j to the variable. Proximity data consist of proximities among objects of the same kind: either proximities among individuals, proximities among variables, proximities among stimuli, or proximities among objects of any single cohesive type. A proximity (following the terminology suggested by Shepard), refers to a similarity, or dissimilarity, or correlation, or overlap measure, or any other variable for measuring closeness or distance between two objects of a single type.

Some readers may be surprised at the inclusion of clustering data among the types of data we deal with, since clustering is what we wish to obtain as the result of the data analysis, rather than data we start with. However, it turns out, particularly in psychological context, that a subjective clustering provided by subjects is a very useful form of data in some circumstances. The most useful way to collect these data appears to be something like this: We present the subject with the stimuli in some manner which makes it easy for him to deal with them, for example, each stimulus may appear on a single index card. These cards are presented to the subject in random order and he is asked initially to form a cluster of cards or stimuli which are very similar in some way which he can choose. We may also ask him to indicate the nature of the

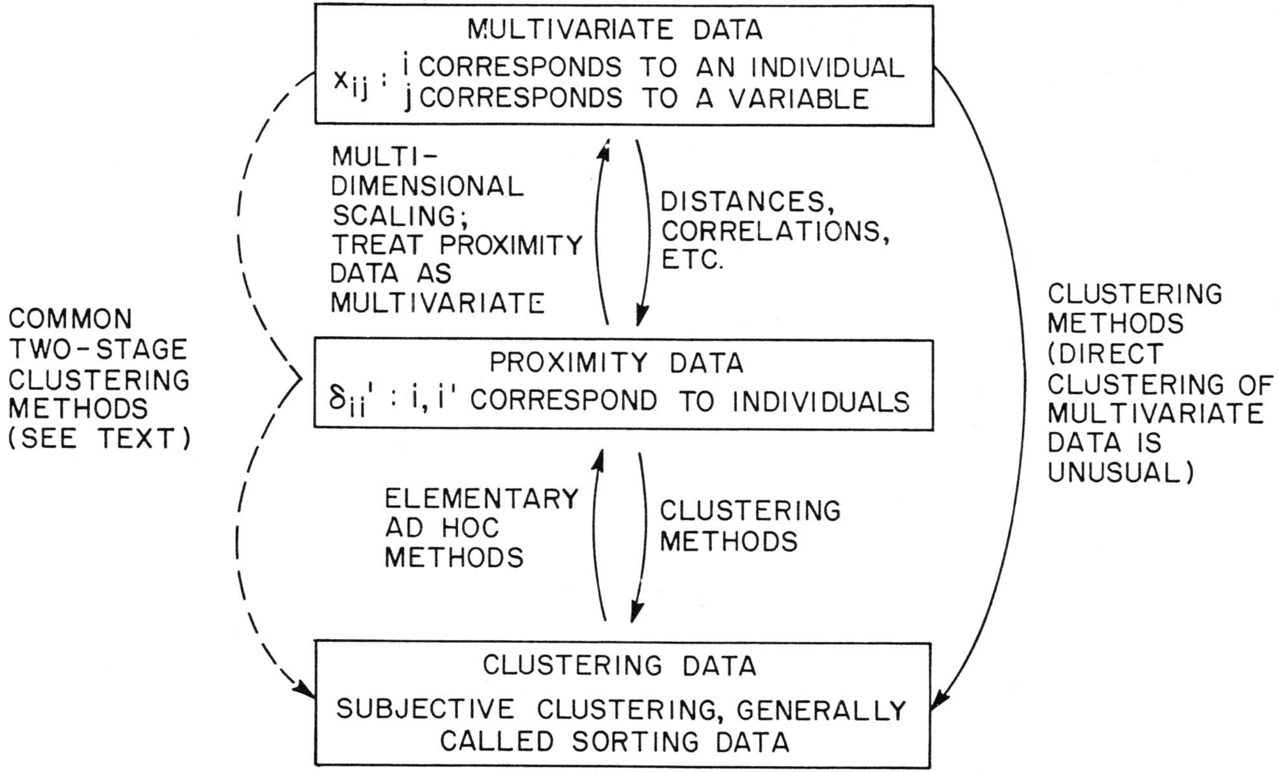
MULTIVARIATE DATA
x_{ij} : i CORRESPONDS TO AN INDIVIDUAL
j CORRESPONDS TO A VARIABLE
MULTI-DIMENSIONAL SCALING; TREAT PROXIMITY DATA AS MULTIVARIATE
DISTANCES, CORRELATIONS, ETC.
PROXIMITY DATA
$\delta_{ii'}$: i, i' CORRESPOND TO INDIVIDUALS
ELEMENTARY AD HOC METHODS
CLUSTERING METHODS
CLUSTERING DATA
SUBJECTIVE CLUSTERING, GENERALLY CALLED SORTING DATA
COMMON TWO-STAGE CLUSTERING METHODS (SEE TEXT)
CLUSTERING METHODS (DIRECT CLUSTERING OF MULTIVARIATE DATA IS UNUSUAL)

Figure 1

similarity that he found among this cluster of objects. Then we ask him to form a second cluster, and we permit the second cluster (if he so desires) to include some members of the first cluster. We repeat the whole procedure again and again until he has formed as many clusters as he wishes, permitting him to reuse stimuli since it has been learned that this permits the formation of cleaner clusters which correspond to his internal concepts in a more accurate way. Generally this kind of data is called sorting data, but for our purposes it is more convenient to use the name clustering data. Of course, in practice we would typically obtain clustering data from many subjects, not just from one.

One basic common type of clustering algorithm takes proximity data as input, and produces a clustering as an output. Another common approach to clustering starts with multivariate data and converts this by a preliminary processing to proximity data, and then further converts the proximity data to clustering data by a procedure of the type just mentioned. However, I feel that it is best to treat such a method as composed of two separable stages, and use the phrase "clustering algorithm" to cover only the second stage. Then we treat the first stage as a preliminary step prior to the clustering. Many writers on clustering refer to this stage as the calculation of the similarity or dissimilarity indices. A less common type of clustering which has been explored occasionally, most notably by Hartigan, starts with multivariate data and produces a clustering without the use of proximities as intermediaries.

Thus we see that clustering can be thought of as a procedure which starts with one or another type of data and converts it to clustering data. Other interconversions are also of interest. We have already noted that conversion of multivariate data to proximity data is sometimes an important intermediate step in clustering. There are a great many different types of conversion of this sort which are used, including the calculation of Euclidean distances among the rows of a multivariate matrix, the calculation of correlations or covariances among the columns of the matrix, the calculation of overlap measures, and many other kinds of coefficients. Multidimensional scaling, which we shall discuss further below, can be thought of as a procedure for converting proximity data into multivariate data. Another important though trivial procedure for converting proximity data into multivariate data is simply to treat the proximity matrix as though it were a multivariate data matrix, in other words to treat the proximities between a single object j and all the other objects as a variable. It is often useful to first convert proximity data into multivariate data by this elementary technique and then to form proximities from this multivariate data matrix. The new proximities are often called secondary proximities or derived proximities.

There is a very simple procedure which is often used for converting clustering data into proximities. It is only necessary to count how many subjects place object i and object j together in a single cluster in order to obtain a similarity between the objects. Such proximity data can then be analyzed by any of the methods that are appropriate for proximity data, including clustering, which then takes us back to clustering data, or by multidimensional scaling which takes us to multivariate data. In fact, there is even good success with more involved procedures. For example, Rosenberg et al (1969) and others have obtained interesting results starting with clustering data and proceeding by the following set of steps: First they use the simple procedure above to obtain a proximities matrix; then they treat this proximities matrix as a multivariate data matrix (which takes us to the top of the diagram) and form Euclidean distances among the rows of this matrix (which takes us back down to a proximity level); then they apply multidimensional scaling (which takes us back up to the top of the diagram again). It would also make good sense to start with clustering data, and from these data directly form a group clustering without the intermediate step of forming proximities or multivariate data.

A second dimension along which different approaches to clustering vary is the purpose for which the clustering is performed. There are two types of purpose, specific and vague. Both are legitimate and valid. First let me illustrate several of the specific purposes. In connection with economic modeling it is frequently necessary to aggregate companies into industries, and districts into geographic regions. Aggregation is of course a form of clustering. As a matter of fact, the United States has a highly developed hierarchical clustering of companies into industries which is called the Standard Industrial Code or SIC. This five digit classification system is widely used by economists. A second specific purpose for clustering occurs in the medical field, namely, the hope of improving treatment and diagnosis. Medical researchers often cluster cases of a single disease into subgroups. If natural subgroups exist, it is plausible to hope that they will react differently to treatment, and will have different prognosis. If this in fact happens, the clustering can be extremely valuable. A third specific purpose occurs in connection with information retrieval. The subject headings in a library, or in any other information retrieval system, form a very valuable application of clustering, although such clusterings classically have not been made by a quantitative approach based on data. A characteristic of these and other specific purposes is that they lend themselves, at least in principle, to measurement of how well a clustering has contributed to the purpose at hand.

In contrast, the vague purposes do not lend themselves to measurement. Clustering for vague purposes may be very valuable, and we may perhaps be well satisfied with the results afterwards, but it is typically very difficult to give objective verification of the value that has resulted. The first vague purpose is that of exploratory analysis, simply "to see what is there". The second purpose is to permit us to comprehend the data more clearly. A third purpose is to aid subsequent analysis. For example, after clustering we may wish to form a stratified subsample of the data by picking a certain number of individuals from each cluster.

A stratified subsample of this kind is often helpful prior to more sophisticated analysis simply to reduce the bulk of data to manageable size. Another way in which clustering can help subsequent analysis is that we may wish to perform separate analysis (for example, separate regressions) within each cluster. If the objects in one cluster really have a different character from the objects in another, it is quite possible that the regression coefficients might vary widely from one cluster to another, so this would be a much more satisfactory way of describing the data. The final vague purpose I wish to mention is clustering as an end in itself. This occurs most notably when we make family trees of languages or family trees of plants and animals. While family trees in biology are sometimes formed for specific purposes, often they are an end in themselves.

Another dimension, which is illustrated in Figure 2, is the distinction between natural and arbitrary clusters. We call clusters natural if the membership is determined fairly well in a natural way by the data,

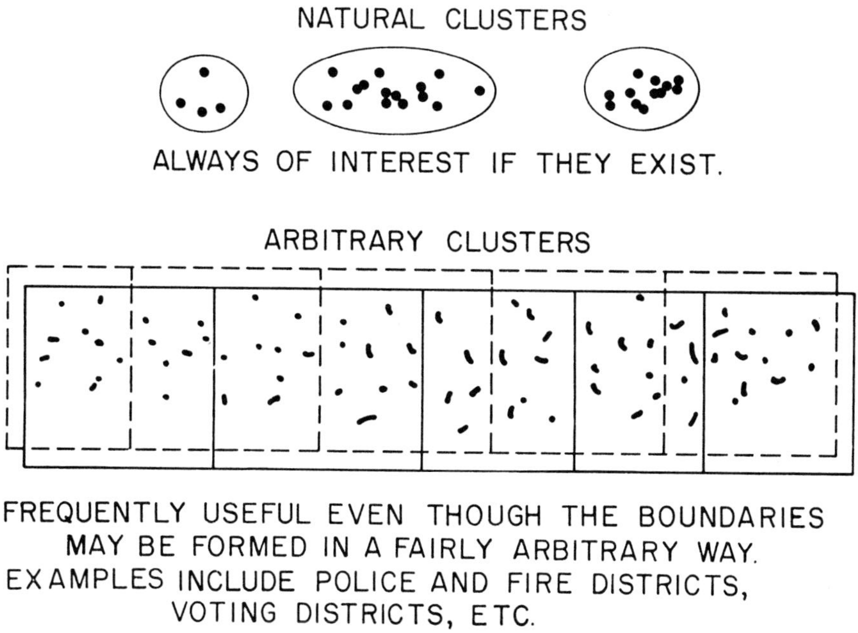
NATURAL CLUSTERS
ALWAYS OF INTEREST IF THEY EXIST.
ARBITRARY CLUSTERS
FREQUENTLY USEFUL EVEN THOUGH THE BOUNDARIES MAY BE FORMED IN A FAIRLY ARBITRARY WAY.
EXAMPLES INCLUDE POLICE AND FIRE DISTRICTS, VOTING DISTRICTS, ETC.

Figure 2

and we call the clusters arbitrary if there is a substantial arbitrary element in the assignment process. When natural clusters exist, they are almost always of interest. Arbitrary clusters are of interest only in some circumstances, but they may be very valuable. Some examples of valuable arbitrary clusters are the divisions of a large city into police precincts, voting districts, fire districts, school districts and so forth. These clusters are arbitrary in the sense that it doesn't make very much difference exactly which street is used to separate two districts. Arbitrary clusters can be of value in connection with most of the purposes mentioned above.

Another dimension applies only to techniques which seek natural clusters, and has to do with the criterion by which we define the natural clusters. In some cases compactness of the cluster is a primary criterion. In other cases, the clarity of separation between the clusters is the main thing. In the multivariate case we may place demands on the shape of the clusters, for example; must they be round ?; do we permit long straggly clusters ? As an illustration of the importance of this dimension, consider Figure 3. This shows a hypothetical density estimate of data in a univariate situation. If the clarity of separation between clusters is the primary criterion, we would probably break these data into two clusters at the deepest valley in the density function. However, if compactness of clusters is of substantial importance, we would probably break them into at least three clusters, and possibly more.

Another dimension has to do with the possibility of overlap among clusters. One possibility is simple clustering in which the clusters are not permitted to overlap at all. A second possibility is simple clustering, but with slight overlap permitted where objects on the boundaries between two clusters are permitted to belong to both. A third possibility is the familiar hierarchical clustering, where one cluster is permitted to completely contain another cluster subject to some rules, but partial overlap is not permitted. Another possibility of overlap among clustering is explored at length in the book by Jardine and Sibson (1972).

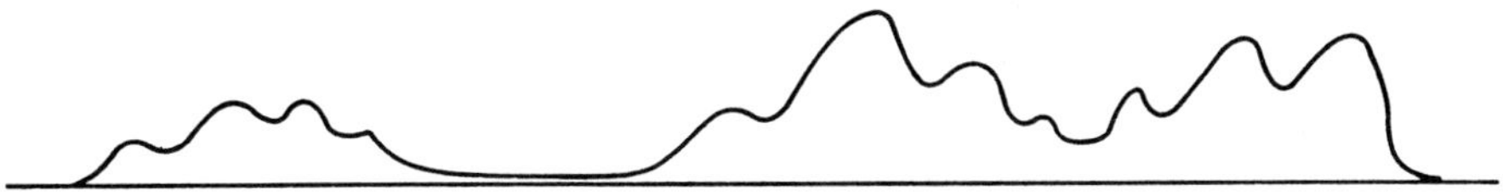

Figure 3

While their ideas are interesting theoretically, it appears to me that their approach is not of value in practice. Another nonhierarchical approach which permits extensive overlap among clusters has been put forth in the last few years by Shepard and Arabie. I believe that their method is of substantial practical interest and I will mention it a little later.

Another dimension of clustering, illustrated in Figure 4, has to do with the statistical model (if any) which underlies the clustering. In some cases no model is explicit or implicit in the approach to clustering, although such approaches are happily less and less common today. Three of the most largely used models and an important new model are illustrated in the figure. The first of these is what I like to call the "time-like tree". This consists of a tree, generally with all the terminal nodes at a single level, and a scale accompaning the vertical dimension of the tree. Each node has a scale value, and these scale values vary monotonically along the tree. Here distance within the tree is given by d_{ij}, which is the scale value associated with the lowest node covering

STATISTICAL MODELS

"TIME-LIKE" TREE

d_{ij} = HEIGHT OF LOWEST NODE COVERING i AND j

MODEL EQUATION:

$$\delta_{ij} = d_{ij} + \text{ERROR}$$

APPLIED TO FAMILY TREES

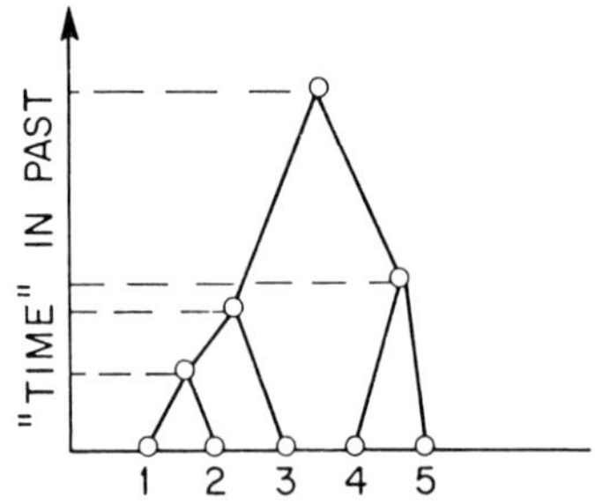

"DISTANCE-LIKE" TREE

d_{ij} = LENGTH OF PATH FROM i TO j

MODEL EQUATION:

$$\delta_{ij} = d_{ij} + \text{ERROR}$$

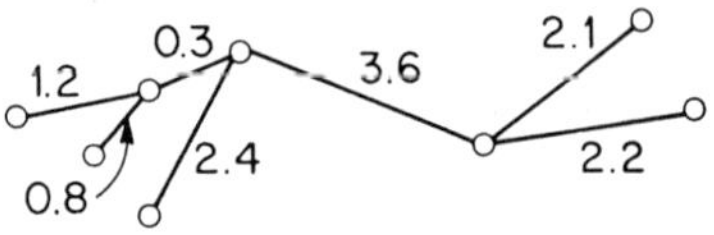

APPLIED TO EVOLUTION OF MACROMOLECULES AND DOCUMENTS

"CLASSICAL MODEL"

x_i = VECTOR OF OBSERVATIONS FOR i-th INDIVIDUAL

MODEL EQUATION:

$$x_i = \mu_k + \text{ERROR, IF } i \text{ IN CLUSTER } k$$

SHEPARD-ARABIE "ADCLUS"

$s_{ij} = \Sigma\, b_k$ + ERROR, WHERE SUM IS OVER ALL CLUSTERS k WHICH CONTAIN i AND j

Figure 4

the two objects i and j . This statistical model applies to data of the proximity type, where the proximities are dissimilarities δ_{ij}. The model equation is $\delta_{ij} = d_{ij}$ + error. This model is often appropriate for family trees of species or languages.

The second model illustrated is what I like to call a distance-like tree. Here there is no scale associated with the tree. Instead, each edge has a length associated with it. The value d_{ij} is length of the path from i to j , and the model equation is again $\delta_{ij} = d_{ij}$ + error. Distance-like trees have turned out to be quite appropriate as models for the development of complex chain-like biological molecules such as cytochrome-c and DNA. It also seems appropriate to describe the evolution of documents which had a long development prior to the many oldest known written copies, such as the Torah (i.e., the Pentateuch), "The Odyssey," "The Iliad," and the "Romance of the Rose". Another model illustrated in the figure may be called the "classical model", and applies to the multivariate data situation. Here we assume that each cluster consists of a sample from a single distribution with a single mean, so the model equation is $S_i = \mu_k$ + error, if object i is in cluster k. Of course other models are used also, but these three models are by far the most common.

Multidimensional Scaling

This is not the appropriate place to give a full exposition of multidimensional scaling, which has received description in many other places. However, for the sake of the reader who may be a little unclear about the nature of this method, we give a brief review.

Multidimensional scaling is a method which is useful for analyzing proximities, typically most often a lower or upper half matrix of them. Very often the proximities are judged similarities or dissimilarities among other objects, though a wide variety of proximity indices are used. Each object is represented by a point x_i in the plane or in space. In the simplest kind of multidimensional scaling, each proximity δ_{ij} is the distance between the points x_i and x_j, $\delta_{ij} = d(x_i, x_j) + \text{error}$. Thus multidimensional scaling is a procedure for describing a matrix of proximities as the distances between points. In general, points need not be in the plane or in three-dimensional space; any low dimensional Euclidean space may be used. (Indeed, even the Euclidean character is not necessary, but I shall not press this point further.) Loosely speaking then, we can say that multidimensional scaling is a method for representing a matrix of proximities by a configuration of points in low dimensional space.

More generally, the model for multidimensional scaling is given by the equation $f(\delta_{ij}) = d_{ij} + \text{error}$. If we know the function f, this doesn't change the procedure very much. If we only know that the function f belongs to a given parametric family, the procedure is not very much more difficult. If we only know that f is monotonic increasing (or we only know that it is monotonic decreasing) then it sounds as if the procedure should be a good deal harder. In fact, it turns out that this is not the case. When multidimensional scaling is done under this monotonic assumption, it is called nonmetric. In actual fact today nonmetric multidimensional scaling is the most common type, though I'm not sure that it should be.

The central fact about multidimensional scaling is that it takes the matrix of proximities as input, and yields a configuration of points as output. Thus, it provides a spatial representation of the proximities. Like any other statistical method which provides a representation or description of the data, the representation may be more or less accurate. Of course the accuracy of the representation is important. When it is too bad, the representation has little value.

When several matrices of proximities among the same projects are available there are special methods of multidimensional scaling available. These are referred to collectively as three-way methods of scaling. One of these methods is the three-mode multidimensional method due to Tucker. Another is the INDSCAL method of multidimensional scaling due to Carroll and Chang, which has an important special advantage: the solutions are not freely rotatable, unlike other methods of scaling. This characteristic turns out to be of great practical importance, and gives this method a substantial advantage over other related methods. Recently a new computer program to do INDSCAL with some significant improvements has been made public by Takane, Young, and De Leeuw. Not only does this computer program generalize the original INDSCAL model in some helpful ways, but it also is computationally quite efficient.

The Mathematical and Statistical Relationship Between Multidimensional Scaling and Clustering.

Since multidimensional scaling deals with proximity matrices and provides a representation of them, it is appropriate to compare multidimensional scaling with clustering methods having these two characteristics. The key difference between multidimensional scaling and such clustering methods is that multidimensional scaling provides a spatial representation for the proximities, while clustering provides a tree representation for them. Of course, this is a slight oversimplification, since clustering does not always provide precisely a tree representation. However, it always provides a discrete combinatorial representation, and in most cases this representation is either a tree or an object very much like a tree. Thus, the mathematical relationship between clustering and multidimensional scaling is quite clear and simple.

The statistical relationship however is more complex. Many people have found it useful to apply both methods to the same proximities, for reasons that we will discuss below. At the same time Eric Holman (1972) proves some theorems which appear to show that there is a competitive relationship between the two models. Roughly speaking one of his main points is that if some data fit a particular cluster model perfectly, then it would require a great many dimensions to be accurately represented by multidimensional scaling. I don't believe that there is really any conflict as I shall now explain.

Figure 5 is an impressionistic diagram which indicates my idea of one relationship between multidimensional scaling and clustering. For any set of data, we can fit both a clustering model and a scaling model. For each model we obtain some residual error. For every set of data (possibly satisfying some constraints) we plot the residual error from clustering on the horizontal axis, and the residual error from scaling on the vertical axis, thus obtaining a point. The set of all possible points forms a region in the plane. The figure shows my impression as to what that region probably looks like. If the cluster model fits perfectly, so that the point lies on the vertical axis, then according to Holman's result the scaling model cannot fit too well, so the point must lie fairly high on the vertical axis. Similarly if the scaling model fits perfectly, a clustering model will presumably not fit too well so that the point will lie fairly well to the right on the horizontal axis. The boundary of the region of possible points includes a negative-sloping curve going from the vertical axis to the horizontal axis. It is the negative slope of this curve which underlies the impression that there is a competitive relationship between the two models. If one model fits better the other model must fit worse and vice versa. How can we reconcile this apparent competitive relationship with the cooperative relationship which occurs in practice?

Figure 5 also shows my subjective conception of the contours of the bivariate frequency of data sets which occur in practice. It is my impression that for data sets encountered in practice, there is a positive relationship between how well one model fits and how well the other model fits. In other words, when one model fits better the other model fits better. In other words, the competitive relationship rests on the boundary of what is possible, while the cooperative relationship rests on the frequency of what I believe occurs in practice. What occurs in practice is typically far from the boundary of best possible fit, so there

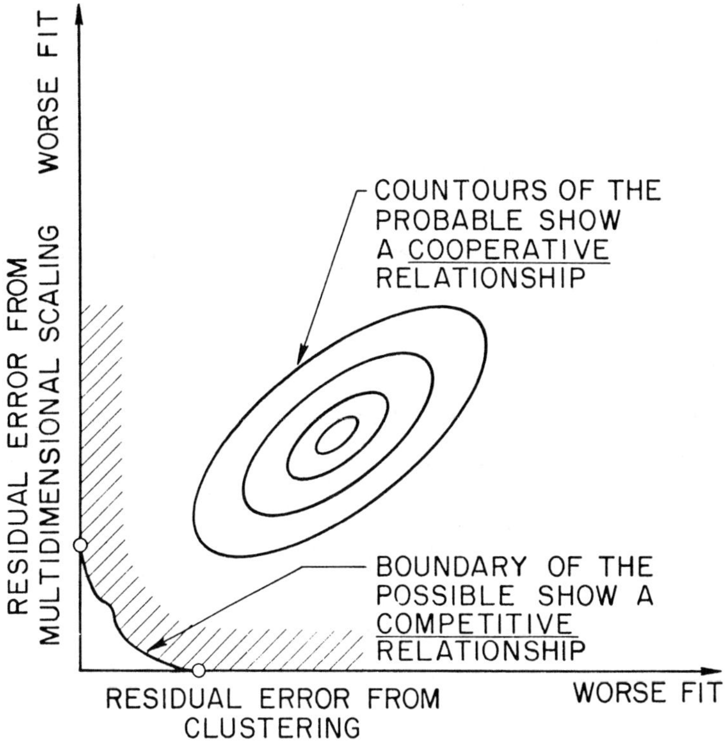

AN IMPRESSIONISTIC DIAGRAM

Figure 5

really is no conflict between the competition pointed out by Holman and the cooperation observed in practice.

It is possible that the reason for the positive correlation between how well one model fits and how well the other model fits is the great effect of random error. I would guess that if we take data which fits one model perfectly and start adding random error, we would discover that after the random error reaches some reasonably small size that the larger the random error, the worse the fit is to the other model. Thus, for random error that is not too small, the smaller the random error gets, the better both models fit. At the same time this is not to say that we cannot distinguish between the models. As Clifford Gillman (1976) shows, the use of F tests applied in the proper manner often distinguishes in an effective manner which model is in fact more valid.

The fact that both clustering and multidimensional scaling may give equally accurate representations in many practical situations does not mean that these two methods give the same kind of information about the data. In fact, quite the contrary is true. It has long been an item of folklore among some of us that scaling gives the information contained in the large dissimilarities while clustering gives the information contained in the small dissimilarities. After explaining what this folklore means I will describe a recent paper which gives partial scientific confirmation.

Consider hierarchical clustering. It is a common experience in many applications to discover that the small clusters fit well, and are often meaningful groups, but that the large clusters high up on the tree fit poorly and do not seem to be meaningful. (The only common exceptions to this which I know of involve situations such as evolution where the family tree model appears to be a truly valid explanation of the development of the groups.) Now the small clusters are based on the small dissimilarities and the large clusters are based on the large dissimilarities. Thus, clustering appears to be extracting meaning from the small dissimilarities but not from the large ones.

On the other hand consider multidimensional scaling. It is notorious that local features of the arrangement are not meaningful. Small changes in the data can cause drastic changes in the local position and arrangement of the points. Indeed different local minima (that is, different solutions which fit the data almost as well as one another) often exist and typically differ just by some local perturbation. On the other hand the general position of the points within the configuration is meaningful. For example, the fact that certain points are near the middle of the configuration will not change, even though the arrangement at the middle will change. Since the local arrangement reflects the small dissimilarities, and the global position reflects the large dissimilarities, we see that multidimensional scaling is extracting information about the large dissimilarities.

A recent paper by Graeff and Spence (1976) provides partial scientific confirmation of this folklore. They did a Monte Carlo study based on a very simple concept. First they would pick a configuration at random in two dimensions. Then they calculated the interpoint distances and added some random error, to yield dissimilarities. They arranged the dissimilarities in order of size and broke them into three equal groups: the large dissimilarities, the medium dissimilarities, and the small dissimilarities. They reconstructed the configuration using only two thirds of the dissimilarities. On each reconstruction they would delete one third of the dissimilarities, either the large ones, the medium ones, or the small ones. For each reconstruction they measured how well it matches the original configuration. The results were very clear cut. Deleting the large dissimilarities badly damages the reconstruction, while deleting either the small dissimilarities or the medium ones has only a slight effect. This clearly shows that the large dissimilarities are especially important in multidimensional scaling and partially confirms the folklore.

The Practical Relationships Between Clustering and Multidimensional Scaling.

Since multidimensional scaling and clustering are sensitive to complementary aspects of the data, the large dissimilarities versus the small ones, it seems appropriate to use them both on the same data in many cases. When the scaling happens to be in two dimensions, which is very common, there is a very happy way to combine both results into a single diagram. To illustrate this I use an analysis by Shepard of some data by Miller and Nicely shown in Figure 6. The position of the points on such a diagram is obtained from the multidimensional scaling, while the loops show the objects which have been grouped together by the clustering process. This representation of data is frequently used and can be very revealing.

Because this method is so useful it may be worth mentioning an important variation of it which can be done without the use of clustering. Again we use the multidimensional scaling configuration, but instead of using loops that indicate clusters we simply use lines between the points to indicate the similarities which are smaller than some threshold values. In some cases, multiple lines or lines of different thickness are used to indicate different degrees of dissimilarity. Figure 7 from a paper by Black shows such a diagram and illustrates a rather common phenomenon in multidimensional scaling. Without the lines the points might appear to form a circle or an annular ring, with some suggestion that the opening at the top is a little bit larger than the spacings at other parts of the ring. However, when we draw in the lines for all dissimilarities below the threshold 0.2, as in the figure, we discover that in fact we have a "horseshoe", that is, a line segment which has been curved around into an arc. This phenomenon has been emphasized by David Kendall, and may also be observed in a paper by Guttman.

Figure 8 shows another application of this method which I introduce for the purpose of describing an important new development in clustering due to Shepard and Arabie. This method, which they call

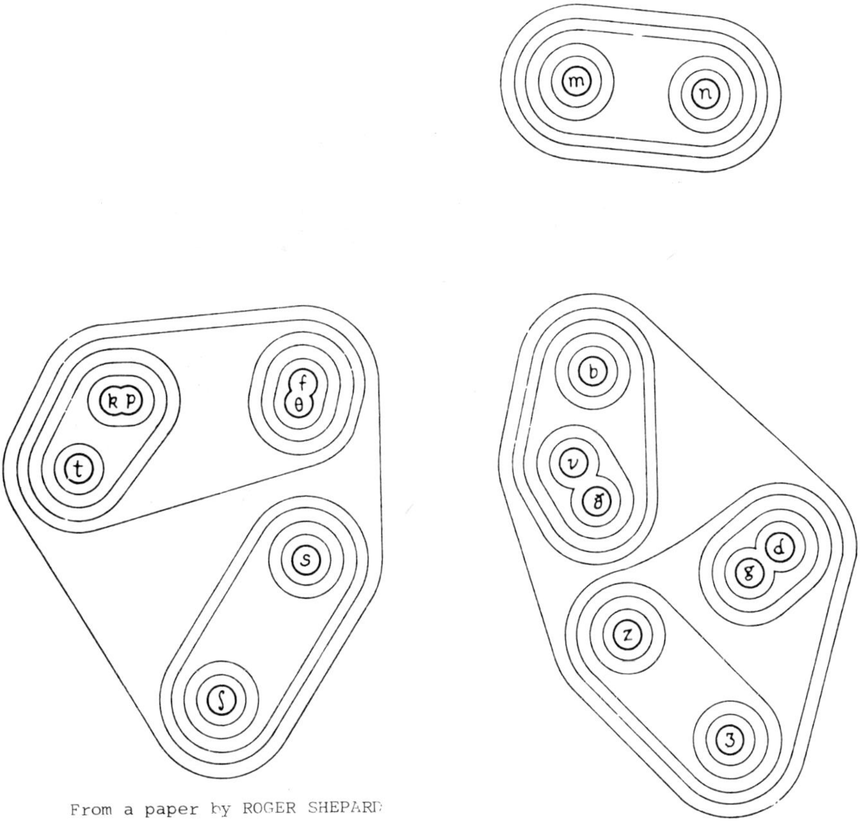

From a paper by ROGER SHEPARD

Figure 6

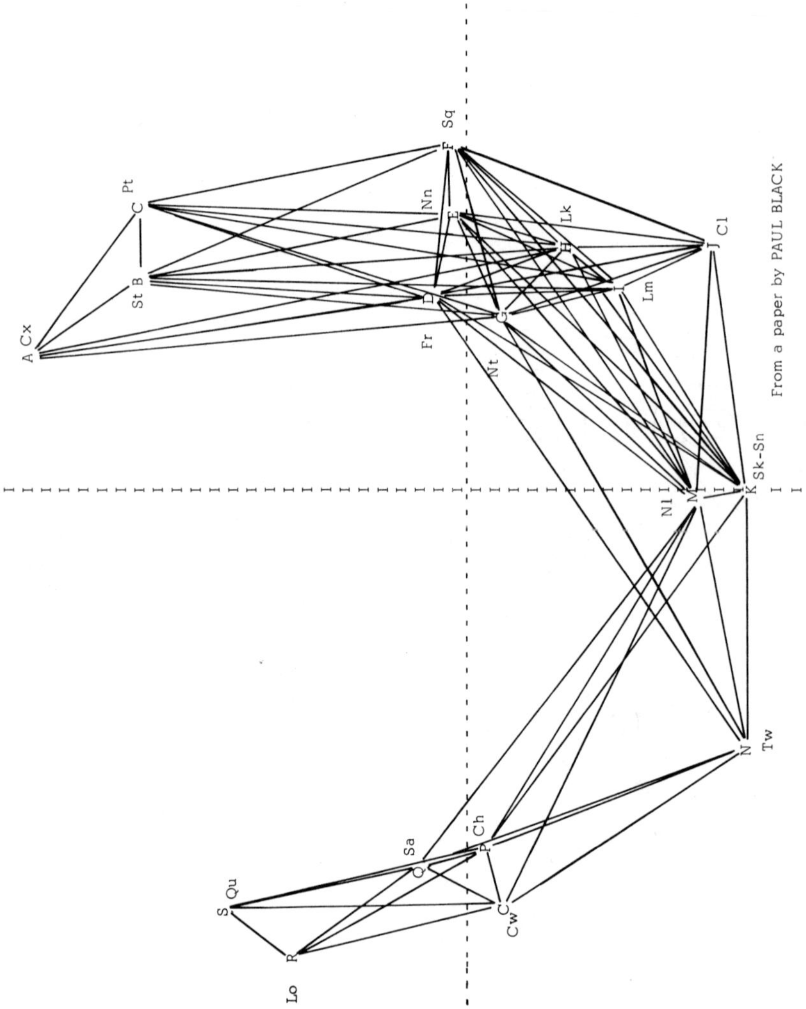

From a paper by PAUL BLACK

Figure 7

CONFUSIONS AMONG 16 CONSONANT PHONEMES

Fraction of variance accounted for with 32 subsets = .991

Plotted: First 16 subsets (embedded in 2-D MDS solution obtained by Shepard, 1972)

From a paper by
Roger Shepard and Phipps Arabie

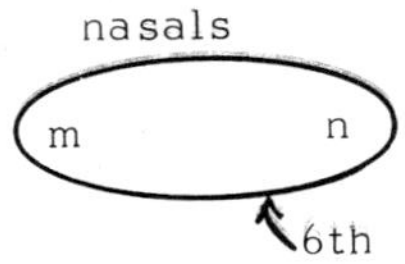

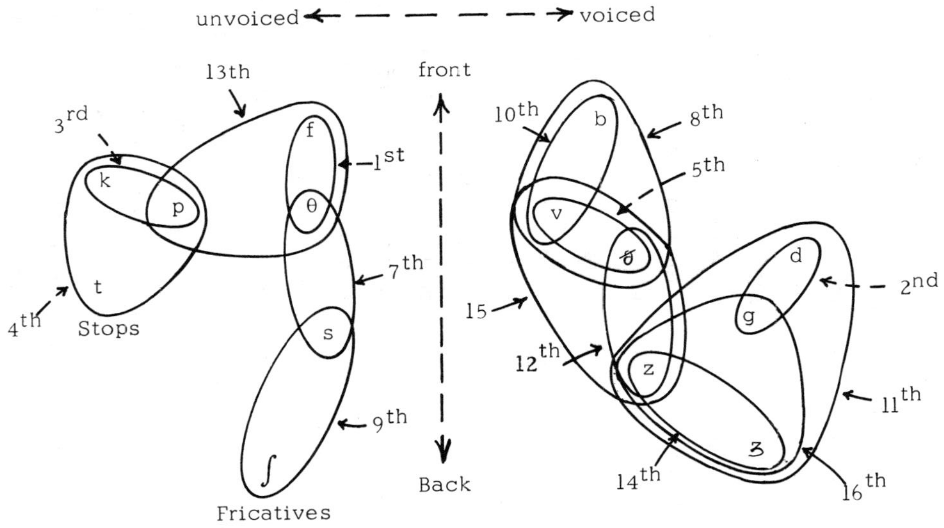

Figure 8

ADCLUS, is based on the last statistical model in Figure 4. Here s_{ij} indicates the observed similarity between objects i and j . This model resembles others used in psychology and is used with direct judged similarities. It assumes that each cluster has an associated parameter b_k, and that the total similarity between two objects i and j is the sum of the parameters associated with the clusters which contain two objects. Conceptually this model is quite simple, but it is very difficult to devise a practical method for fitting the model to real data. Shepard and Arabie have devised a method which works quite well, and have had successful results with several sets of data.

A second practical relationship between clustering and multidimentional scaling is quite similar to the first. Suppose the multidimensional scaling configuration appropriate for the data is not two dimensional, but is three dimensional or higher. Alternatively, suppose that multidimensional scaling just doesn't seem useful or appropriate for the data you are dealing with. It may nevertheless make sense to use the two-dimensional scaling configuration, and to portray the clustering by drawing loops on it, simply as a way of presenting the clustering. Thus, two-dimensional scaling configurations are useful as an auxiliary tool to aid the presentation. The same is true to some extent for one-dimensional scaling solutions, but I shall not pursue this point.

A third practical relationship is quite different. Here multidimensional scaling is the technique of central interest, and the question is how we are going to interpret the spatial configuration. By far the most popular type of interpretation is a dimensional interpretation. However, neighborhood interpretations also make sense and can be helpful. Here we take clusters of objects from a clustering (or directly from the scaling solution, although the former method is better). Then we interpret the neighborhood where each cluster lies in terms of what is common to the elements of the cluster. One example of this may be found in Kruskal and Hart, where 10,000 objects were scaled in 6-dimensional space.

(The objects were deliberately introduced malfunctions in a large digital computer. For example, a particular transistor might be stuck at the high or low voltage, or a particular diode might have a short circuit.) Our hope ultimately was to obtain a dimensional interpretation. However, it was so difficult to work with such a large number of objects, each of which was so complex to understand, that we started initially with the interpretation of regions in space. It turned out that there was no essential difficulty of interpreting regions, although it did require some effort, and we gave clear meaning to more than 20 regions. However, time ran out and we never did get to the stage where it was reasonable to attempt a direct dimensional interpretation with a minor partial exception.

There is one more "practical" relationship between clustering and scaling, though to me it seems more of an impractical relationship. Some people have clustered proximity data by first subjecting it to multidimensional scaling, generally in two dimensions, and then using the configuration to pick out clusters visually. This is not recommended, for reasons which are obvious at this point: the scaling configuration reflects the large dissimilarities, and is quite irregular in regard to the local arrangement of the points, so the configuration does not give a good grasp on which points are nearest to which other points. In other words, this is a poor way to cluster for precisely the same reason it is useful to apply both clustering and multidimensional scaling to the same proximity matrix.

Some Applications From Astronomy Which Should Be Famous In Clustering.

Astronomy may well be the oldest science using systematic analysis of numerical data. It has led the way in several parts of statistics. Let us not forget that Gauss invented the Gaussian (i.e., normal) distribution in connection with statistical problems of astronomy. In the field of clustering also, astronomy provides a major application which should be famous in the field of clustering, although in fact it is little known. This application has provided the basis for major portions of astronomy in the last six or seven decades.

It was in the 1840's that Father Secchi first produced photographic spectra from starlight. Each spectrum is a continuous band of light, interrupted by occasional black lines. He noticed that the spectra of different stars vary substantially in the position and darkness of the lines, and he grouped these spectra into four types which remained of great importance for many decades afterwards. (This clustering of the stars into four clusters is simply the first stage of this application.) The reason for the dark lines in the spectrum was not known and excited a great deal of speculation. Naturally, the reason for the four types was still more mysterious. It should be noted that his clustering was based purely on perceived similarity and had no theoretical basis, just like many clusterings today in the social sciences.

Some decades later, with better photographic and optical techniques available, other astronomers began to extend and refine his work. During the course of this process a very large number of clusters were proposed and labelled with letters of the alphabet. In the process almost the whole alphabet was used up, and in some cases subscripts were used to distinguish subclasses within these tentative classes. The work went forward most rapidly during the period from 1890 to 1915, with the Harvard Observatory playing a very notable role. At the same time that the observational knowledge about the stellar spectra was improving, the photographic and optical techniques were also improving very rapidly, and it is difficult now historically to sort out exactly what advances can be attributed to what sources. Such a study would require a detailed examination of the records at the Harvard Observatory.

However, as the classes became better defined and spurious classes based on artifacts were discarded, the astronomers began to notice that between certain clusters there were intermediate cases, while between other pairs of clusters such intermediate cases did not occur. Somewhere during this process they discovered that the clusters could be placed in a linear order, with intermediate types of spectra occurring only between clusters which were adjacent in the linear order. In other words, the clustering in this case led to the discovery that there was an underlying one-dimensional scale, and that the clusters reflected neighborhoods on that scale. At this time the meaning of the scale was still very mysterious and there was much speculation as to what it might correspond to. Some of the variables which were suggested included the age of the star and its mass. The reasons for the dark lines was still not well understood. Today it is known that the dark lines are due to absorption, by the outer cooler layers of gas on the star, of certain frequencies of the light which is emitted from the hot inner core. It is also known that the one-dimensional scale corresponds in a very direct way with the temperature of this outer layer of absorbing gas. To this day, however, those spectral classes discovered before 1900 continue to live in daily astronomical terminology: every beginning astronomy student has to learn the apparently haphazard arrangement of letters which form the sequence of spectral classes: O, B, A, F, G, K, M, R, N. Our sun is in spectral class G2, where the 2 indicates that the spectrum is two-tenths of the way from G to the next cluster, namely K. (Originally, the division of each interval into tenths was done entirely on a subjective basis.)

When we consider this application of clustering, the first thing we realize is that this clustering was done subjectively, without the kind of systematic methods that are used today. It is impossible to say whether this work would have benefited by the application of modern clustering methods, but it might have. The work was carried out over several decades, with the quality of the data improving at a rapid rate. Thus, while the field of clustering by systematic methods cannot take credit for this great step forward in astronomy, I do think that we can look on this example as an illustration of the great value that can occur from successful data analysis through the grouping of objects into clusters.

The second application in astronomy which I wish to discuss rests directly on the results of the first. In 1913 an astronomer named Russell considered a substantial group of stars belonging to a small group of adjacent spectral classes. He discovered that the intrinsic brightness of these stars falls into two very clearly distinguishable clusters: brighter stars, which today would be called "red giants", and the less bright stars, which today would be called "main sequence" stars. A year later the Swedish astronomer Hertzsprung was led by this observation to plot a scatter diagram of intrinsic brightness versus spectral class. This diagram reveals a great amount of structure, and has been extremely important in astronomy since that time. The life cycle of a star may be plotted on such a diagram, and such diagrams are frequently referred to today merely as H-R diagrams. Thus, Russell's discovery of univariate clustering in the brightness variable was a second application of great importance. I have not seen Russell's original paper, so I do not know what motivated him, nor do I know exactly what means he used to discover this clustering. I hope someday to study further the historical development of both these applications.

Another example from astronomy may be worth mentioning, although it has rather different character. In the 1940's and early 1950's Walter Baade discovered a clustering of the cepheid variable stars. He referred to the two types as population 1 and population 2. His discovery of two different types led to a very important reinterpretation of astronomical distances. As a result it was realized that the larger astronomical distances are approximately 2.5 as large as they had been thought prior to that time, and the age of the universe was increased by the same factor. This resolved a major discrepancy of long standing, since up to that time the astronomical age of the universe was substantially smaller than the well established geological age of the earth. Thus the discovery of a clustering once again had major implications in astronomy.

References

Black, P. W. (1977) Multidimensional Scaling Applied to Linguistic Relationships, in Lexico Statistics in Genetic Linguistics II, Proceedings of the Montreal Conference, I. Dyen, ed. l'Institut de Linguistique: Louvain, Belgium.

Carroll J. D. & Chang, J. J. (1970) Analysis of Individual Differences in Multidimensional Scaling via an N-Way Generalization of Eckart-Young Decomposition, Psychometrika 35 (3): 283-319.

Gillman, C. (1976) Empirical Considerations in Euclidean Nonmetric Multidimensional Scaling vs. Hierarchical Clustering, Talk presented at the Psychometric Society meeting, April 1-3.

Graef, Jed & Spence, Ian (1976) Using Prior Distance Information in Multidimensional Scaling. Paper presented at Joint Meeting of the Psychometric Society and Mathematical Society Group, Bell Laboratories, Murray Hill, April.

Hartigan, J. A. (1975) Clustering Algorithms, John Wiley & Sons, New York.

Holman, E. W. (1972) The Relation Between Hierarchical and Euclidean Models for Psychological Distances, Psychometrika, vol. 37, No. 4, December.

Jardine, N. and Sibson, R. (1971) Mathematical Taxomony, John Wiley & Sons, London.

Kendall, D. G. (1975) The Recovery of Structure from Fragmentary Information, Philosophical Transactions of the Royal Society of London, vol. 279, No. 1291, 547-582.

Kruskal, J. B. & Hart, R. E. (1966) Geometric Interpretation of Diagnostic Data from a Digital Machine, based on a study of the Morris, Illinois Electronic Central Office, Bell System Technical Journal, Vol. 45, 1299-1338.

Rosenberg, S., Nelson, C. & Vivekananthan, P.S (1968) Multidimensional Approach to Structure of Personality Impressions, J. Person. Social Psychol. 9(4), 283-294.

Shepard, R. N. (1974) Psychological Representation of Speech Sounds, in Human Communication: A Unified View, E. E. David & P. B. Denes, Eds. McGraw-Hill, New York.

Shepard, R. N. (1974) Representation of Structures in Similarities: Problems and Prospects, Psychometrika, Vol. 39, 373-421.

Shepard, R. N. & Arabie, P. (1974) Representation of Similarities as Additive Combinations of Discrete Overlapping Properties, unpublished, but see also Shepard (1974).

Takane, Y., Young, F. W., and De Leeuw, J. (1977) Nonmetric Individual Differences Multidimensional Scaling: An Alternating Least Squares Method with Optimal Scaling Features, Psychometrika, (in press).

Tucker, L. R. (1972) Relations Between Multidimensional Scaling and Three-Mode Factor Analysis, Psychometrika, Vol. 37, No. 1, March, 3-27.

Mathematics & Statistics Research Center
600 Mountain Avenue
Bell Telephone Laboratories
Murray Hill, New Jersey 07974

Distribution Problems in Clustering

J. A. Hartigan

1. Introduction.

The very large growth in clustering techniques and applications is not yet supported by development of statistical theory by which the clustering results may be evaluated. A number of branches of statistics are relevant to clustering: discriminant analysis, eigenvector analysis, analysis of variance, multiple comparisons, density estimation, contingency tables, piecewise fitting, and regression. These are all areas where the techniques may be used in evaluating clusters, or where clustering operations occur.

The statistical problem considered in this paper is that of deciding which of the many clusters presented to us by algorithms are "real". There is no easy generally applicable definition of "real". Our approach is to assume that the data consist of a sample from a distribution P on a population. A data cluster is "real" if it corresponds to one of the population clusters.

Mixture techniques, k-means, single linkage, complete linkage and other common algorithms are examined to give measures of the "reality" of their clusters. Most of the statistical problems are stated as asymptotic distribution theory problems, because finite and exact theory is almost always out of the question. A reasonable significance testing procedure requires the asymptotic theory to be validated by Monte Carlo experiments. There are many guesses, conjectures, analogies, and hopes, and only a few hard results.

2. Two clusters, one dimension, mixture model.

Suppose observations are drawn from a mixture of two normal distributions with the same variance. There are "real" clusters in the population if the normal means are different. The statistical question is how to decide this given the sample.

P1: Let $x_1, \ldots, x_n$ be observations from the mixture $N(\mu_1, \sigma^2)$ with probability p and $N(\mu_2, \sigma^2)$ with probability $(1-p)$. Let $L_0(\underset{\sim}{x})$ be the maximum log likelihood under the assumption $\mu_1 = \mu_2$ and let $L_1(\underset{\sim}{x})$ be the unconstrained maximum log likelihood. What is the asymptotic distribution of $L_1(\underset{\sim}{x}) - L_0(\underset{\sim}{x})$?

The distribution could be used to test the reality of two clusters in $x_1, \ldots, x_n$ - i.e., whether or not the difference in the likelihoods is unusually large for samples from an unmixed normal. Unfortunately $L_1(\underset{\sim}{x}) - L_0(\underset{\sim}{x})$ is not asymptotically $\frac{1}{2}\chi_d^2$ which might be anticipated from standard theory; the unconstrained problem has 4 parameters and the constrained problem has 2 so one would think $d = 2$; but precisely when $\mu_1 = \mu_2$ the value of p is irrelevant, suggesting $d = 1$. Simple interpolation suggests an asymptotic distribution between $\frac{1}{2}\chi_1^2$ and $\frac{1}{2}\chi_2^2$.

The problem may be graphically represented (Figure 1) in the simplified case $x \sim p\, N(\Delta, 1) + (1-p)\, N(\frac{-\Delta p}{1-p}, 1)$, when the asymptotic distribution of the log likelihood ratio for testing $\Delta = 0$ is to be obtained. Assume $0 \leq p \leq \frac{1}{2}$ so that parameters are identifiable. For n large, the distribution of $L(\underset{\sim}{x}, \Delta, p) - L(\underset{\sim}{x}, 0, p)$ for each fixed Δ, p is approximately normal, and it takes its maximum in expectation along the lines $\Delta = 0$, $p = 0$. [Here observations are assumed to be normal, according to the null hypothesis $\Delta = 0$.] For large n, the maximum of $L(\underset{\sim}{x}, \Delta, p) - L(\underset{\sim}{x}, 0, p)$ will be close to the lines $\Delta = 0$, $p = 0$. The locus of maximizing (Δ, p) for p fixed will be close to these lines; for each fixed p, $\sup_{\Delta} [L(\underset{\sim}{x}, \Delta, p) - L(\underset{\sim}{x}, 0, p)]$ is approximately $\frac{1}{2}\chi_1^2$. Thus asymptotically $\sup_{\Delta, p} [L(\underset{\sim}{x}, \Delta, p) - L(\underset{\sim}{x}, 0, p)]$ will be greater than $\frac{1}{2}\chi_1^2$, but its exact

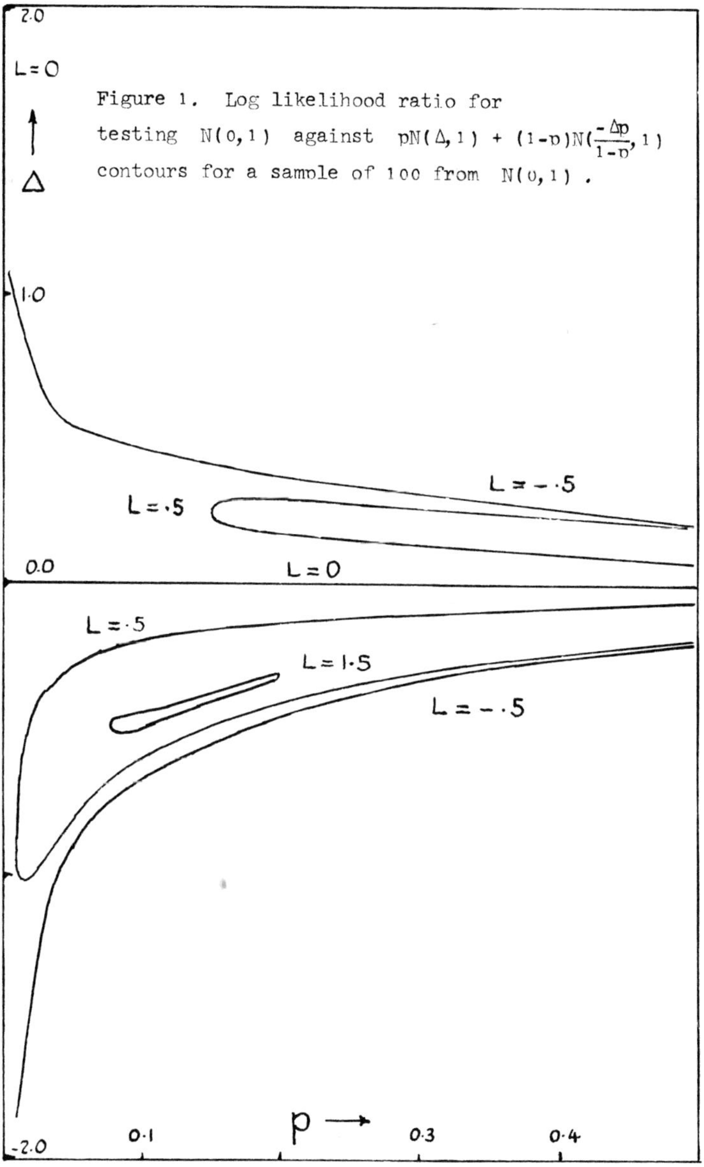

Figure 1. Log likelihood ratio for testing N(0,1) against $pN(\Delta,1) + (1-p)N(\frac{-\Delta p}{1-v},1)$ contours for a samnle of 100 from N(0,1) .

distribution is the maximum of a family of correlated $\frac{1}{2}\chi_1^2$.

3. 2 clusters, 1 dimension, F-ratio.

Especially in the null case, the maximum likelihood for the mixture model is difficult to compute. See, for example, Dick and Bowden (1973). A simpler criterion, which is equivalent to maximum likelihood when μ_1 and μ_2 are well separated, computes the maximum F-ratio over all possible divisions of the sample into two sets. This criterion may be considered for use with more general models.

P2. Let $x_1, \ldots, x_n$ be a sample from a distribution function F. Let R_n denote the maximum F-ratio over all possible divisions of the sample into two sets. What is the asymptotic distribution of R_n?

Suppose that the maximum F-ratio for the population occurs when the population is split into two sets $x \leq x_0$ and $x > x_0$. Suppose that x_0 is unique, that the distribution F has finite variance, and that F^{-1} is differentiable at x_0. Then R_n is asymptotically normal and equivalent to the F-ratio computed on the sample split:

$$S_1 = \{x_i | x_i \leq x_0\}$$
$$S_2 = \{x_i | x_i > x_0\}.$$

The actual sample split will occur at a value x_0^n which converges to x_0, but the difference between x_0^n and x_0 is negligible in computing the asymptotic behaviour of R_n. There isn't much left in this problem except weakening the already weak conditions on F.

Suppose we wish to conclude from a large value of R_n that the population is bimodal; the largest unimodal population value of R, given that the population is optimally split into two clusters in proportions p_0 and $(1-p_0)$, occurs for the uniform-spike distribution: the random variable is uniform over $[0,1]$ except for an atom $(1-2p_0)$ at $\frac{3}{4}$. The observed value of R_n should be tested against the theoretical asymptotic normal distribution based on this worst unimodal case. For $p_0 = \frac{1}{2}$, R_n is approximately $N(3n, 19.2n)$; thus in a sample of size 100, we need an F of about 372 for significance at the 5% level. These results appear in part in Hartigan (1975), but are proven in an unpublished paper, Hartigan (1976). The distribution of R_n for small n, in samples from the normal, was estimated in Engelman and Hartigan (1969). Scott and Knott (1974) apply this asymptotic distribution in a multiple comparisons problem.

4. Two modes.

If our aim is to discover the presence of two modes, we might consider the likelihood ratio test for the model that the density has two modes, against the model that the density has one mode. Maximum likelihood estimates for densities under these two models may be computed using the monotone fitting techniques of Barlow et al. [1972]. The distribution theory is very difficult; the techniques generalize to many modes in one dimension but not to many dimensions. Still, the F-ratio discussed in §3 is affected markedly by the tails of the distribution which should not affect decisions about the presence of more than one mode.

A good test for the presence of two modes should be simple to compute, should have accessible distributional properties, and should consistently distinguish between one and two modes as the data increase.

P3: Define $R_n(a,b,c)$ to be the F ratio between the samples $\{x_i \mid a \le x_i < b\}$ and $\{x_i \mid b \le x_i < c\}$. What is the distribution of $\sup_{a,b,c} R_n(a,b,c)$ as $n \to \infty$?

This is a robustified F with the tails omitted. For consistency, it will be necessary to constrain a,c so that as the total number of observations approaches ∞, the number of observations in (a,c) approaches ∞ (as in kernel estimates of densities, Wegman (1972), for example). I would expect that $\sup R_n(a,b,c)$ is asymptotically normal and that a,b,c can be chosen to maximize $R(a,b,c)$ for the population in order to compute the parameters of the asymptotic distribution.

Suppose that a histogram with equal cell sizes yields counts $n_1, n_2, \ldots, n_k$. The value of i maximizing n_i is the sample mode, say i_0; if the population is unimodal n_i should decrease approximately monotonically for $i > i_0$, and increase for $i < i_0$. An indication of bimodality occurs if there is an n_i, $i < i_0$, such that $n_i \geq n_{i+1}, \ldots, n_j$ for j large. Similarly bimodality may be indicated to the right of i_0 .

P4: Let $n_1, n_2, \ldots, n_k$ be sampled from Poisson $[\lambda]$. What is the distribution of the maximum length between ladder points, $V_k = \sup(j-i)$ where $n_i \geq n_{i+1}, \ldots, n_j$?

If the $\{n_i\}$ were sampled from a continuous distribution, V_k would depend in distribution only on k . Suppose in the histogram m_1 is the maximum length for $i < i_0$ and m_2 is the maximum length for $i > i_0$. Reject unimodality at significance level α if $\min P[V_{i_0-1} \geq m_1]$, $P[V_{k-i_0} \geq m_2] \leq \frac{1}{2}\alpha$. Given that the true mode occurs in the interval i_0, I would guess that this test has true significance level $\leq \alpha$. Another test for bimodality will be described in the discussion of single linkage techniques.

5. Many clusters, one dimension.

These problems generalize to many clusters in one dimension. The mixture problem will be to decide how many components are in the mixture, a difficult compound decision problem. A simpler case is to decide whether k or $k + 1$ components are necessary; an additional component adds two parameters to the model, but one would speculate that the log likelihood ratio lies between $\frac{1}{2}\chi_1^2$ and $\frac{1}{2}\chi_2^2$ as before.

The F-ratio, R_n , generalizes easily to many clusters; the optimal partition of $x_1, \ldots, x_n$ into k clusters can be computed in $0(n^2k)$ computations using dynamic programming techniques, W. D. Fisher (1958). The asymptotic distribution of R_n is normal under weak assumptions on the parent populations, and it is the same as the asymptotic distribution of the F-ratio computed for the optimum population division, Hartigan (1976). To decide whether k or $k + 1$ clusters are necessary, the distribution of R_n for $(k+1)$ clusters must be assessed for a population with k modes. Relevant information about such a null population is contained in the optimal partition into k clusters.

P5: <u>Let $R_n(k+1)$ denote the maximum F-ratio over all partitions of $x_1,\ldots,x_n$ into (k+1) clusters. Let $\{n_i, \bar{x}_i, s_i^2\}$ denote the number, mean, and sample variance of the observations in the ith cluster of the optimal partition of $x_1,\ldots,x_n$ into k-clusters. What is the asymptotic distribution of $R_n(k+1)$ given $\{n_i, \bar{x}_i, s_i^2, i = 1,\ldots,k\}^2$?</u>

One would expect this asymptotic distribution to be normal, and to depend on the unknown parent population F which in practice would be chosen to be least favourable given $\{n_i, \bar{x}_i, s_i^2, i = 1,\ldots,k\}$. It may happen, if k clusters are well established, that the (k+1) clusters are obtained by splitting one of the k-clusters, and the significance of this split might be evaluated by the F for splitting this cluster into two clusters.

6. k Clusters, many dimensions, mixture model.

Consider first the normal mixture model

$$x \sim p_1 N[\mu_1, \Sigma] + p_2 N[\mu_2, \Sigma] + \ldots \; p_k N[\mu_k, \Sigma]$$

where μ_j are p-dimensional vectors, and Σ is a $p \times p$ covariance matrix. Maximum likelihood estimation of parameters of this model has been considered by Wolfe (1970), Day (1969), Scott and Symons (1971), Hartigan (1975) and others.

P6: <u>Let $x_1,\ldots,x_n$ be sampled from a multivariate normal mixture,</u> k = 2. <u>Define $L(x_1,\ldots,x_n)$ to be the log likelihood ratio for the hypothesis $\mu_1 = \mu_2$ against $\mu_1 \neq \mu_2$. What is the asymptotic distribution of</u> L ?

The statistic L gives a test for two clusters. By analogy with the one-dimensional case, one might guess $\frac{1}{2}\chi_p^2 \leq L \leq \frac{1}{2}\chi_{p+1}^2$ (which is probably an adequate approximation for p large).

Suppose that x_i comes from $N(\mu_j, \Sigma)$ with probability p_{ij}. If $p_{ij} = p_j$, so that every observation has the same probability of coming from $N(\mu_j, \Sigma)$, the above mixture model obtains. If the p_{ij} are unconstrained, the maximum likelihood estimates maximize

$$\sum_i \log(\sum_j p_{ij} f[x_i | \mu_j, \Sigma]) \text{ subject to } p_{ij} \geq 0 \,, \; \sum_j p_{ij} = 1 \,.$$

For a particular i , the optimal $\{p_{ij}\}$ will be $p_{ij} = 1$ for j maximizing $f(x_i|\mu_j,\Sigma)$ and 0 otherwise. Thus $\Sigma\Sigma\, p_{ij} \log f(x_i|\mu_j,\Sigma)$ is to be maximized. An iterative technique for finding the maximum likelihood estimates is to select $p_{ij} = 1$ if j maximizes $\log f(x_i|\mu_j,\Sigma)$, to select μ_j to be the mean of those x_i's allocated to the jth population, and to select Σ to be the within cluster covariance matrix of the x_i's [using divisor n rather than $(n-k)$]. This process is repeated, increasing the log likelihood at each stage, until no further reallocation of the x's occurs. This generalization of the k-means technique may be used with any probability model for the distribution of observations within clusters. The relation between the k-means technique and the mixture model is discussed by Scott and Symons (1971). If the components are well separated the two techniques give similar estimates of parameters; every observation can be unequivocally assigned to one of the components. In general, the k-means technique is easier computationally, but gives estimates of parameters μ_i which are more widely separated than the mixtures model; the mixture model is therefore more conservative. A very bad property of the k-means technique is that it cuts up large clusters while failing to detect small reasonably distinct clusters. I suspect that the mixtures model, being asymptotically consistent when there are k distinct components, would not have this failing. The usual likelihood asymptotics is inapplicable for k-means because of the infinitely many parameters p_{ij} estimated as $n \to \infty$.

A simpler version of the k-means algorithm assumes $\Sigma = \sigma^2 I$. Thus all variables have equal weight, whereas weighting is done iteratively if Σ is unconstrained. (Use of arbritrary Σ , when the data contains discrete variables, can lead to clusters within which one of the discrete variables has zero variance, and so the likelihood becomes infinite. Clustering is according to an arbitrary choice among the discrete variables.) MacQueen (1967) studies asymptotic consistency of an algorithm of the k-means type.

7. 2 clusters, p dimensions, k-means.

Consider now the case $k = 2$, p arbitrary, $\Sigma = \sigma^2 I$. The optimal clustering (maximum likelihood according to the above model) is obtained by splitting the data into two sets by a hyperplane to minimize within cluster sum of squares.

P7: Let $x_1, \ldots, x_n$ be points in p dimensions. Let R_n be the maximum ratio of between cluster sum of squares to within cluster sum of squares over all divisions of the data into two clusters. If $x_1, \ldots, x_n$ are sampled from F, what is the asymptotic distribution of R_n?

The asymptotic normality that holds for $p = 1$ does not always generalize. I conjecture the following asymptotics: suppose that F is such that $E|x|^2$ is finite and that the population is divided by a unique hyperplane optimizing the within cluster sum of squares. Then the asymptotic distribution of R_n is the same as the asymptotic distribution of the between-within ratio for clusters specified by the population hyperplane, and this distribution will be normal. If F is spherically symmetric, then there will not be a unique hyperplane splitting the population, and R_n will be distributed as the maximum of a normal process on a sphere; this is the case when F is spherical normal.

8. 2 clusters, p dimensions, projected F-ratio.

In testing for bimodality, it seems plausible to use as the null distribution F the worst case: F is chosen to maximize the population value of R, given that F is unimodal and the population is optimally divided in the proportions p and $1-p$. The worst case has F concentrated on the line $(0,1)$; uniform over the line except for an atom of size $(1-2p)$ at $\frac{3}{4}$. Thus the worst case asymptotic distribution of R_n is just the one dimensional distribution specified after P2. It will usually be evident that F is not concentrated on a line, and so it might be desirable to consider a modified F-ratio which ignores that component of the within sum of squares which is orthogonal to the line between the cluster means.

P8: Let $x_1, \ldots, x_n$ be points in p dimensions sampled from some distribution F. Let $\bar{x}_1, \bar{x}_2$ be the means of the two clusters obtained by dividing $x_1, \ldots, x_n$ by a hyperplane chosen to maximize between cluster sum of squares. Let R_n^* be the maximum F-ratio for the projections of $x_1, \ldots, x_n$ onto the line between $\bar{x}_1$ and $\bar{x}_2$. What is the asymptotic distribution of R_n^*?

I would conjecture that the asymptotic distribution is normal, the same as for the one dimensional case with the distribution F projected onto the line between $\bar{x}_1$ and $\bar{x}_2$. Thus the one dimensional F-ratio may also be used as a test for bimodality. The division of points to maximize between cluster sum of squares is the same as the division of points to maximize the projected F-ratio, given $\bar{x}_1, \bar{x}_2$. It might be thought that the points $x_1, \ldots, x_n$ should be divided into two clusters to maximize the projected F-ratio; this procedure is invariant under linear transformations of the data, unlike k-means with $\Sigma = \sigma^2 I$. However, this procedure is very susceptible to the presence of 0 - 1 variables - frequently, a single variable with marked bimodality will decide the division. It is recommended that k-means be used to decide where to split, and then that the projected F-ratio be used to decide whether or not the split indicates bimodality.

In a study led by Scott Henderson, Australian National University, Canberra, 300 attempted suicides were questioned about personal and family background, previous medical history, immediate circumstances and method of the attempt, and psychological state at the time of the attempt. A principal aim of the study was to identify different types of suicide requiring different treatment. After considerable preselection and rescaling of variables, the k-means algorithm generated the projected values in Figure 2, with a projected F-ratio 1366, 60% of the patients in one group, and 40% of the patients in the other. A uniform-spike distribution with an atom of size .2 at $\frac{1}{4}$ is the worst unimodal distribution; the distribution of R_n is asymptotically normal N(3.6n, 31.8n) from

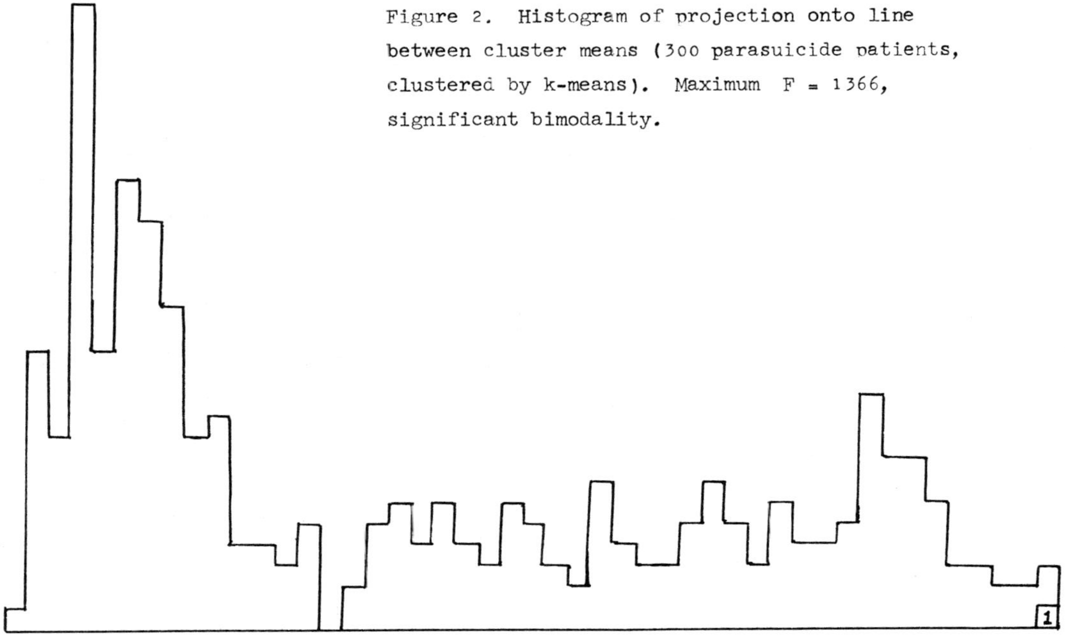

Figure 2. Histogram of projection onto line between cluster means (300 parasuicide patients, clustered by k-means). Maximum F = 1366, significant bimodality.

Hartigan (1976). Thus $R_n = 1366$ corresponds to the normal deviate $[1366 - 3.6 \times 300]/[31.8 \times n]^{\frac{1}{2}} = 2.93$, quite significant, but not astoundingly so. The variable which was the most important in distinguishing the two groups was the diagnosis of OPERANT; an OPERANT person is one who is attempting to affect others through the attempt at suicide. Approximately 40% of the patients were classified as operant, 60% as non-operant.

9. k-clusters, p dimensions, projection.

The same projection technique is useful with more than two clusters. For example with three clusters, the points in p-dimensions are projected into the plane passing through the three cluster means, and this projection is examined for evidence of trimodality. In the attempted suicides data, the projected values showed two modes not three (Figure 3) so it was concluded that 3 types of suicide could not be justified. To formalize this procedure, it would be necessary to test for 3 rather than 2 modes in 2-dimensions. The difficulty lies in selecting an appropriate null distribution F . Possibly, the distribution of R_n should be considered conditionally on the cluster parameters for the optimal 2-split, as in P5.

P9: Suppose that a distribution F in two-dimensions has two modes, but is divided into 3 clusters, in proportions p_1, p_2, p_3 to maximize R, the ratio of between to within sum of squares. For fixed p_1, p_2, p_3, which distribution F maximizes R ?

Just guessing, I would suggest the equilateral spike distribution, uniform over the three lines from the centroid to the vertices of an equilateral triangle, with atoms at the midpoint of two of the lines. This problem, and this guess, generalize to p dimensions.

After application of the k-means algorithm, the sample points are divided into k convex clusters; the centers of the clusters are the mean of points in the cluster, and every point is closer to its cluster center than to any other. The k-means procedure will not consistently identify

Figure 3: Projection of 300 parasuicide patients onto plane through means of 3 clusters (k-means).

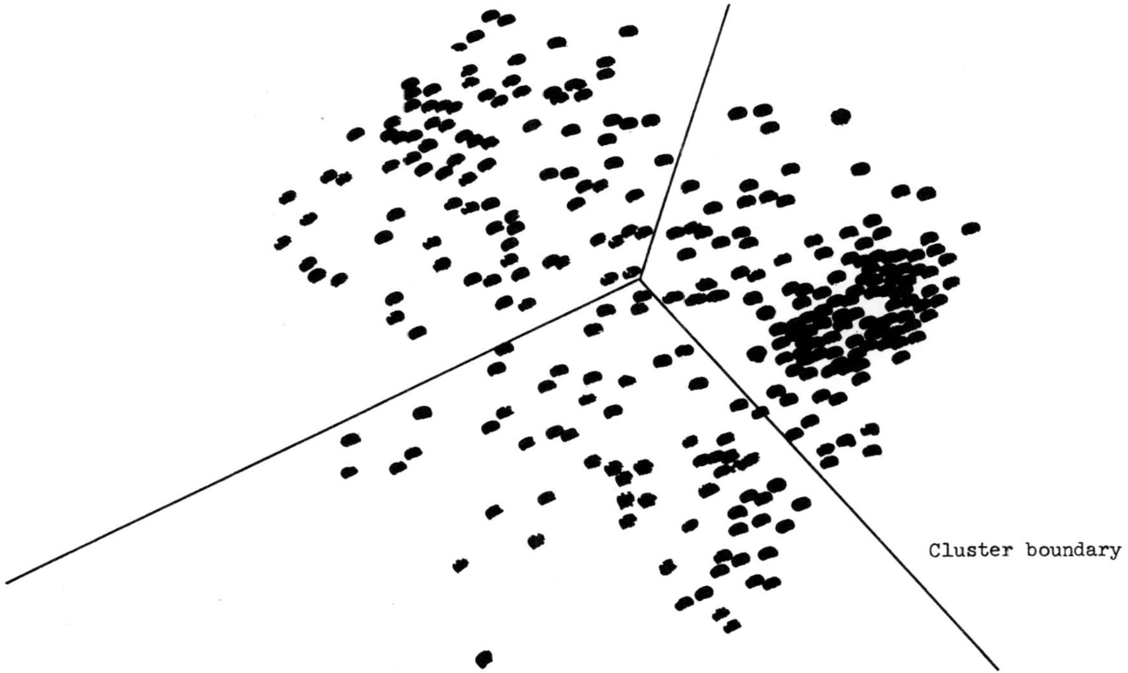

k distinct modes. The reason that the k-means procedure fails to identify modes, is that all points within a cluster have the same weight in computing the cluster center, whereas the outlying points should have no weight at all in determining the mode.

10. Identification of modes.

A mode is usually defined as a maximum of a density. Thus in estimating modes from a sample one quickly becomes involved in density estimation, as in Eddy (1976). A different definition is adopted here, because it is applicable to general observation spaces, and to computation of modes on samples.

Let P be a probability distribution on a general space $\mathcal{X}$. Assume given a distance d on $\mathcal{X}$, satisfying $d(x,y) = d(y,x) \geq 0$, each $x,y \in \mathcal{X}$. Define

$$a(y,\varepsilon) = P[d(X,y) \leq \varepsilon].$$

Then y is an ε-mode of width c, if for each $\delta > 0$

$$a(y,\varepsilon) \geq a(x,\varepsilon) \text{ whenever } d(x,y) < c$$

$$a(y,\varepsilon) < a(x,\varepsilon) \quad \text{some } d(x,y) < c + \delta .$$

Finally y is a mode of width c, if there is a sequence of ε-modes of width approaching c which approach y as $\varepsilon \to 0$.

P10: Let $x_1, \ldots, x_n$ be a sample from P. Let $y(\varepsilon, n)$ denote an ε-mode of $x_1, \ldots, x_n$ (i.e., with respect to the empirical distribution on $x_1, \ldots, x_n$.) Suppose $\varepsilon \to 0$ as $n \to \infty$. When do the sample ε-modes of widths approaching c have as limit points the population modes of width c? What is the asymptotic distribution of a sample ε-mode?

By analogy with asymptotic results for kernel estimation, one would expect, under general conditions, that the sample ε-modes are consistent and asymptotic normal. Choice of ε is a difficult practical problem; in general, every x_i is an ε-mode, so it is also necessary to select a width c below which observations are not considered as modes. The observation y_1 maximizing $a(y,\varepsilon)$ for the sample has maximum width $\sup_x d(x, y_1)$. The mode of next greatest width, say y_2 might be used in testing for the presence of two clusters.

P11: Let y_1 and y_2 be the sample ε-modes of largest and second largest width. For samples from a unimodal P , what is the asymptotic distribution of the second largest width?

I speculate that this second largest width approaches zero in probability as $n \to \infty$, $\varepsilon \to 0$ for a unimodal P but approaches a positive number for a bimodal P . In selecting a worst case P , the values of $a(y_1, \varepsilon)$ and $a(y_2, \varepsilon)$, which indicate the density at y_1 and y_2 , should be used. For example in the univariate case, the worst case for euclidean distance is probably a density taking a constant value in an interval including y_1 , and taking another constant value in an interval including y_2. In the univariate case, use of the second largest width is similar to the test for bimodality proposed in Section 4.

11. Trees and ultrametrics.

A tree τ on a set of points $\mathcal{X}$ is a family of subsets (clusters) such that

(i) $\mathcal{X} \in \tau$

(ii) $S_1, S_2 \in \tau \Rightarrow S_1 \supset S_2$ or $S_2 \supset S_1$ or $S_1 \cap S_2 = \varphi$.

A graph theoretic tree may be used to represent the clusters, one nodc corresponding to each cluster and $\mathcal{X}$ corresponding to the root. The widespread use of trees in clustering derives from the success in biology of a tree classification, which also has evolutionary significance. But one wonders when 100 innocent cases are passed into one of the standard programs to produce 99 clusters; perhaps the statistician can be of assistance in pruning back the 99 to the 1, 2, or 3 usually justified by the data.

Many of the standard algorithms assume a distance matrix given. This is a big assumption. In this context, a tree may be viewed as defining an ultrametric distance d^* , satisfying $d^*(x,y) = \sup[d^*(x,z), d^*(y,z)]$, and the clustering technique is viewed as a function from an input distance matrix to an output ultrametric or tree. The ultrametric idea was advanced simultaneously in Jardine and Sibson (1967), Johnson (1967) and Hartigan (1967). Jardine and Sibson (1971) show that the clustering function has certain continuity properties only if it corresponds to the single linkage technique. Single linkage clusters are, for each $\delta > 0$, the minimal sets S such that $x \in S$ whenever $d(x,y) \leq \delta$ some y in S. (Thus $x \in S$ if there is a single link to an element of S).

12. Joining algorithms.

A variety of joining algorithms may be considered other than single linkage. In all of these distances are assumed given between objects, and the closest pair are joined to form a cluster, which is treated as a single object in further joins by an amalgamation rule which defines its distances to other objects. Suppose that i and j are joined to form ij.

(i) Single linkage: $d(ij,k) = \min d(i,k), d(j,k)$

(ii) Complete linkage: $d(ij,k) = \max d(i,k), d(j,k)$

(iii) Average linkage: $d(ij,k) = \frac{1}{2}[d(i,k) + d(j,k)]$
(unweighted pair group)

(iv) Weighted average linkage $d(ij,k) = \dfrac{n_i d(i,k) + n_j d(j,k)}{n_i + n_j}$

where n_i is the number of original objects in the cluster i.

The single linkage tree is related to the minimum spanning tree, the tree of minimum length connecting the objects, Gower and Ross (1970). Given the minimum spanning tree, the single linkage clusters are obtained by deleting links from the MST in order of decreasing length; the connected sets after each deletion are single linkage clusters. Despite its nice mathematical properties, the single linkage tree is widely regarded with disfavour for "chaining", tending to produce no clearcut divisions into clusters. Perhaps there are no clusters there, perhaps single linkage is not an effective means of discovering them. These algorithms and also the k-means algorithms are compared in Fisher and Van Ness (1971); asymptotic properties are not considered. Monte-Carlo comparisons are given in Kuiper and Fisher (1975). Baker (1974), Baker and Hubert (1975), and Hubert (1974) compare single linkage with complete linkage in a number of Monte Carlo studies; in general single linkage seemed less effective in recovering an ultrametric contaminated by errors. The measure of effectiveness used was the rank correlation between the true and fitted ultrametric. This measure is similar to one proposed by Sokal and Rohlf (1962): the correlation coefficient between the input distances and output ultrametric is to be used as the measure of effectiveness of the clustering. Farris (1969) suggests that the "unweighted pair group" clustering technique should give the largest value to this coefficient. I disagree. Suppose that the objects have been grouped into three clusters 1, 2, 3, containing n_1, n_2, n_3 objects, and suppose that the average distance between objects in clusters i and j is d_{ij}. Then the optimal grouping of clusters 1, 2, 3 into two clusters joins i,j to minimize $(d_{ik} - d_{jk})^2/(n_i+n_j)$ where i, j, k is a permutation of 1, 2, 3. This corresponds to none of the standard joining techniques.

In any case, the choice of the correlation coefficient as a criterion makes a particular technique look favourable, and this may be happening in the Hubert and Baker studies; it may be that choice of the rank correlation makes complete linkage look superior to single linkage.

An exact distribution theory, under the null hypothesis that the $n(n-1)/2$ distances between n objects are equally likely to be in any of $[n(n-1)/2]!$ orders, is given by Ling (1973). This theory is related to work by Erdos and Renyi (1961) and others on the evolution of random graphs as links are added randomly. Anything exact in clustering is rare and valuable but one should be uncomfortable about applying these distributions in testing. When no clustering is present in the data, such as for observations from a multivariate normal, the distances will be far from uniformly distributed over all permutations.

13. Density contour models.

Let $\mathcal{X}$ be p-dimensional euclidean space, and let a probability distribution P on $\mathcal{X}$ have density f with respect to Lebesgue measure on $\mathcal{X}$.

A c-cluster S is a maximal connected set such that $f(x) \geq c$, $x \in S$. It is easy to show that the set of all c-clusters forms a tree. Also each mode of the density f is the limit of a decreasing sequence of such clusters. These density contour clusters are discussed in Hartigan (1975). A c-cluster S is said to be rigid if for every $c' > c$, there is no unique c'-cluster contained in S. The rigid clusters form a tree, and there is a single point rigid cluster for each mode.

For a hierarchical clustering τ, and any set of points A, let $\tau(A)$ denote the member of τ which is the minimal cluster including A. Consider a hierarchical clustering τ_n defined on a sample of points $x_1, \ldots, x_n$ from $\mathcal{X}$, and let τ be a hierarchical clustering on $\mathcal{X}$. For any subset A of $\mathcal{X}$, let $A_n = A \cap \{x_1, \ldots, x_n\}$. Say that τ_n is consistent for τ if for any A, B in τ, $A \cap B = \varphi$, $P(\tau_n(A_n) \cap \tau_n(B_n) = \varphi) \to 1$ as $n \to \infty$. Of course $A \subset B \Rightarrow A_n \subset B_n \Rightarrow \tau_n(A_n) \subset \tau_n(B_n)$, so the limit result means that the tree relationships in τ_n converge to the tree relationships in τ. For example, one statement of τ is of the form: x and y are more similar to each other than to z. If x,y,z appear in the sample, eventually τ_n will make the same similarity judgment.

P12: For what clustering models τ, and clustering algorithms τ_n, is τ_n consistent for τ?

I suspect that the single linkage tree τ_n is consistent for the density contour tree τ for quite general densities. Possibly a related result is that of Cover and Hart (1967) who show that classifying a new observation by the nearest neighbour rule (the observation is classified into the same group as the closest already classified observation) leads to a classification error rate at most twice the optimum error rate.

14. One dimensional data, single linkage.

Suppose $x_1,\ldots,x_n$ are real observations from a density f. Let $g_1,g_2,\ldots,g_{n-1}$ be the lengths of intervals or gaps, between the order statistics $g_i = x_{(i+1)} - x_{(i)}$. The single linkage algorithm may be expressed as follows: divide the observations into two clusters $x_{(1)},\ldots,x_{(i)}$ and $x_{(i+1)},\ldots,x_{(n)}$ where g_i is the largest gap. Divide these clusters into further clusters, again at the largest gaps. The clusters obtained this way are the single linkage clusters. The essential element is thus the largest gap.

P13: Let $x_1,\ldots,x_n$ be observations from a density f on $[0,1]$. What is the distribution of the largest gap, $g_i = x_{(i+1)} - x_{(i)}$, $1 \leq i \leq n-1$?

This is a familiar problem in the theory of order statistics. If $f = 1$, $\sup g_i$ is distributed as

$$\sum_{i=1}^{n-1} (e_i/i) \Big/ \sum_{i=1}^{n+1} e_i = Z_n ,$$

where the e_i are independent exponentials (using the representations $e_j/\sum_{i=1}^{n+1} e_i$ for the gaps g_j).

If $a \leq f \leq b$, $0 \leq x \leq 1$, $x_i = F^{-1}(u_i)$ where the u_i are from the uniform and F^{-1} has a derivative between $1/b$ and $1/a$. Thus $Z_n/b \leq \sup g_i \leq Z_n/a$ in distribution. ($X \leq Y$ means that $P[X \leq x] \geq P[Y \leq x]$ for every x.) As $n \to \infty$, Z_n has the extreme value distribution given by $P[nZ_n - \log n < z] \to e^{-e^{-z}}$. Thus asymptotically

$$P[\sup g_i \leq [z + \log n]/bn] \leq e^{-e^{-z}} \leq P[\sup g_i \leq [z + \log n]/an].$$

P14: Let $x_1,\ldots,x_n$ be observations from a continuous density f. Suppose that the minimum of the density f in $[\alpha,\beta]$ is at x_0. Let $g_n[\alpha,\beta]$ be the largest gap of $x_1,\ldots,x_n$ occurring in the interval $[\alpha,\beta]$. What is the asymptotic location and distribution of $g_n[\alpha,\beta]$?

By a similar argument to that following P13, it may be shown that asymptotically

$$P[g_n(\alpha,\beta) \le (z + \log n)/bn] \le e^{-e^{-z}} \le P[g_n(\alpha,\beta) \le (z + \log n)/an]$$

where

$$b = \sup_{\alpha \le x \le \beta} f(x), \qquad a = \inf_{\alpha \le x \le \beta} f(x) .$$

If $[\alpha_1, \beta_1]$ and $[\alpha_2, \beta_2]$ are two intervals such that

$$\sup_{\alpha_1 < x < \beta_1} f(x) < \inf_{\alpha_2 < x < \beta_2} f(x) ,$$

it follows that $g_n[\alpha_1, \beta_1] > g_n[\alpha_2, \beta_2]$ in probability as $n \to \infty$. Thus for each $\varepsilon > 0$, the maximum gap $g_n(\alpha, \beta)$ is located eventually in the interval $[x_0 - \varepsilon, x_0 + \varepsilon]$ - <u>the location of the maximum gap converges to the location of the minimum density.</u>

There is a crude argument that $ng_n(\alpha,\beta) f(x_0) - \log n + \frac{1}{2} \log \log n + \frac{1}{2} \log [\sqrt{2\pi}\, f''(x_0)/f^3(x_0)]$ has asymptotic c.d.f. $e^{-e^{-z}}$, provided $f(x_0) \neq 0$, $f''(x_0) \neq 0$. There is another crude argument that $g_n(\alpha,\beta)$ is located $0(\log n)^{\frac{1}{2}}$ from x_0, so that the convergence is not very rapid.

P15: <u>Let</u> $x_1, \ldots, x_n$ <u>be observations from a continuous density</u> f. <u>Let</u> $\alpha_1 < \beta_1 < \alpha_2 < \beta_2$ <u>be such that</u> $[\alpha_1, \beta_1]$ <u>and</u> $[\alpha_2, \beta_2]$ <u>form disjoint density contour clusters for</u> f. <u>Let</u> A_n <u>be the smallest single linkage clusters containing all</u> x_i, $\alpha_1 < x_i < \beta_1$ <u>and let</u> B_n <u>be the smallest single linkage cluster containing all</u> x_i, $\alpha_2 \le x_i \le \beta_2$. <u>Then</u> A_n <u>and</u> B_n <u>are eventually disjoint in probability.</u>

This shows that the single linkage algorithm is consistent for density contour clusters.

To prove the result, note that for some (α, β)

$$\beta_1 < \alpha < \beta < \alpha_2, \quad \sup_{\alpha \le x \le \beta} f(x) < \inf_{\alpha_1 < x < \beta_1} f(x), \quad \inf_{\alpha_2 \le x \le \beta_2} f(x) ,$$

for otherwise $[\alpha_1, \beta_1]$ and $[\alpha_2, \beta_2]$ would not be disjoint contour clusters.

Now

$$n\, g_n(\alpha_i, \beta_i)/\log n \le (1+\varepsilon) \sup_{\alpha_i \le x \le \beta_i} f$$

$$n\, g_n(\alpha, \beta)/\log n > (1-\varepsilon)/ \sup_{\alpha \le x \le \beta} f$$

with probability 1 as $n \to \infty$. Thus the largest gap in (α, β) exceeds the gaps in (α_1, β_1) and (α_2, β_2) in probability as $n \to \infty$. Thus, in constructing single linkage clusters, a split will occur in (α, β) before (α_1, β_1) or (α_2, β_2) so that they will be disjoint, in probability as $n \to \infty$.

Intuitively, bimodality occurs in one dimension if there is a large gap with numerous small gaps on either side of it. Define $G_i^+ = [x_{(i+1)} - x_{(i)}]/[x_{(j)} - x_{(i)}]$ where $x_{(j)} - x_{(j-1)}, x_{(j-1)} - x_{(j-2)}, \ldots, x_{(i+2)} - x_{(i+1)} \le x_{(i+1)} - x_{(i)}$ but $x_{(j+1)} - x_{(j)} > x_{(i+1)} - x_{(i)}$. In the uniform case, G_i^+ has distribution function, given $k = j - i$,

$$F^+(a) = P[G_i^+ < a] = \sum_{r=0}^{k} (-1)^{r+1} r\binom{k+1}{r+1} \{1 - [\frac{1-ra}{1+a}]^{+(k-1)}\}.$$

This is established by standard arguments for uniform order statistics, such as in David (1970). A ratio G_i^- for gaps to the left of $x_{(i+1)} - x_{(i)}$ is similarly defined. The gap $[x_{(i+1)} - x_{(i)}]$ is chosen which maximizes $p_i = \min[F^+(G_i^+), F^-(G_i^-)]$ and $\max p_i$ is used as a test statistic for bimodality. What is the asymptotic distribution of $\max p_i$ in the uniform case? Another test based on gaps has been proposed by J. B. Kruskal in Giacomelli et al. (1971).

15. <u>One dimensional data, complete linkage.</u>

After complete linkage joining has operated on $x_1, \ldots, x_n$ for a certain time, the clusters will consist of intervals $[\alpha_1, \beta_1], [\alpha_2, \beta_2], \ldots, [\alpha_k, \beta_k]$ where $\alpha_i \le \beta_i \le \alpha_{i+1}$. If $\ell = \sup[\beta_i - \alpha_i]$, then necessarily $[\beta_i - \alpha_{i-1}] \ge \ell$, otherwise $[\alpha_{i-1}, \beta_{i-1}]$ and $[\alpha_i, \beta_i]$ would already be joined. If f has positive density, asymptotically the distance $\alpha_{i+1} - \beta_i$ will be negligible; the joining process may be modelled as follows: let $\ell_1, \ell_2, \ldots, \ell_k$ be a sequence of lengths; join ℓ_i to ℓ_{i+1} if $\ell_i + \ell_{i+1}$ is a minimum; continue until a single length remains.

P16: <u>Let</u> $\ell_1, \ell_2, \ldots, \ell_n$ <u>be a random sample from some distribution</u> P. <u>Join the</u> ℓ_i <u>as specified; the final join will combine</u> $\ell_1 + \ldots \ell_j$ <u>to</u> $\ell_{j+1} + \ldots + \ell_k$. <u>What is the asymptotic distribution of</u> j <u>as</u> $n \to \infty$?

It would also be interesting to know what the distribution of the ℓ_i is after a large amount of joining. I suspect that it does not depend asymptotically on P, nor does the asymptotic distribution of j.

After much joining, we know that for every ℓ_i, ℓ_{i+1}, $\ell_i + \ell_{i+1} \geq \sup \ell_i$; thus the sequence ℓ_i consists of a run of elements of length $\geq \frac{1}{2} \sup \ell_i$, an element of length $\leq \frac{1}{2} \sup \ell_i$, another run of elements of length $\geq \frac{1}{2} \sup \ell_i$ and so on. This suggests that the lengths ℓ_i will all have about the same value asymptotically, and that j will be approximately $n/2$.

P17: Let $x_1, \ldots, x_n$ be sampled from a distribution with continuous positive density f on [0,1]. Let $[x_{(1)}, x_{(j)}]$, $[x_{(j+1)}, x_{(n)}]$ be the two complete linkage clusters which are last to join. What is the asymptotic distribution of $x_{(j)}$?

I suspect that $x_{(j)}$ has an asymptotic distribution which does not depend on f. For large n, I suspect more generally that the last k clusters to be joined by complete linkage do not depend on f. Thus the large complete linkage clusters are uninformative about the density f. Complete linkage is not completely worthless, because it will detect two intervals of positive density sufficiently widely separated by an interval of zero density. However consider the case $f = 1$ for $0 < x < .9$, $f = 1$ for $1.0 < x < 1.1$, $f = 0$ elsewhere. There is at least a 50% chance that the final join will occur for $0 < x < .9$, rather than for $.9 \leq x \leq 1.0$ as indicated by the density contour tree. See Figure 4.

Complete linkage is, I think, favoured over single linkage because it splits the data up more neatly. Single linkage is a neglected Cassandra, telling us there are no clusters there; we prefer the good news of complete linkage which shows fine even splits regardless of the true density.

Figure 4. Complete linkage fails to split in interval of zero density for sample of size 20 from uniform over (0,.9), (1,1.1).

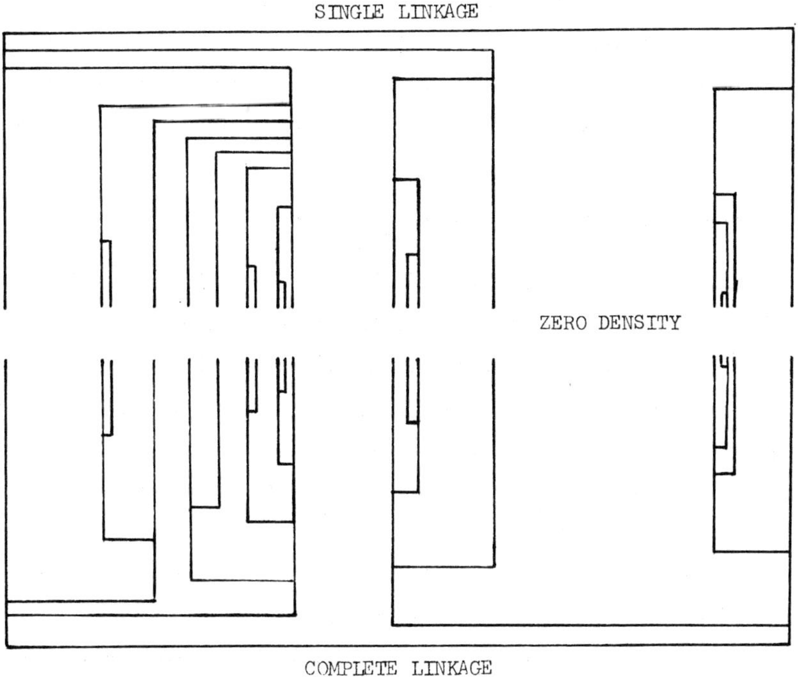

16. <u>One dimensional data, average linkage.</u>

The unweighted average joining technique, in the one dimensional case, may be rephrased:

(i) find the closest pair of points

(ii) delete points and replace by the average

(iii) continue until a single join remains.

P18: <u>Let</u> $x_1,\ldots,x_n$ <u>be sampled from a continuous positive density</u> f <u>in</u> [0,1]. <u>Let</u> $[x_{(1)},x_{(j)}]$ <u>and</u> $[x_{(j+1)},x_{(n)}]$ <u>be the average linkage clusters before the final join. What is the asymptotic distribution of</u> $x_{(j)}$?

After a certain amount of joining, one expects that the cluster centers will be approximately uniformly distributed over (0,1].

The large average linkage clusters do not depend on the density f.

Thus average linkage acts about the same as complete linkage; it can detect intervals of positive density well separated by intervals of zero density. But as with complete linkage it will not surely separate $f = 1$ for $0 < x < .9$, $f = 1$ for $1 < x < 1.1$, $f = 0$ elsewhere into clusters $0 < x < .9$ and $1 < x < 1.1$. K-means is similarly inconsistent for density contour clusters.

18. p-dimensional data, minimal spanning tree.

P19: Let $x_1, \ldots, x_n$ be sampled from a continuous density f in R^p. Let $g(x_0)$ be the link in the minimum spanning tree which is closest to some point x_0. What is the asymptotic distribution of $g(x_0)$?

By analogy with the one dimensional case, one would expect that $f(x_0)g^p$ is exponential with parameter independent of x_0, $f(x_0) > 0$. One would also expect that $g(x_1)$ and $g(x_2)$ are asymptotically independent for $x_1 \neq x_2$. Rohlf (1975) suggests that the distribution of the ordered minimum spanning tree links is like that of an ordered sample from a gamma with suitably selected parameters.

The gap test for bimodality described in §14 is extended to the minimum spanning tree as follows:

(1) Let g be an arbitrary link in the MST connecting say x_1 and x_2.

(2) Let $g_{i_1}, \ldots, g_{i_r}$ be the set of links connected to x_1 by links less than g.

(3) Let $G_1^p = \Sigma g_j^p$, where j runs over $i_1, \ldots, i_r$ such that $g_{i_k} \leq g$.

(4) Let $P_1(g/G_1)$ denote the probability of observing a gap ratio less than g/G if g^p and $g_{i_1}^p, \ldots, g_{i_r}^p$ are independent exponentials; $P_1(g/G_1)$ will depend on r and on the number of $g_{i_k} > g$.

(5) Define $P_2(g/G_2)$ analogously.

(6) Accept g as a real splitting point if $P_1(g/G_1)$ and $P_2(g/G_2)$ are both sufficiently high.

The test statistic $\max_g \min[P_1(g/G_1), P_2(g/G_2)]$ has an unknown distribution which must be explored, for uniform populations.

19. Density estimates.

The single linkage algorithm may be viewed as a density estimation procedure as follows - construct the minimum spanning tree, and estimate the distribution to lie entirely on the minimum spanning tree, distributed uniformly over its length! The contour clusters for this distribution are the single linkage clusters.

A smoothed density estimate might lead to improved and fewer clusters. For example, estimate the density at each point by the number of points within ε of the point. Estimate the density on the MST links by interpolation between points, and construct the contour clusters for this density. Procedures similar to this are discussed by Wishart [1974] and Hartigan [1975].

References

Barlow, R. E., Bartholomew, D. J., Bremmer, J. M., and Brunk, H. D., Statistical Inference under Order Restrictions, John Wiley and Sons, New York, 1972.

Baker, F. B., Stability of Two Hierarchical Grouping Techniques; Case I: Sensitivity to Data Errors, Journal of the American Statistical Association, 69 (1974), 440-445.

Baker, F. B., and Hubert, L. J., Measuring the Power of Hierarchical Cluster Analysis, Journal of the American Statistical Association, 70 (1975), 31-38.

Cover, T. M., and Hart, P. E., Nearest Neighbour Pattern Classification, IEEE Transactions on Information Theory, IT-13 (1967), 21-27.

David, H. A., Order Statistics, John Wiley and Sons, New York, 1970.

Day, N. E., Estimating the Components of a Mixture of Normal Distributions, Biometrika 56 (1969), 463-474.

Dick, N. P., and Bowden, D. C., Maximum Likelihood Estimation for Mixtures of Two Normal Distributions, Biometrics, 29 (1973), 781-790.

Eddy, William F., Optimum Kernel Estimators of the Mode, Unpublished Ph. D. dissertation, Department of Statistics, Yale University, May 1976.

Engelman, L., and Hartigan, J. A., Percentage Points of a Test for Clusters, Journal of the American Statistical Association, 64 (1969), 1647-1648.

Erdos, P., and Renyi, A., On the Evolution of Random Graphs, Bulletin de l'Institut Internationale de Statistique Tokyo, 38 (1961), 343-347.

Farris, J. S., On the Cophenetic Correlation Coefficient, Systematic Zoology, 18 (1969), 279-285.

Fisher, Walter D., On Grouping for Maximum Homogeneity, Journal of the American Statistical Association 53 (1958), 789-798.

Fisher, L., and Van Ness, J. W., Admissable Clustering Procedures, Biometrika, 58 (1971), 91-104.

Giacomelli, F., Wiener, J., Kruskal, J. B., Pomeran, J. W., and Loud, A. V., Subpopulations of blood lymphocytes and demonstrated by quantitative cytochemistry, Journal of Histochemistry and Cytochemistry, 19 (1971), 426-433.

Hartigan, J. A., Representation of similarity matrices by trees, Journal of the American Statistical Association, 62 (1967), 1140-1158.

Hartigan, J. A., Clustering Algorithms, John Wiley and Sons, New York, 1975.

Hartigan, J. A., Asymptotic Distributions for Clustering Criteria, Unpublished, 1976.

Hubert, L., Approximate Evaluation Techniques for the Single Link and Complete Link Hierarchical Clustering Procedures, Journal of the American Statistical Association, 69 (1974), 698-704.

Jardine, C. J., Jardine, N., and Sibson, R., The Structure and Construction of Taxonomic Hierarchies, Mathematical Biosciences, 1 (1967), 173-179.

Jardine, N., and Sibson, R., Mathematical Taxonomy, John Wiley and Sons, London, 1971.

Johnson, S. C., Hierarchical Clustering Schemes, Psychometrika, 32 (1967), 241-254.

Kuiper, F. K., and Fisher, L., A Monte Carlo Comparison for Six Clustering Procedures, Biometrics, 31 (1975), 777-784.

Ling, R. F., A Probability Theory of Cluster Analysis, Journal of the American Statistical Association, 68 (1973), 159-169.

MacQueen, J., Some Methods for Classification and Analysis of Multivariate Observations, in Proceedings of the Fifth Berkeley Symposium on Mathematical Statistics and Probability, L. Le Cam and J. Neyman, eds., University of California Press, Berkeley and Los Angeles, 1967, 281-297.

Rohlf, F. J., Generalization of the Gap Test for Multivariate Outliers, Biometrics, 31 (1975), 93-101.

Scott, A. J., and Knott, M., Cluster-analysis Method for Grouping Means in Analysis of Variance, Biometrics 30 (1974), 507-512.

Scott, A. J., and Symons, M. J., Clustering Methods Based on the Likelihood Ratio Criteria, Biometrics, 27 (1971), 387-397.

Sokal, R. R., and Rohlf, F. J., The Comparison of Dendrograms by Objective Methods, Taxonomy 11 (1962), 33-39.

Wegman, E. J., Nonparametric Probability Estimation, I, Technometrics, 14 (1972), 533-546.

Wishart, D., A Generalization of Nearest Neighbor which Reduces Chaining Effects, in Numerical Taxonomy, A. J. Cole, ed., Academic Press, London, 1969.

Wolfe, J. H., Pattern Clustering by Multivariate Mixture Analysis, Multivariate Behavioural Research, 5 (1970), 329-350.

This research was supported in part by National Science Foundation Grant DCR75-08374.

Department of Statistics
Yale University
New Haven, Connecticut 06520

The Botryology of Botryology

I. J. Good

Introduction.

The Greek word βοτρυς means a cluster of grapes. It is the origin of the English prefix botryo- as in the word botryoidal which means resembling a cluster of grapes; for example, the pancreas is botryoidal. Thus the word "botryology", meaning the theory of clusters, is hardly a neologism. It has occurred in a few places in the literature of clustering (e.g. Good, 1962; Needham, 1966; Sneath and Sokal, 1973), but is probably not yet in any English dictionary. It seems to me that the subject of clustering is now wide enough and respectable enough to deserve a name like those of other disciplines, and the existence of such a name enables one to form adjectives and so on. For example, one can use expressions such as "a botryological analysis" or "a well-known botryologist said so and so". There is another word that serves much the same purpose, namely "taxonomy", but this usually refers to biological applications whereas "botryology" is intended to refer to the entire field, provided that mathematical methods are used. The subject is so large that it might not be long before there are professors and departments of botryology. Another possible name would be aciniformics, but it sounds inelegant. On the other hand "agminatics" is a good contender, forming "agminaticist", etc.

Most studies of clustering do not make much reference to the time dimension in a genuinely dynamic manner, although there have been applications in historical and archealogical research which refer to time. One could imagine an investigation of the behavior of an artificial neural network in which the clusters were assemblies and subassemblies of reverberating neurons. Here of course the clusters vary with time. Similarly a sociologist might study the formation of groups of people by using some general theory from the field of botryology. I mention these two examples because they are very far from the classical taxonomic problem of classifying plants or animals into a logical tree.

Bertrand Russell's definition of a thing was botryological. He said "Thus 'matter' is not part of the ultimate material of the world, but merely a convenient way of collecting events into bundles". (Russell, 1946, p. 861.)

Botryology can be regarded as a contribution to the subject of hypothesis formulation. Whenever we say that an object belongs to a cluster or a clump we are in effect stating a hypothesis, especially if the clump has a clear interpretation.

I shall quote here partly verbatim what I have published before on the definition of a definition because I think it reveals the comparisons and contrasts between most of the work published under the name of clustering or numerical taxonomy on the one hand, and the concept of botryology in general. (Good, 1962, pp. 124-5; 1965b, p. 42.)

Whenever one introduces a new word or definition one is attempting to identify a new clump. The philosopher G. E. Moore emphasized that the meanings of words are much less clear-cut than is sometimes imagined. Later John Wisdom emphasized that we call an object a cow if it has enough of the properties of a cow, with perhaps no single property being essential. Thus the notion of a cow can be regarded as "polythetic". (For the history of this word see Sneath and Sokal, 1973, p. 20.) It has seemed to me to be worthwhile to convert "Wisdom's cow" into a probabilistic form, both for its philosophical interest and for future elaborate information retrieval systems. An object is said to have credibility π of belonging to class C (such as the class of cows) if some function $f(p_1, p_2, \ldots, p_m) = \pi$, where the p's are the credibilities (logical probabilities) that the object has qualities $Q_1, Q_2, \ldots, Q_m$. These probabilities depend on further functions related to other qualities, on the whole more elementary, and so on. A certain amount of circularity is typical. For example, a connected brown patch on the retina is more likely to be caused by the presence of a cow if it has four protuberances that look like biological legs than if it has six; but each protuberance is more likely to be a biological leg if it is connected to something that resembles a cow rather than a table. In view of the circularity in this interpretation of 'definition', the stratification in the structure of the cerebral cortex can be only a first approximation to the truth. (See also Hayek, 1952, p. 70.)

The function f here is unconstrained and this is what makes the definition of "definition" qualitatively distinct from the work on numerical taxonomy where the functions f tend to be linear or quadratic. The qualitative distinction between linear and quadratic functions on the one hand and more complicated ones on the other corresponds roughly to the distinction between numerical taxonomy and general concept formation which includes the more difficult aspects of pattern recongition. If the problem could be solved of automatically determining the function f in any given situation then I believe the problem of pattern recognition would be solved, and this would be a major breakthrough in the field of machine intelligence also, since any description can be expressed in functional notation.

This definition of a definition is both iterative and circular. Obviously it is not possible to define all words in terms of no words at all without reference to the real world and therefore circularity of definitions is inevitable. (In practice subjective probabilities must be used and they are only partially ordered, but I have ignored these further complications.)

When a person recognizes a cow he presumably does it without being conscious of his methods. It seems quite reasonable to suppose that the brain carries out an iterative calculation of the kind just described. It is possible, although it may not be very likely in most situations, for the brain to temporarily lock in to a local solution and then to switch to another solution. This happens, for example, in the well known psychological experiment in perception involving Necker's cube (see Figure 1). A similar phenomenon of temporary locking in occurs if one listens to a word being repeated again and again: after a time one hears words that are not objectively there, a fact also well known to the

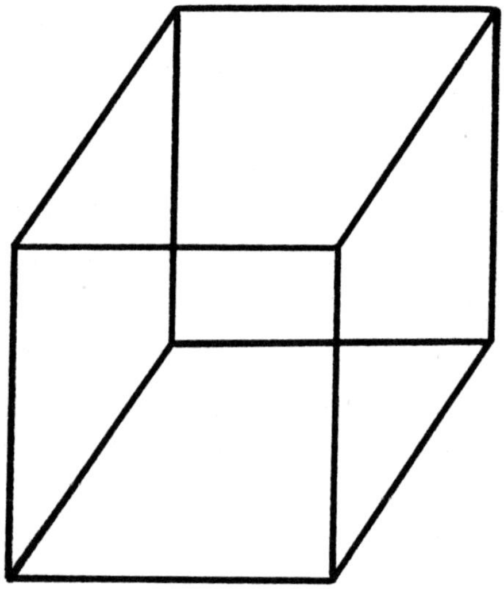

Figure 1. Necker's cube.

experimental psychologist. The reason for the switching is presumably that subassemblies of neurons can suffer from fatigue and break up, thus allowing a distinct solution to be obtained. (Cf. Good, 1965b.) It would be interesting to experiment with an analogous procedure in a numerical iterative calculation. It would be possible to have certain nodes in the calculation where some branches are preferred to others at certain times but in which the use of a branch leads to a decrease in the probability of its use soon afterwards. This would enable the calculation to branch out in different directions on different occasions, and thus to reach more than one point of metastability.

If the function f is chosen in some arbitrarily complicated manner, it is unlikely to enter into the definitions of other things or concepts previously regarded as interesting. Only if variations in the value of f do cause variations in the probabilities of other interesting things or concepts do we regard f as a candidate for defining an interesting thing or concept itself. When we decide that f is a candidate we attempt to modify f so that its value has almost its maximum effect on the probabilities of other interesting things or concepts. At this stage we regard f as itself defining an interesting thing or concept.

A distinction is sometimes made between a cluster and a clump: a cluster is usually spherical or ellipsoidal in attribute space, whereas a clump can be of any shape. Thus the general definition of a definition is apt to lead to a clump rather than a cluster if this terminology is adopted.

If a search is to be made for "interesting" clumps or functions f we need a test for deciding whether the clump obtained is statistically significant. This necessarily depends on the prior probability that the function f will define a thing or concept and this is related to some extent to its complexity. (Only to some extent: "0 = 1" is simple but improbable: see Good, 1974.) Thus an important unsolved problem is to find ways of choosing prior probabilities of functions or of hypotheses or theories. This problem of choosing prior probabilities of hypotheses and theories is the basic difficulty in the Bayesian approach to statistics and in the philosophy of scientific induction. The fact that it is difficult does not mean that there is any other satisfactory solution to these problems.

These remarks should be sufficient to show how wide the subject is, and even within the somewhat narrower but still wide field of numerical taxonomy there are many botryological techniques. Techniques are apt to be invented in different fields of application without people noticing the duplication of the work. Since one of the purposes of botryology is to overcome this kind of duplication by detecting clusters it is surprising that more people have not said "Botryologist, clump thyself!" My purpose in this talk is to propose this as a research problem rather than to solve it. A natural plan is (i) to make a list of all the different attributes or qualities that a clustering technique might have and the list at the end of this paper leans in this direction; (ii) to determine for each published technique which of these qualities it has; (iii) to apply each of the techniques to the entire set of information thus obtained so as to put the techniques into clusters; (iv) to use some careful judgment to decide which of these clusters are useful. Perhaps a number of distinct solutions would be obtained each of which could be the basis of yet another book on botryology. Each botryological technique T would lead to a collection of clusters $C_1(T)$, $C_2(T)$, $C_3(T)$, ... Interesting questions would then arise; for example, if the techniques T and U both belong to $C_i(T)$ would they usually both belong to $C_j(U)$ for some j?

Purposes of Botryology.

To give a short list of some of the applications and purposes of botryology I shall lean heavily on Good (1965a).[†]

(i) For avoidance of fear of the unknown. This application is so familiar that it is apt to be overlooked. It would be terrifying if you suddenly could not classify your sensory input. If someone in a turban and white cloak then told you it was merely the Korbensky effect, the experience would join the familiar clump of named experiences so it would become less terrifying.

(ii) For mental clarification and communication.

(iii) For discovering new fields of research.

(iv) For planning an organizational structure, such as a university.

(v) For planning the structure of a machine. For example, in designing the elementary instructions of a computer one should first

---

[†] Permission to quote extensively from this paper has been kindly granted by the British Medical Research Council.

categorize the elementary arithmetic and other operations, the attributes being micro-operations. One could imagine a computer carrying out a botryological program for this purpose and then microprogramming itself! In effect it would change itself into a distinct machine. Perhaps in a few decades machines will modify themselves in this manner as a matter of course.

(vi) For concept formation in ordinary life (clustering) and for recognition (identification and diagnosis).

(vii) For probability estimation from effectively small samples. For example, consider the problem of determining the probability that a patient with a certain disease D has a certain collection of indicants $I_1, I_2, \ldots$. The number of indicants may be so large that, for any reasonable sample, there will be no previous example of a patient having that particular set of indicants. One way of solving the problem of probability estimation in this case is to cluster the patients within the sample. Then one could regard a patient with a certain set of indicants as belonging to a cluster containing a reasonable number of previous patients. One could then associate a probability of having the disease with each cluster. This method could be refined by allowing for the probability of a patient's belonging to each of a set of clusters, and also by allowing overlapping clusters. Much work needs to be done in this area.

(viii) For information retrieval.

(ix) For pattern recognition, e.g. optical character recognition. Here the need for speed has provoked research in automatic methods.

(x) For word clustering for the understanding of linguistics. The methods suggested for information retrieval apply here also.

(xi) For biological taxonomy. Here the size of the problem and the need for objectivity led to research on automatic methods.

(xii) For medical research.

(xiii) For every other discipline.

The Facets of Botryological Techniques.

Attempts have previously been made to classify the facets of classification, in fact every book on classification must make such an attempt so as to break the subject into chapters and paragraphs. An explicit attempt was made by Good (1965a).

There does not appear to be much discussion on how to categorize the various attributes of botryological techniques. The paper I just mentioned contains some material on this and also section 5.4 of Sneath and Sokal (1973) discusses eight such facets. I shall base the beginning of my discussion on the attributes mentioned in (Good, 1965a), then I shall list the ones not published there that were given by Sokal and Sneath and finally I shall make a list of facets which will include those mentioned in these two references.

The following categorization of categorization is obtained from Good (1965a). The labelling of the various headings is as in the references: more than one "alphabet" is used, to allow for cross-categorizations,

A. Mutually exclusive clumps. In the classification of species of animals or plants, one usually aims at mutually exclusive classes. This is so often possible, owing to so-called 'isolating mechanisms', that one is liable to assume (as did Plato in his Theory of Ideas) that it is always possible, but this would be a mistake. (See, for example, Grant, 1957, p. 58.)

B. Overlapping clumps. For example, when documents are indexed by means of index terms, the clumps of documents corresponding to two index terms will often overlap. One also gets some overlap when two clumps in Euclidean space are specified by, say, two normal distributions. Any cross-categorization, such as a contingency table, involves overlapping clumps if each row and each column is regarded as corresponding to a clump.

1. General-purpose or 'natural'. Such categorization is possible if the clumps are well isolated once enough properties are given. One can then determine the clumps without taking account of all the properties. It is even possible for distinct sets of properties to lead to the same clumps. This often happens with animal species. In such cases there is more scope for objective methods of categorization than when the clumps are not well isolated. It is then possible, as Sneath has emphasized (Sneath, 1965), to ignore phylogenetic evidence, and later to use the categorization into species as a firm basis for phylogenetic research. But when the features of the animals or organisms do not provide overwhelming evidence for the purposes of classification, it is necessary to fall back on mature human judgment to a greater extent. A fair degree of objectivity is possible when one can afford to throw away some evidence, as in other statistical problems.
2. Special-purpose, for example the classification of books by size.

a. Qualitative (intuitive categorization).

b. Quantitative, i.e. 'botryology' .

    b.1 Classical multivariate statistics, assuming normal distributions, and using product-moment correlations. Items are placed in a multidimensional Euclidean space in which the coordinates of an item are taken as equal to measurements of features of that item. The assumption of normal distributions is mathematically convenient but of course not necessarily valid, and is especially awkward when a feature (or facet) requires a discrete measure, such as the presence or absence of an attribute.

b.11 Cluster analysis. Usually cluster analysis starts from a table or matrix of intercorrelations. (See, for example, Tryon, 1938, and Thurstone, 1947.)

b.12 Factor analysis and principal components. The reduction of the dimensionality of the space, and the search for simple structure. These techniques can be used as a prelude to the search for clusters, and also have the following geometrical relationship to cluster analysis. Suppose that a number of points in multidimensional Euclidean space are all found to be close to some hyperplane, then they will form a cluster if they are projected into a subspace orthogonal to this hyperplane. It is geometrically natural to try to find all the subspaces such that if we project all the points into one of these subspaces we get a significant separation into clusters. There will generally be an infinity of such subspaces, but they can be reduced to a finite set by insisting that the measure of significance is a local maximum, i.e. the significance is decreased when the subspace is slightly rotated. (Cf. Rao, 1952, and Healy, 1965, p. 93.)

b.2 Rough-and-ready methods. When the number of items is large, as, for example, in the categorization of animals into species, factor analysis involves a great deal of arithmetic, and there is a premium on rough-and-ready methods, at any rate when there is a great redundancy of evidence. This redundancy of evidence is apt to occur more for higher animals than for plants. The reason is that such animals are living vehicles, and are more complicated than plants. Consequently the constraints on animals are more elaborate. Thus the so-called isolating mechanisms are more effective, in other words species are more easily definable. It can therefore be expected that rough-and-ready methods will be more effective for animals than for plants. Such methods have been discussed especially by Sokal and Sneath (1963), Parker-Rodes (1959) and Needham (1965, p. 111). A simple example of a rough-and-ready method is to replace all correlations by 1, -1, or 0 before picking out the clusters.

b. 3 Space of items not Euclidean. Sometimes it is more natural and convenient to estimate the relatedness of pairs of items than to estimate their individual properties. The items can then be thought of as nodes of a linear graph in which the edges are labelled with measures of relatedness, similarity or relevance. There are many possible measures of relatedness (see, for example, Good, 1958). A categorization into clumps will depend to some extent on what measure of relatedness is used, and also on the definition of a clump, but one hopes that this dependence will not be too sensitive. It was shown by R. N. Shepard (1962) and Kruskal (1964) how one can force the space to be Euclidean by defining a suitable nearly monotonic function of the relatedness measure. They reconstruct the metric so as to embed the points in Euclidean space of small dimensionality. ("Multi-dimensional scaling".) The technique is visually useful if the dimensionality is small enough: and in any case can be used as a prelude to a clumping procedure just like the method of principal components.

b.4 Ad hoc methods. Some methods seem worth trying although they are neither clearly tied to any mathematical model nor rough-and-ready. I shall give an example from information retrieval (Good, 1965b). Suppose that we have n abstracts of documents and w index terms (w for "words"). Let f_{ij} be the frequency with which index term i occurs in document j, and consider the w by n matrix $F = (f_{ij})$. Various botryological computations with F have been proposed: the present one is closest to one used by Needham (1965); however, he was concerned with a square symmetric matrix of frequencies of co-occurrence of index terms, and he did not use logarithms or 'balancing' as described below. One unusual feature of the present method is that it does not make use of measures of relatedness, a feature shared with an independent proposal by Hartigan (1975).

First replace the matrix F by the matrix $(\log(f_{ij} + k))$, where k is a small constant (less than unity). A reason for using logarithms is that the sum of log-frequencies is approximately a log-likelihood. The constant k is introduced to avoid taking the logarithm of zero. The modified matrix is now 'balanced' - that is, we add $a_i + b_j$

to cell (i,j) $(i,j = 1,2,...)$ in such a manner that each row and column adds up to zero. It is easy to evaluate the constants a_i and b_j, and to show that the modified matrix is unique. Let B be the balanced matrix. By means of a simple iterative process we can now find vectors x and y consisting exclusively of 1's , 0's and -1's, in such a manner as to maximize the bilinear form $x'By$, where the prime indicates transposition. (At any stage of the iteration x is updated by computing By and then taking the sign of each non-small component, or y is updated by computing $B'x$, etc. No multiplications are required, only additions and subtractions. At the final stages of the iteration "non-small" is interpreted as "non-zero".) The effect is to separate the words into two large clumps, and the documents into two clumps conjugate to these, as we might say. The words and documents corresponding to small components can be omitted. (With somewhat more arithmetic we can allow the components of x and y to range over all real numbers with $x'x = y'y = 1$.)

Consider one of the two small matrices obtained by extracting from B the rows and columns corresponding to a clump and its conjugate. Balance this matrix and apply the same procedure to it. This will split our clump into two smaller clumps, and will simultaneously split the conjugate clump. (For a significance test, see the Appendix.) In this manner we can continue to dichotomize the clumps until they are of any desired size. The whole collection of clumps would form a tree.

The procedure can easily be modified in order to obtain overlapping clumps if desired, and also we can allow low-scoring index terms and documents to lie outside clumps if we wish.

A similar procedure could be applied with index terms replaced by indicants and documents replaced by people. A disease would correspond to a clump of people and the conjugate clump to the relevant indicants.

The application to information retrieval is as follows: for each document, D_j , and each index term, W_i, we should like to know the Bayes 'factor in favour' of the hypothesis that the document is wanted when the index term is used. (For the terminology see Good, 1950.) But, owing to the large number of index terms and documents, the sample will never be large enough to estimate this factor directly. We therefore propose to estimate the factor by replacing the index term by a whole clump of terms and the document by the conjugate clump of documents. (Cf. Maron and Kuhns, 1960.)

b.a Adaptive categorization. When a procedure is modified in the light of human or machine experience we say that it is adaptive. A simple kind of adaptive procedure is one whose description involves some parameters whose values are gradually optimized. (See, for example, Good, 1959, and Samuel, 1959.) During this modification of the parameters it would be advisable to give more weight to the more recent past, as, for example, in the diagnosis of a disease during an epidemic.

b. b Non-adaptive categorization.

b.α The study of the dependence between two or more categorizations. For example, somatic type and personality.

The following six facets of clustering methods are mentioned by Sneath and Sokal and were not covered in Good (1965a):

Agglomerative versus devisive. In an agglomerative method clusters are built up by appending items to them whereas in a devisive method the clusters are gradually broken up into subclusters. It is like the distinction in sculpturing between working in clay and in marble.

Hierarchical versus nonhierarchical. In a hierarchical technique the different clusters or clumps form a "tree" whereas in a nonhierarchical method all the clumps are "siblings" so to speak.

Sequential versus simultaneous. If in the course of the calculation each stage of reasonable size treats the taxonomic items simultaneously and symmetrically then the calculation is "simultaneous". Most methods of clustering are sequential.

Local versus global criteria. Imagine the various items embedded in some abstract space. If the distance function varies from one part of the space to another then one can describe the clustering process as a local one, otherwise as global.

Direct versus iterative solutions. It is usually intuitively obvious whether a calculation should be regarded as iterative.

Weighted versus unweighted clustering. At various stages in a technique in which items are added sequentially to clusters, the different clusters or different directions can be given different weights.

I shall now make a list of facets which include those already mentioned. I have not succeeded in finding any nice logical order for these facets, but the order is not entirely random. Presumably many further facets could be found.

Apart from the proposed application to putting the various techniques into clumps, the list may be of independent interest and may suggest new ideas.

Some Facets of Botryological Techniques.

Although these facets are often expressed as a disjunction between two or three alternatives, the choice between the alternatives will often not be clear-cut. Hence the components of the vector corresponding to a specific botryological technique will often be more naturally taken as real numbers rather than as say -1, 0, or 1. All facets are fuzzy but some are fuzzier than others.

(i) Based on a "philosophy" or ad hoc.

(ii) General-purpose (or natural) versus special-purpose.

(iii) Uses phylogenetic (evolutionary) knowledge or not (in biological taxonomy).

(iv) Interpretable or not.

(v) "Descriptive", that is, the definition of a clump is analytically explicit but complex; or on the other hand not descriptive. ("Clumps versus clusters".)

(vi) Uses classical multivariate analysis or not.

(vii) Involves the relation between two or more categorizations, versus not doing so.

(viii) Agglomerative versus devisive.

(ix) Permits "chaining" or not. When chaining is permitted, items can be added to a clump so that it might crawl all over the place.

(x) Hierarchical versus nonhierarchical.

(xi) Number of clusters fixed or number flexible.

(xii) With a significance test or not. The notion of "substantialism" is relevant here: an example of it is given in the Appendix. Also we'd like a test for determining the number of clumps.

(xiii) Has mutually exclusive clumps, versus overlapping clumps.

(xiv) Subjective (personal), or objective (by explicit rules), or mixed. Explicit rules are usually selected subjectively! Most methods are mixed even when they are said to be either subjective or objective; it is a matter of degree.

(xv) Non-numerical versus numerical.

(xvi) Qualitative (intuitive) versus mathematical (botryological). (Not quite the same as xv.)

(xvii) Corrigible versus incorrigible. All methods can be made corrigible.

(xviii) Adaptive versus non-adaptive.

(xix) Logical versus probabilistic. If logical, the logic might be fuzzy.

(xx) Weighted facets versus "unweighted". When the attributes are assigned to the items they can be given weights or not. In the latter case the technique is called Adansonian. Really even the Adansonian method is weighted because the attributes that are not used are given a weight of zero.

(xxi) Weighted versus unweighted clustering.

(xxii) Direct versus iterative solutions.

(xxiii) Sequential versus simultaneous.

(xxiv) "Dynamic" (as if gravitational, so that clusters condense like galaxies) or not.

(xxv) Estimates probability densities versus not doing so. Density estimation methods have a large literature and could be categorized; for example, there are parametric methods including the mixing of distributions, window methods (Rosenblatt, 1956; Parzen, 1962), methods based on penalized likelihood (maximization of log-likelihood minus a roughness penalty; Good and Gaskins, 1971, 1972) and histospline methods (Boneva, Kendall and Stefanov, 1971).

(xxvi) If probability densities are estimated, the technique uses modes versus "bumps". A bump on a density curve is a part between two points of inflexion. In more dimensions it can be defined as the part encircled by a curve or surface etc. on which the Gaussian curvature vanishes.

(xxvii) Exhaustive search or not. The number of partitions of a set into subsets is given by the Bell or exponential numbers which increase at an alarming rate. (See, for example, Good, 1975.)

(xxviii) Rough-and-ready versus "accurate".

(xxix) Monothetic or polythetic. (Polythetic means "like Wisdom's cow".)

(xxx) Based on distances or on similarities. The similarities might be judged or might be "objective" (correlations).

(xxxi) Uses only resemblances between items or uses only attributes, or uses both.

(xxxii) "Resemblances" between two items symmetric or not.

(xxxiii) Uses the metric $(\sum x_i^2)^{\frac{1}{2}}$ or the Mahalanobis metric $(x'C^{-1}x)^{\frac{1}{2}}$ (invariant under a linear transformation), where C is an average of the within-cluster covariance matrices, estimated iteratively.

(xxxiv) Metric based on Euclidean geometry or not.

(xxxv) Local versus global metric.

(xxxvi) Genuine metric or not. For example, the "informational" or "evidential" distances between multinomial distributions, $\sum(p_i - q_i)\log\frac{p_i}{q_i}$ (divergence) and $\sum p_i \log\frac{p_i}{q_i}$ (dinegentropy, or Gibbs-Szilard-Shannon-Watanabe-Turing-Good-Jeffreys-Kullback-Liebler dinegentropy, expected weight of evidence, or directed divergence), do not satisfy the triangle inequality.

(xxxvii) Objective function "informational" (related to entropy) or not.

(xxxviii) Uses measurable or countable qualities (e.g. "f_{ij}") or only nominal qualities (attributes).

(xxxix) Uses matrix algebra in a more or less inevitable manner or not. (This is an especially fuzzy facet.)

(xl) Visual display or not, e.g. using the "dendrogram".

(xli) Makes allowance for hierarchical structure of the qualities (or attributes) or not. I don't know how to do this.

(xlii) Clustering of qualities (or attributes) also used or not.

(xliii) One clump being merely a "ragbag" or not.

(xliv) Preceeded by multidimensional scaling or not.

(xlv) Uses graph theory or not.

The use of these 45 facets for the clustering of all published botryological techniques is left as an exercise for the reader.

Appendix

A test of significance for the ad hoc clumping procedure b.4.

If x and y are random vectors whose components are all 1 or -1, then $E(z) = 0$, where $z = x'By$, and $\mathrm{var}(z) = \sum_{ij} b_{ij}^2 = b$, say. The tail-area probability corresponding to a value of z, if x and y are random is close to

$$z^{-1} e^{-\frac{1}{2}z^2/b} (b/2\pi)^{\frac{1}{2}} . \tag{1}$$

If this is much smaller than 2^{-n-w+1}, then the (first stage of the) clumping procedure is statistically significant. (Later stages can be discussed in the same manner, with appropriate changes in the values of n and w.) This is a sufficient condition for significance, but is not a necessary one owing to a phenomenon that may well be called 'substantialism'. This means that a large number of vectors 'close' to x (differing from it only in a small number of components) can be regarded as substantially the same as x, and give rise to large values of z. (The similarity to, and the difference from, either the 'substantialization' of sign sequences or error-correcting codes may be noted: see Good, 1954.) To allow for substantialism it seems reasonable to me to replace expression (1) by its harmonic mean when x and y range over all possible vectors whose components are 1 and -1, when we wish to test the significance of a pair of vectors x, y that maximize $x'By$ (cf. Good, 1958a). The effect of taking the harmonic mean can be shown to lead approximately to the same result as does multiplying expression (1) by

$$(2) \qquad \prod_r \{1 + \exp(- \frac{z}{b} \operatorname{sgn}(x'B)_r)^{-1} \prod_s \{1 + \exp(- \frac{z}{b} \operatorname{sgn}(By)_s)\}^{-1} .$$

There is a factor in expression (2) corresponding to each component of x and each component of y . If a component of x or y can be changed in sign without affecting the value of z , then the corresponding factor in (2) is $\frac{1}{2}$, as it clearly ought to be (because it is as if n or w were decreased by 1).

The notion of substantialism may be of more general value than this exemplification of it.

References

Boneva, L., D. G. Kendall and I. Stefanov (1971), Spline transformations; three new diagnostic aids for the statistical data analyst, J. Roy. Statist. Soc. Ser. B , 33, 1-70 (including discussion).

Good, I. J. (1950), Probability and the Weighing of Evidence, (London: Charles Griffin; New York: Hafners).

Good, I. J. (1954), The substantialization of sign sequences, Acta cryst. 7 , 603.

Good, I. J. (1958a), Significance tests in parallel and in series, J. Amer. Statistic. Assoc. 53 , 799-813.

Good, I. J. (1958b), Speculations concerning information retrieval, Res. Rep. RC-78, IBM Res. Center, Yorktown Heights, New York, pp. 14.

Good, I. J. (1959), Could a machine make probability judgments ?, in Computers and Automation 8, 14-16 and 24-26.

Good, I. J. (1962), Botryological speculations in The Scientist Speculates: An Anthology of Partly-Baked Ideas, (ed. I. J. Good, A. J. Mayne, and J. Maynard Smith; paper back edn., New York: Putnam, 1965), 120-132.

Good, I. J. (1965a), Categorization of classification in Mathematics and Computer Science in Biology and Medicine (London, HMSO and Medical Research Council), 115-125; discussion 126-128.

Good, I. J. (1965b), Speculations concerning the first ultra-intelligent machine in Advances in Computers 6, 31-88 .

Good, I. J. (1974), A correction concerning complexity, British J. Philosophy Science 25, 289.

Good, I. J. (1975), The number of hypotheses of independence for a random vector or for a multidimensional contingency table, and the Bell numbers, Iranian J. Sc. Technology 4, 77-83.

Good, I. J. and R. A. Gaskins (1971), Non-parametric roughness penalties for probability densities, Biometrika 58, 255-277.

Good, I. J. and R. A. Gaskins (1972), Global nonparametric estimation of probability densities, Virginia J. of Science 23, 171-193.

Grant, V. (1957), The plant species in The Species Problem (ed. E. Mayr; American Assoc. Adv. Pub. no. 50), 46.

Hartigan, J. A. (1975), Clustering Algorithms (New York: Wiley).

Hayek, F. A. (1952), The Sensory Order (Chicago: University Press).

Healy, M. J. R. (1965), Descriptive uses of discriminant functions, in Mathematics and Computer Science in Biology and Medicine (London, HMSO), 93-102.

Kruskal, J. B. (1964), Nonmetric multidimensional scaling: a numerical method, Psychometrika 29, 115-129.

Maron, M. E. and J. L. Kuhns (1960), On relevance, probabilistic indexing and information retrieval, J. Assoc. Comp. Mach. 7, 216-244.

Needham, R. M. (1965), Automatic Classification: models and problems in Mathematics and Computer Science in Biology and Medicine (London, HMSO), 111-114.

Needham, R. M. (1966), The termination of certain iterative processes, Memo. RM-5188-Pr, The Rand Corporation, California, pp. 7.

Parker-Rhodes, A. G. (1959), Notes for a prodomus to the theory of clumps (Cambridge Language Research Unit).

Parzen, E. (1962), On estimation of a probability density and mode, Ann. Math. Statist. 33, 1065-1076.

Rao, E. R. (1952), Advanced Statistical Methods in Biometric Research, (New York: Wiley).

Rosenblatt, M. (1956), Remarks on some nonparametric estimates of a density function, Ann. Math. Statist. 27, 832-837.

Russell, Bertrand (1946), History of Western Philosophy (London).

Samuel, A. L. (1959), Some studies in machine learning, using the game of checkers, IBM J. Res. Dev. 3, 210-229.

Shephard, R. N. (1962), The analysis of proximities: multidimensional scaling with an unknown distance function, Psychmetrika 27, 125 and 219.

Sneath, P. H. A. (1965), The application of numerical taxonomy to medical problems in Mathematics and Computer Science in Biology and Medicine (London, HMSO), 81-91.

Sneath, P. H. A. and R. R. Sokal (1973), Numerical Taxonomy (San Francisco: W. H. Freeman).

Sokal, R. R. and P. H. A. Sneath (1963), Principles of Numerical Taxonomy (San Francisco and London: W. H. Freeman).

Thurstone, L. L. (1947), Multiple-factor Analysis (Chicago: University Press).

Tyron, R. C. (1939), Cluster Analysis (Berkeley: University of California Press).

This work was partly supported by a grant, No. NIH-ROI GM18770, from the Dept. of Health, Education and Welfare (U.S.)

Department of Statistics
Virginia Polytechnic Institute and
State University
Blacksburg, Virginia 24061

Graph Theoretic Techniques for Cluster Analysis Algorithms

David W. Matula

1. Introduction and Summary.

Following numerous authors [2,12,25] we take as available input to a cluster analysis method a set of n objects to be clustered about which the raw attribute and/or association data from empirical measurements has been simplified to a set of $n(n-1)/2$ proximity values on the object pairs. The output of a cluster analysis method will be a collection of subsets of the object set termed clusters characterized in some manner by relative internal coherence and/or external isolation, along with a natural stratification of these identified clusters by levels of cohesive intensity.

In formalizing a model of such cluster analysis methods it is essential to consider the nature and inherent reliability of the proximity data that will constitute the input in substantive clustering applications. Proximity value scales may be simply dichotomous, e.g. the object pair associations {likes, dislikes}, or finite, e.g. the Hamming distance between a sequence of binary valued attributes on the objects, or continuous, e.g. the Euclidean distance between objects plotted in an n-dimensional space with axes given by n different numerical attribute variables on the objects.

It is the practice of most authors of cluster methods to assume that the proximity values are available in the form of a real symmetric matrix where any unjustified structure implicit in these real values is either to be ignored or axiomatically disallowed, the formalism of Jardine and Sibson [12] being the most complete instance of the latter. Such models can be antithetical to that required by the practitioner whose proximity data is available in elemental dichotomous or ordinal form and whose desired tool is a straightforward efficient method to achieve a clustering explicitly consistent with the admittedly weak structure of his input data.

It is our contention that the most desirable cluster analysis models for substantive applications should have the input proximity data expressible in a manner faithfully representing only the reliable information content of the empirically measured data. Biological and social science measurements are generally not sufficiently structured to allow proximity values to be real numbers possessing all the mathematical structure inherent in the real number field. Practitioners generally agree that attempts to ascribe more meaning to the proximity data than simply the rank order relation of proximity values is not justified in the substantive application, and the authors of cluster analysis monographs [2,12,25] generally stress the importance of characterizing those clustering methods which actually depend only on the rank order of the proximity values. An order theoretic formulation of the cluster analysis model of [12] appears in [10, 11].

The intent of this article is to demonstrate that a formalism for cluster analysis methods founded on simply an assumed ordinal relation amongst the proximity values is readily accessible utilizing the theory of graphs, where specifically the input data from the substantive application is assumed to be available in the form of a proximity graph. Extant results from graph theory are then utilized to characterize many of the well known clustering methods and to suggest new methods with important properties. In all resulting methods the practitioner can be assured that the output of such methods is a faithful realization of the reliable information content of his empirically measured input data.

A decided computational advantage is obtained by the formulation of proximity data input as a proximity graph rather than a symmetric matrix of real numbers. Proximity graphs may be represented utilizing the recent theory [1: ch. 5,17,26] of space conserving graphical data structures along with the efficient algorithmic techniques for manipulating these structures. Thus our formal graph theoretic characterizations of cluster methods can often be complemented by efficient algorithmic realizations of these methods.

In section II some elements of graph theory are described and the notion of a proximity graph $P = (V, E)$ is introduced where V is the set of objects to be clustered and E is the ordered set of links (link = object pair) reflecting the ordinal relation of the proximity data on the object pairs. Level, stratified and hierarchical clusterings of the objects are defined and a cluster method is formalized as a mapping from proximity graphs to stratified clusterings.

Ordinal stratified clustering methods are characterized by the stratification levels being directly related to the rank order of the proximity values. Our main result is theorem 1 of section III which shows that any graph theoretic function satisfying four specified properties characterizes a threshold ordinal stratified clustering method on proximity graphs. Standard connectivity properties of graphs are described and shown to characterize the single-linkage, weak and strong k-linkage, and k-overlap clustering methods. Complete-linkage and Jardine and Sibson's [12] $\underline{B}_k$ clustering methods are also characterized graph theoretically. Although a technical difficulty in Jardine and Sibson's formulation of the $\underline{B}_k$ method is avoidable in our formalism, a shortcoming of the complete-linkage method is shown to be intrinsic to the method and makes that procedure suspect for general application.

In section IV we show that stratified clusterings may be obtained from dichotomous proximity data utilizing the intensity of connectivity within the proximity graph to stratify the cluster levels.

Data structures for representing proximity graphs are described in section V. The computational efficiencies [7,16,17,20] of various graph connectivity algorithms utilizing such data structures are employed to assess the efficiencies achievable with stratified clustering algorithms incorporating these concepts.

Finally, in section VI, the question of random proximity data and cluster significance is considered. The presentation is limited to a brief assessment of the implications for cluster validity of a result of our work [15,19] on random graphs indicative in the following numeric example. Suppose one thousand objects are to be clustered and each object pair independently has a proximity type of either "like" or "dislike" with equal probability. Several clustering methods applied to this data would determine subsets of objects for which all object pair relations of the subset must be of type "like". The surprising result is that the distribution of the random variable N, $1 \leq N \leq 1000$, giving the largest object subset with all pairwise relations of type "like" is not broad but rather so highly spiked as to yield the subset size $N = 15$ with probability over .8.

II. Proximity Graphs and Stratified Clustering.

A graph $G = (V,E)$ is composed of a non-void vertex set $V = \{o_1, o_2, \ldots, o_n\}$ and an edge set $E = \{\ell_1, \ell_2, \ldots, \ell_m\}$, $m \leq n(n-1)/2$, where each edge ℓ_k is a distinct pair $o_i, o_j \in V$, denoted $\ell_k = o_i o_j$. If $m = n(n-1)/2$, then G is a complete graph, and if the set E possesses an order relation $\ell_1 \leq \ell_2 \leq \ldots \leq \ell_m$, then G is an ordered graph. For graph theoretic models of cluster analysis it is convenient to refer to the members of V as objects and the members (objects pairs) of E as links.

A proximity graph $P = (V,E)$ is then an ordered graph where $V = \{o_1, o_2, \ldots, o_n\}$ is a set of objects to be clustered, and $E = \{\ell_1, \ell_2, \ldots, \ell_m\}$ is a set of object pairs termed links, and the order relation on the links is determined by the proximity data on the object pairs. Specifically for $\ell_i = o_p o_q$, $\ell_j = o_r o_s$, $\ell_i \leq \ell_j$ denotes that the objects o_p and o_q are no less similar than the objects o_r and o_s. Notationally $\ell_i \equiv \ell_j$ denotes that the possibly distinct links ℓ_i and ℓ_j have the same order with the equality $\ell_i = \ell_j$ denoting that ℓ_i and ℓ_j are the same links, i.e. $i = j$. A complete proximity graph thus succinctly represents the ordinal proximity relation over the object pairs.

The splitting levels of the proximity graph $P = (V, E)$ are the levels $s = 0$, $s = m = |E|$, and all s, $1 \leq s \leq m-1$, for which $\ell_s < \ell_{s+1}$. For each splitting level $0 \leq s \leq m$, with $E_s = \{\ell_1, \ell_2, \ldots, \ell_s\}$, the ordered graph $P_s = (V, E_s)$, where E_s possesses the order relation of E restricted to E_s, is the s th order proximity subgraph of P, and the graph $T_s = (V, E_s)$, where E_s is not assumed to be ordered, is the s th order threshold subgraph of P.

The various formal terms of graph theory are generally intuitively clear in the graph diagram, which is an informal drawing where the vertices (objects) of the graph are represented by nodes and the edges (links) are represented by lines. Figure 1 illustrates a diagram of a complete proximity graph P on five objects where the order relation on the ten links,

$$o_2o_3 < o_3o_4 < o_1o_2 < o_2o_4 < o_1o_3 < o_4o_5 < o_3o_5 \equiv o_2o_5 < o_1o_5 < o_1o_4,$$

corresponds to the distance between objects in the diagram. The splitting levels of P are $0,1,2,3,4,5,6,8,9$ and 10. The 4-th and 8-th order threshold subgraphs T_4, T_8 of the proximity graph P are also shown, where only the adjacencies and not the distance between objects are considered relevant in the diagrams of T_4 and T_8.

A level clustering of the object set $V = \{o_1, o_2, \ldots, o_n\}$ is a set L of non-nested subsets of V termed clusters that cover V, i.e. C, $C' \in L$, $C \subset C'$ implies $C = C'$ and $\bigcup L = V$. A stratified clustering of the object set $V = \{o_1, o_2, \ldots, o_n\}$ is a sequence $S = (L_0, L_1, \ldots, L_k)$ of level clusterings of V where

(1)

i) $L_0 = \{\{o_1\}, \{o_2\}, \ldots, \{o_n\}\}$ [all objects are single object clusters at level 0],

ii) $C \in L_i$, $i \leq k-1$ implies $C \subset C'$ for some $C' \in L_{i+1}$ [sequential refinement: each cluster at level i is part of a cluster at level $i+1$]

In substantive applications it is generally desired that all objects are in a single cluster at the final cluster level of a stratified clustering.

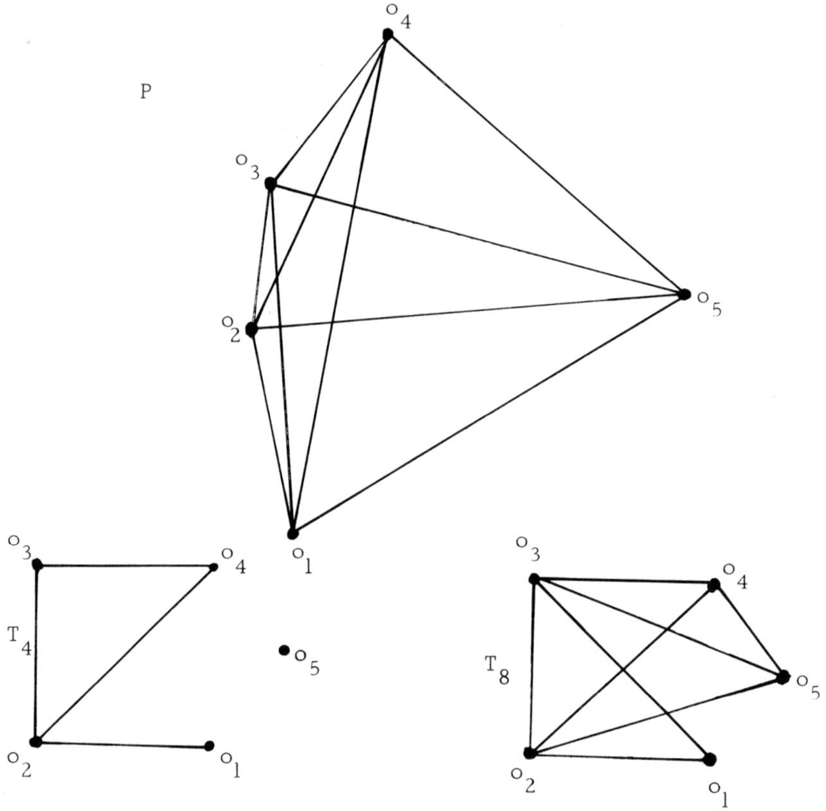

Figure 1. Diagrams of a complete proximity graph P and its 4th and 8th order threshold subgraphs T_4 and T_8

However, for technical reasons, we do not require $L_k = \{V\}$ for stratified clusterings of arbitrary object sets V. A stratified clustering $S = (L_o, L_1, \ldots, L_k)$ is hierarchical if each level clustering L_i, $0 \leq i \leq k$, constitutes a partition of V. The clusters of L_i are termed clusters of S at level i. As a consequence of (1), any cluster C of S has a formation level $i_1 = \min\{j | C \in L_j\}$. Furthermore, for $C \notin L_k$, C also has an absorption level $i_2 = 1 + \max\{j | C \in L_j\}$ such that (i) C cannot be a subset of any cluster of level less than i_1, (ii) C is a cluster at level j for $i_1 \leq j \leq i_2 - 1$, and (iii) C is always a proper subset of at least one cluster of level j for $j \geq i_2$.

Pursuing our graph theoretic model of cluster analysis, a <u>level clustering method</u> is a mapping $\gamma : \mathcal{P} \to \mathcal{L}$ of the set, $\mathcal{P}$, of proximity graphs into the set, $\mathcal{L}$, of level clusterings, and a <u>stratified clustering method</u> is a mapping $\sigma : \mathcal{P} \to \mathcal{S}$ of proximity graphs into the set, $\mathcal{S}$, of stratified clusterings where both $\gamma(P)$ and $\sigma(P)$ are clusterings of the object set V of the proximity graph $P = (V,E)$. To assure that γ and σ depend only on the ordinal proximity relation between objects and not on the nature of the objects being clustered, we add the further technical condition that the application of these clustering methods to order isomorphic proximity graphs must yield isomorphic clusterings. Specifically, when there is a one-to-one correspondence $\phi : V_1 \to V_2$ which takes the links $o_{i_1} o_{j_1} \le o_{i_2} o_{j_2} \le \ldots \le o_{i_m} o_{j_m}$ of $P_1 = (V_1, E_1)$ into the correspondingly ordered links $\phi(o_{i_1})\phi(o_{j_1}) \le \phi(o_{i_2})\phi(o_{j_2}) \le \ldots \le \phi(o_{i_m})\phi(o_{j_m})$ of $P_2 = (V_2, E_2)$ preserving all strict inequalities, then for any level clustering method $\gamma : \mathcal{P} \to \mathcal{L}$, $C \in \gamma(P_1)$ if and only if $\phi(C) \in \gamma(P_2)$, denoted $\gamma(P_2) = \phi(\gamma(P_1))$, and for any stratified clustering method $\sigma : \mathcal{P} \to \mathcal{S}$ with $\sigma(P_1) = (L_0, L_1, \ldots, L_k)$, then $\sigma(P_2) = \phi(\sigma(P_1)) = (\phi(L_0), \phi(L_1), \ldots, \phi(L_k))$. Finally, if the level clustering method $\gamma : \mathcal{P} \to \mathcal{L}$ has the property that for all $P = (V,E) \in \mathcal{P}$, $\gamma(P)$ is independent of the order of the set E, then γ is a <u>graphical</u> level clustering method and we may write $\gamma : \mathcal{G} \to \mathcal{L}$, where $\mathcal{G}$ denotes the set of all graphs.

This graph theoretic model assures that any resulting stratified clustering method is inherently dependent only on the ordinal relation of the initial proximity data. Lest these assumptions appear too restrictive, it should be noted that single-linkage, complete-linkage, and many other popular cluster methods typically formulated in terms of real valued proximity matrices actually depend only on the ordinal relation of the proximity values and therefore are straightforward instances of stratified clustering methods in our formalism.

III. Ordinal and Threshold Stratified Clustering Methods.

Single-linkage and complete-linkage are two of the most prevalent hierarchical clustering procedures in the applied substantive literature. In both of these methods the order of the proximity data corresponds directly to the stratification level order in the resulting hierarchical clustering and this correspondence is indicative of a whole class of stratified clustering procedures.

A stratified clustering method $\sigma : \mathcal{P} \to \mathcal{S}$ shall be termed ordinal if there is a level clustering method $\gamma : \mathcal{P} \to \mathcal{L}$ defined on proximity graphs such that if $0 = s_0 < s_1 < \ldots < s_k = |E|$ are the splitting levels of the proximity graph $P = (V,E)$, then utilizing the s_ith order proximity subgraphs of P,

$$\sigma(P) = (\gamma(P_{s_0}), \gamma(P_{s_1}), \ldots, \gamma(P_{s_k})). \tag{2}$$

Furthermore, if γ is a graphical level clustering method so that $\gamma : \mathcal{G} \to \mathcal{L}$ then σ is termed a threshold stratified clustering method and we may write

$$\sigma(P) = (\gamma(T_{s_0}), \gamma(T_{s_1}), \ldots, \gamma(T_{s_k})) , \tag{3}$$

and say that $\gamma : \mathcal{G} \to \mathcal{L}$ characterizes $\sigma : \mathcal{P} \to \mathcal{S}$.

An important consequence of this formulation is that a threshold stratified clustering method may be characterized by a graph theoretic function, as we now illustrate for the standard single-linkage procedure.

The induced (proximity) subgraph $\langle V' \rangle$ of the (proximity) graph $P = (V,E)$ is the (proximity) graph having object set $V' \subset V$ and all links $o_i o_j$ of E where o_i and o_j are both in V'. For the (proximity) graph $P = (V,E)$, a sequence of distinct objects $(o_1, o_2, \ldots, o_p)$ with $o_i o_{i+1} \in E$ for $1 \leq i \leq p-1$ denotes a path joining o_1 to o_p in P, and an induced subgraph $\langle V' \rangle$ of P is a component if V' is maximal such that every pair of objects in V' is joined by a path. The component function α defined for all $P \in \mathcal{P}$ by

$$\alpha(P) = \{V' \mid \langle V' \rangle \text{ is a component of } P = (V,E)\} \tag{4}$$

is then a graphical level clustering method $\alpha : \mathcal{G} \to \mathfrak{L}$ where $\alpha(P)$ is a partition of V.

Let α^* be defined over P by

(5) $$\alpha^*(P) = (\alpha(T_{s_0}), \alpha(T_{s_1}), \ldots, \alpha(T_{s_k}))$$

for $s_o < s_1 < \ldots < s_k$ the splitting levels of $P = (V,E)$. It is immediately verified that α^* satisfies conditions (i) and (ii) of (1), so $\alpha^* : \mathcal{P} \to \mathcal{S}$ is a threshold hierarchical clustering method which we recognize as the single-linkage clustering method. The hierarchical clustering generated by α^* is conveniently described by a dendrogram [12, p. 46] as illustrated in Figure 2 for the complete proximity graph P of Figure 1.

A graph function $\gamma : \mathcal{G} \to \mathcal{V}$ is a function from the set $\mathcal{G}$, of all graphs to the set, $\mathcal{V}$, of all subsets of object sets where specifically for $G = (V,E) \in \mathcal{G}$, $\gamma(G)$ denotes a set of subsets of V. A graphical level clustering $\gamma : \mathcal{G} \to \mathfrak{L} \subset \mathcal{V}$ is a graph function, and if γ also characterizes a threshold stratified clustering method $\gamma^* : \mathcal{P} \to \mathcal{S}$ by the equation

(6) $$\gamma^*(P) = (\gamma(T_{s_0}), \gamma(T_{s_1}), \ldots, \gamma(T_{s_k})) \qquad \text{for all } P \in \mathcal{P}$$

with $s_0 < s_1 < \ldots < s_k$ the splitting levels of P, then γ is termed a stratifying graph function. Thus the component function α given by equation (4) is a stratifying graph function. Our principal theorem is the following characterization of stratifying graph functions in terms of graph properties.

Theorem 1: The graph function $\gamma : \mathcal{G} \to \mathcal{V}$ is a stratifying graph function if and only if for all $G = (V,E) \in \mathcal{G}$,

(i) Invariance property: $\gamma(G')$ is isomorphic to $\gamma(G)$ if G' is isomorphic to G,

(ii) Non-nesting property: $C, C' \in \gamma(G)$, $C \subset C'$ implies $C = C'$,

(iii) Isolation property: $\gamma(G) = \{\{o\} \mid o \in V\}$ if $E = \phi$,

(iv) Absorption property: $G = (V,E)$ a subgraph of $G' = (V,E')$ and $C \in \gamma(G)$ implies $C \subset C'$ for some $C' \in \gamma(G')$.

Splitting Level	Adjoined Links	Threshold Graph T_{s_i}	Component Function $\alpha(T_{s_i})$
10	o_1o_4		(o_1,o_2,o_3,o_4,o_5)
9	o_1o_5		(o_1,o_2,o_3,o_4,o_5)
8	o_2o_5, o_3o_5		(o_1,o_2,o_3,o_4,o_5)
6	o_4o_5		(o_1,o_2,o_3,o_4,o_5)
5	o_1o_3		$(o_1,o_2,o_3,o_4 \mid o_5)$
4	o_2o_4		$(o_1,o_2,o_3,o_4 \mid o_5)$
3	o_1o_2		$(o_1,o_2,o_3,o_4 \mid o_5)$
2	o_3o_4		$(o_1 \mid o_2,o_3,o_4 \mid o_5)$
1	o_2o_3		$(o_1 \mid o_2,o_3 \mid o_4 \mid o_5)$
0	-		$(o_1 \mid o_2 \mid o_3 \mid o_4 \mid o_5)$

Single-Linkage Dendogram

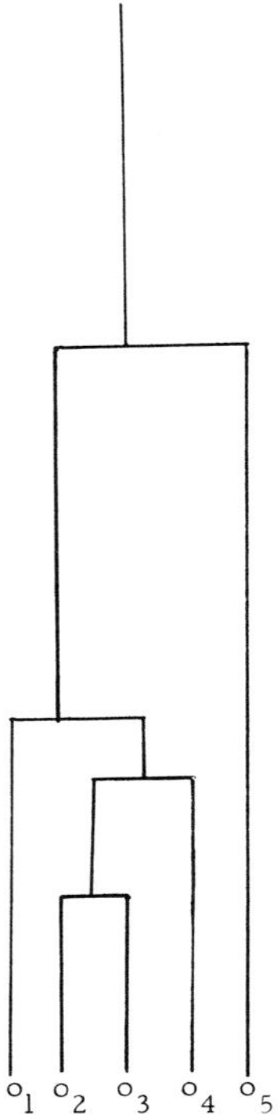

Figure 2: Graph component interpretation of single-linkage clustering.

Proof: The stratifying graph function $\gamma: \mathcal{G} \to \mathcal{S} \subset \mathcal{V}$ is readily shown to satisfy properties (i)-(iv). For the converse, let $\gamma: \mathcal{G} \to \mathcal{V}$ be any graph function satisfying properties (i)-(iv). The isolation and absorption properties then imply that $\gamma(G)$ covers V for any $G = (V, E) \in \mathcal{G}$. Along with the invariance and non-nesting properties it then follows that $\gamma: \mathcal{G} \to \mathcal{S} \subset \mathcal{V}$ is a graphical level clustering method. Let $s_0 < s_1 < \ldots < s_k$ be the splitting levels of $P = (V, E) \in \mathcal{P}$. The threshold subgraph T_{s_0} of P has no links so that the isolation property yields $\gamma(T_{s_0}) = \{\{o\} \mid o \in V\}$. The threshold subgraph T_{s_i} of P is a subgraph of the threshold subgraph $T_{s_{i+1}}$ of P with the same object set for $0 \leq i \leq k-1$, so by the absorption property and equation (1), $\gamma^*(P) = (\gamma(T_{s_0}), \gamma(T_{s_1}), \ldots, \gamma(T_{s_k}))$ is a stratified clustering of V. Hence $\gamma^*: \mathcal{P} \to \mathbb{S}$ is a threshold stratified clustering method, confirming that $\gamma: \mathcal{G} \to \mathcal{S} \subset \mathcal{V}$ is a stratifying graph function.

Thus theorem 1 provides graph theoretic conditions for determining a stratifying graph function $\gamma: \mathcal{G} \to \mathcal{V}$ and equation (6) then defines the threshold stratified clustering method $\gamma^*: \mathcal{P} \to \mathbb{S}$ characterized by γ. Together these results provide a tool which we now employ to characterize other familiar cluster procedures from the literature.

An often stated defect of single linkage clustering is that a single non-representative proximity value can force the union of two otherwise dissimilar clusters at a premature level in the hierarchy. A proposed resolution of this defect is provided by "k-linkage clustering", where two distinct interpretations of k-linkage have appeared in the literature. Ling [14] requires that each object of a singly linked object set be related at a specified level to at least k other objects of that object set to constitute a k-linkage cluster at that level, whereas Sneath [24] and Hubert [9] require the stronger condition that k-linkages of suitable proximity are required to effect merger of two clusters at a specified level in the hierarchy. The distinction between these methods is evident in the threshold graph of Figure 3 which yields one cluster by

Ling's weak 2-linkage criterion and two clusters $\{\{1,2,3,4\},\{5,6,7\}\}$ by the latter strong 2-linkage criterion.

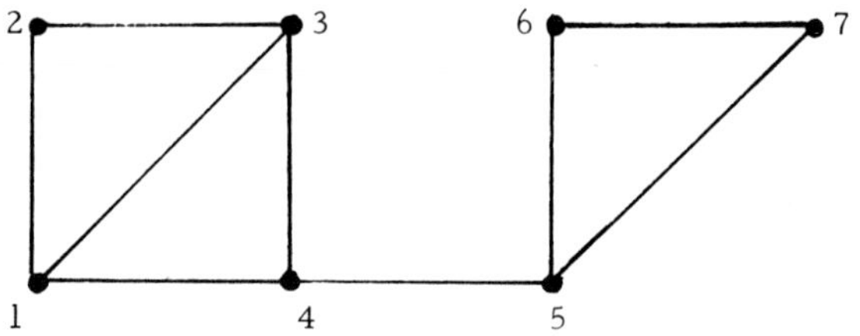

Figure 3: A threshold graph having one cluster by weak 2-linkage two clusters by strong 2-linkage.

These k-linkage methods can both be formalized utilizing well known connectivity properties of graphs and theorem 1. The degree of an object $o_i \in V$ in the graph G is the number of links $o_i o_j \in E$ incident to o_i. For $k \geq 1$, a k-bond of $G = (V,E)$ is a maximal connected induced subgraph $\langle A \rangle$ where every object of A has degree at least k in the graph $\langle A \rangle$. For $k \geq 1$, a k-component of $G = (V,E)$ is a maximal induced subgraph $\langle A \rangle$ with the property that for every partition of A into two parts A_1, A_2, there are at least k links of E each incident with some object of A_1 and some object of A_2.

For $k \geq 1$, the k-bond function $\delta_k : \mathcal{G} \rightarrow \mathcal{V}$ and the k-component function $\lambda_k : \mathcal{G} \rightarrow \mathcal{V}$ are given by

(7) $\delta_k(G) = \{A \mid \langle A \rangle \text{ is a k-bond of } G\} \cup \{\{o\} \mid o \in V \text{ is in no k-bond of } G\}$,

(8) $\lambda_k(G) = \{A \mid \langle A \rangle \text{ is a k-component of } G\} \cup \{\{o\} \mid o \in V \text{ is in no k-component of } G\}$.

The invariance, non-nesting, isolation and absorption properties are readily verified for δ_k and λ_k for $k \geq 1$, so by theorem 1, δ_k and λ_k are stratifying graph functions for each $k \geq 1$. The threshold stratified clustering method $\delta_k^* : \mathcal{P} \rightarrow \mathcal{S}$ characterized by δ_k is termed the weak k-linkage cluster method, and $\lambda_k^* : \mathcal{P} \rightarrow \mathcal{S}$ characterized by λ_k is then the strong k-linkage cluster method. The k-components are disjoint subgraphs [16] as are the k-bonds, so weak and strong k-linkage are both hierarchical clustering methods. Figure 4 shows the dendrogram for the weak 2-linkage cluster method applied to the proximity graph P of Figure 1, where it is noted that strong 2-linkage would yield the identical result in this case.

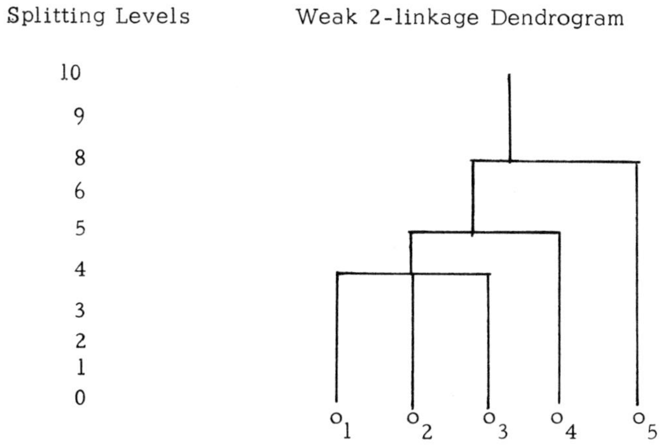

Figure 4: The dendrogram for $\delta_2^*(P)$ for the complete proximity graph P of Figure 1.

k-bonds and k-components are two of a family of four classes of subgraphs of a graph determined by four corresponding well known measures of graph connectivity. The other two measures of graph connectivity deal respectively with object connectivity and complete subgraphs and are now shown to yield overlapping threshold stratified clustering methods.

For $k \geq 1$, the induced subgraph $\langle A \rangle$ of $G = (V,E)$ is <u>k-connected</u> if $|A| \geq k+1$ and the removal of any $k-1$ objects of A still leaves the subgraph connected, i.e. $\langle A' \rangle$ is connected for any $A' \subset A$, $|A'| \geq |A| - k+1$. A <u>k-block</u> $\langle A \rangle$ is a maximal k-connected induced subgraph of $G = (V,E)$. A <u>clique</u> of $G = (V,E)$ is a maximal induced subgraph $\langle A \rangle$ such that every object pair $o_i, o_j \in A$ denotes a link

$o_i o_j$ of E. For $k \geq 1$, the k-block function $\kappa_k: \mathcal{G} \to \mathcal{V}$ is given by

(9) $\kappa_k(G) = \{A \mid \langle A \rangle$ is a k-block of $G\} \cup \{\{o\} \mid o \in V$ is in no k-block of $G\}$

and the k-clique function $\omega_k : \mathcal{G} \to \mathcal{V}$ is given by

(10) $\omega_k(G) = \{A \mid \langle A \rangle$ is a clique of G with $|A| \geq k+1\} \cup$

$\{\{o\} \mid o \in V$ is in no clique of G having over k objects$\}$.

It is readily shown utilizing theorem 1 that for every $k \geq 1$, $\kappa_k: \mathcal{G} \to \mathcal{V}$ and $\omega_k : \mathcal{G} \to \mathcal{V}$ are stratifing graph functions. The threshold stratified clustering method $\kappa_k^*: \mathcal{P} \to \mathcal{S}$ characterized by κ_k is non-hierarchic for $k \geq 2$ and is termed the k-overlap cluster method. $\omega_k^* : \mathcal{P} \to \mathcal{S}$ as characterized by $\omega_k: \mathcal{G} \to \mathcal{V}$ is termed the k-clique cluster method and is non-hierarchic for $k \geq 1$. Application of the 2-overlap method to the proximity graph P of Figure 1 results in the same clustering as the weak and strong 2-linkage method on P shown in Figure 4. Figure 5 shows the stratified clusterings resulting from applying the 2-clique and 1-clique cluster methods to the proximity graph P of Figure 1.

For any graph $G \in \mathcal{G}$, it follows that every subgraph which is a clique on at least k+1 objects is k-connected and hence is contained in a k-block of G. Furthermore, every k-block of G is a subgraph of some k-component of G which in turn is a subgraph of some k-bond of G which in turn is a subgraph of some component of G. These graph connectivity relations impart a relation of successive refinement on the sequence $(\omega_k^*, \kappa_k^*, \lambda_k^*, \delta_k^*, \alpha^*)$ of cluster methods which we now formalize.

The level clustering L' of V is a refinement of the level clustering L of V, denoted $L' \leq L$, if for any $C' \in L'$, $C' \subset C$ for some $C \in L$. The stratified clustering $S' = (L'_0, L'_1, \ldots, L'_k)$ of V is a refinement of the stratified clustering $S = (L_0, L_1, \ldots, L_k)$ of V, denoted $S' \leq S$, if $L'_i \leq L_i$ for $0 \leq i \leq k$. For the level clustering methods γ', $\gamma: \mathcal{P} \to \mathcal{S}$, γ' is a refinement of γ if $\gamma'(P) \leq \gamma(P)$ for all $P \in \mathcal{P}$, and for the stratified clustering methods $\sigma', \sigma: \mathcal{P} \to \mathcal{S}$, σ' is a refinement of σ if $\sigma'(P) \leq \sigma(P)$ for all $P \in \mathcal{P}$. The aforementioned graph connectivity properties yield in a straightforward manner the following result.

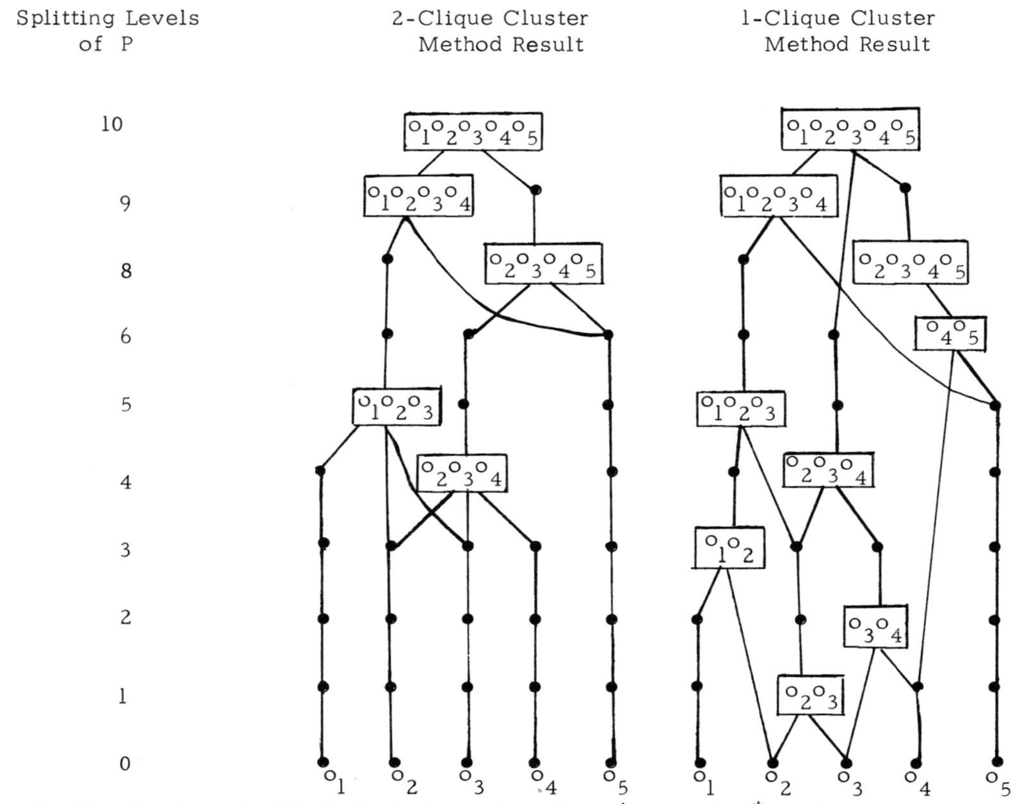

Figure 5: Overlapping stratified clusterings given by $\omega_2^*(P)$ and $\omega_1^*(P)$ for the complete proximity graph P of Figure 1.

Theorem 2: For every $k \geq 1$, each member of the sequence $(\omega_k^*, \kappa_k^*, \lambda_k^*, \delta_k^*, \alpha)$ of threshold stratified clustering methods is a refinement of each of the successive members of the sequence.

In other words the k-clique clustering method is a refinement of the k-overlap method which is a refinement of the strong k-linkage method which is a refinement of the weak k-linkage method which is a refinement of the single-linkage method. In general, a refinement of a hierarchic clustering method always yields at least as many clusters at each level in the hierarchy, however, a refinement of a non-hierarchic method may yield more or fewer clusters at each level. Of specific interest are the numbers of non-trivial clusters, i.e. the clusters with at least two objects, admitted by the methods $\delta_k^*, \lambda_k^*, \kappa_k^*$ and ω_k^*.

Weak and strong k-linkage are hierarchic methods and clearly admit at most $n/(k+1)$ non-trivial clusters at any level in the resulting hierarchic clustering of an n-object set. From deeper results in graph theory [18,23] it follows for the resulting clustering of an n-object set that the k-overlap method allows at most $(n-k+1)/2$ non-trivial clusters at any level, whereas the k-clique method can allow as many as $3^{n/3}$ non-trivial clusters at some level. The exponential growth in this latter case suggests that the k-clique method is impractical as a general clustering strategy, and its application should be limited to properly suited special cases.

The value of graph theory for unifying and categorizing clustering methods is strongly evident in these results. In addition to our proximity graph model of clustering, a graph theoretic interpretation of the theoretical clustering model of Jardine and Sibson [12] has been extensively investigated in our Center by Day [3 ,4], and a unified treatment of the above noted family of cluster methods based on graph connectivity properties is similarly developed in so far as possible within the Jardine and Sibson framework.

The general Jardine and Sibson clustering model [12] may be informally characterized as a two-step procedure proceeding first from an "input" dissimilarity matrix to a "target" dissimilarity matrix in a manner satisfying a host of prescribed axioms, and then from the target dissimilarity matrix to a stratified clustering in a manner equivalent to the 1-clique cluster method applied to the proximity graph corresponding to the target dissimilarity matrix. We feel the conditions imposed by Jardine and Sibson in both of these steps are too restrictive. For $k \geq 2$, the k-linkage methods (weak and strong) are not acceptable hierarchic methods in the Jardine and Sibson model simply because a clique on fewer than $k+1$ objects in a threshold graph is still deemed sufficient to force these objects to be together in some cluster by any acceptable method of the Jardine and Sibson model at that level. A slight relaxation of this condition, essentially ignoring cliques of size less than $k+1$ in a k parameterized family of methods, would allow both weak and strong k-linkage to enjoy the balance of the Jardine and Sibson framework and thus represent additional well founded hierarchic methods. Our objection to Jardine and Sibson's second step is more substantive and is best illustrated by characterizing their $\underline{B}_k$ method in our proximity graph model. As in Harary [8], let $K_n - \ell$ denote a graph on n objects having all but one of the $n(n-1)/2$ possible links. Let $B_k : \mathcal{G} \to \mathcal{G}$ be defined so that for $G \in \mathcal{G}$, $B_k(G)$ is the minimal graph which contains G and has no subgraph isomorphic to $K_{k+2} - \ell$, and let the graph function $\beta_k : \mathcal{G} \to \mathcal{V}$ be defined by $\beta_k(G) = \omega_1(B_k(G))$ for all $G \in \mathcal{G}$. It is readily verified that $B_k : \mathcal{G} \to \mathcal{G}$ is uniquely defined and that β_k is a stratifying graph function, so $\beta_k^* : \mathcal{P} \to \mathcal{S}$ is a threshold stratified clustering method which, with the tagging of the obvious numeric values to the levels of $\beta_k(G)$, is the $\underline{B}_k$-method of Jardine and Sibson [12: p. 65-69]. The $\underline{B}_k$ method has a superficial resemblance to our k-overlap method in that any two distinct clusters at a given level can overlap in at most $k-1$ objects. However, they are quite different methods and we have found no convenient graph theoretic interpretation for a cluster determined by the $\underline{B}_k$ method analogous to the k-block interpretation of the k-overlap method. Now consider the $\underline{B}_2$ method as it applies to a threshold graph G with links corresponding to the solid lines in Figure 6.

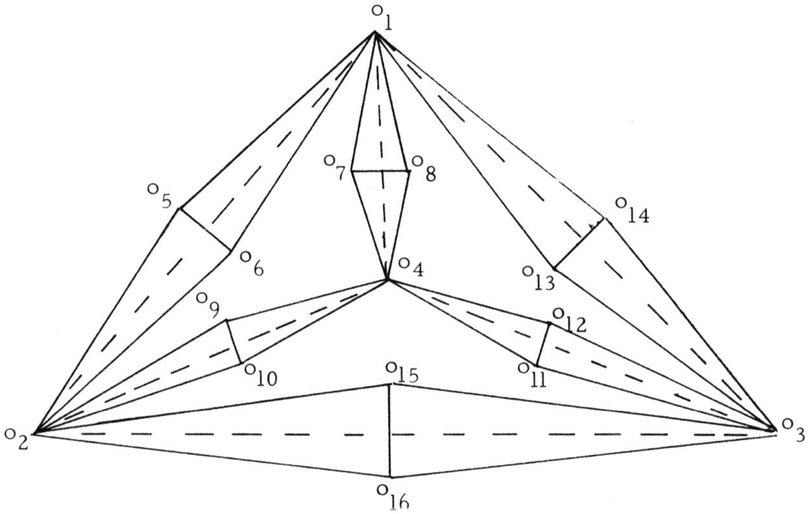

Figure 6

$B_2(G)$ then has as additional links the dashed lines of figure 6, and the function

$$\beta_2(G) = \omega_1(B_2(G)) = L_1 \cup L_2 ,$$
$$L_1 = \{\{o_1,o_2,o_5,o_6\}, \{o_2,o_3,o_{15},o_{16}\}, \{o_1,o_3,o_{13},o_{14}\}, \{o_1,o_4,o_7,o_8\}, \{o_2,o_4,o_9,o_{10}\}, \{o_3,o_4,o_{11},o_{12}\}\},$$
$$L_2 = \{\{o_1,o_2,o_3,o_4\}\},$$

therefore identifies seven four object clusters for this level of the stratified clustering. The induced subgraph $\langle C\rangle$ of G formed by any of the six clusters $C \in L_1$ has the property that $\beta_k(\langle C\rangle) = \{C\}$, and one has no quarrel with the identification of these six clusters. However, one would expect practitioners to have substantive objections to identifying $\{o_1,o_2,o_3,o_4\}$ as an additional distinct cluster, as these four objects are without a single link in the original threshold graph G. Day [4] utilizes a generalized version of the condition $\beta_k(\langle C \rangle) = \{C\}$ to characterize "authentic" clusters, other clusters being termed "specious", and he then pursues an extensive treatment of authentic and specious clusters and methods within the Jardine and Sibson framework.

We note here that by appropriately modifying the $\underline{B}_k$ method to yield only "maximal authentic clusters" of at least k+1 objects and trivial clusters, a threshold cluster method constituting a refinement of the k-overlap method and representing the essence of the $\underline{B}_k$ method can be determined. We shall not pursue the details of this modification in this paper.

It is important to realize that the determination of "specious" clusters, such as $\{o_1, o_2, o_3, o_4\}$ in the preceding example, is not a phenomenon peculiar to the $\underline{B}_k$ method. Rather this is a natural consequence of the fundamental requirement of the Jardine and Sibson model that the clusters at a given level must be the set of cliques of some graph. Hence the identification of exactly the six object sets of L_1 as clusters at a level characterized by the threshold graph G of Figure 6 is an inadmissible result of any "legitimate method" in the Jardine and Sibson axiomatic model. It is primarily this unnecessarily restrictive condition that has motivated us to depart more significantly from the Jardine and Sibson model and prefer a definition of a cluster method as a mapping of proximity graphs directly to stratified clusterings.

The characterization of the complete-linkage method has been deferred until now as it is a non-threshold ordinal stratified clustering method. That is, the level clustering at a given splitting level determined by the complete-linkage procedure is not necessarily independent of the ordering of the links in the proximity subgraph for that level. For the proximity graph $P = (V, E)$, a partition $V_1, V_2, \ldots, V_q$ of V is said to induce a complete subgraph cover of P if $\langle V_i \rangle$ is a complete subgraph of P for $1 \le i \le q$, i.e. oo' is a link of $\langle V_i \rangle$ for all $o, o' \in V_i$, $o \neq o'$, for all i. The range $\mathfrak{L} \cup \{I\}$ is used for the definition of the complete-linkage method ρ on $\mathcal{P}$, with $\rho(P) = I$ denoting that the complete-linkage method is intederminate (not well defined) for P. The complete cover function $\rho : \mathcal{P} \to \mathfrak{L} \cup \{I\}$ is defined for $P = (V,E) \in \mathcal{P}$, where $(V, \emptyset) = P_{s_0}, P_{s_1}, \ldots, P_{s_k} = P$ are the proximity subgraphs of P, such that

$\rho(P) \in \mathfrak{L}$ if and only if

1. $\rho(P_{s_0}) = \{\{o\} | o \in V\}$,
2. $\rho(P_{s_i})$ is the unique partition containing $\rho(P_{s_{i-1}})$ as a subpartition where $\rho(P_{s_i})$ induces a complete subgraph cover of P_{s_i} and is a subpartition of no other partition inducing a complete subgraph cover of P_{s_i} for all $1 \le i \le k$,

and $\rho(P) = I$ if and only if for some $1 \le i \le k$, $\rho(P_{s_{i-1}}) \in \mathfrak{L}$ and condition 2 above is not satistied by any partition of V for the proximity subgraph P_{s_i}.

Let $\mathcal{P}' \subset \mathcal{P}$ be the set of proximity graphs on which ρ is not indeterminant, so $\rho : \mathcal{P}' \to \mathfrak{L}$. For $P \in \mathcal{P}'$ with proximity subgraphs $P_{s_0}, P_{s_1}, \ldots, P_{s_k}$, the sequence $(\rho(P_{s_0}), \rho(P_{s_1}), \ldots, \rho(P_{s_k}))$ is then a hierarchical stratified clustering of V. Hence we define the complete-linkage method $\rho^* : \mathcal{P}' \to \mathfrak{S}$ by

$$\rho^*(P) = (\rho(P_{s_0}), \rho(P_{s_1}), \ldots, \rho(P_{s_k})) . \tag{11}$$

It is readily shown that if a proximity graph $P = (V,E)$ has a strict ordering of E, i.e. $\ell_1 < \ell_2 < \ldots < \ell_m$, then $P \in \mathcal{P}'$ and the complete-linkage procedure may be applied to P. In practical applications where link proximity value ties cause the complete-linkage method to be undefined for P, the ties are broken arbitrarily creating a proximity graph P' with a strict link proximity ordering, so then $P' \in \mathcal{P}'$ and the complete linkage method is simply applied to P'. It is evident that with many ties there may be many different resulting stratified clusterings depending on how the ties are broken, and any such resulting clustering should be treated with due caution. A more annoying property of complete linkage is the overall sensitivity of the resulting clustering to a single transposition of the rank order of the proximity data. This difficulty is clearly illustrated in the following example.

Example: Comparison of Complete-Linkage Method.

Let the complete proximity graph Q on five objects have the following order relation on its links,

$$o_2o_3 < o_3o_4 < o_1o_2 < o_1o_3 < o_2o_4 < o_4o_5 < o_3o_5 \equiv o_2o_5 < o_1o_5 < o_1o_4.$$

Note that this ordering for Q differs by a single transposition of the links o_2o_4 and o_1o_3 from the proximity graph P of Figure 1. The complete-linkage dendrograms for Q and P are shown in Figure 7. Note that although the threshold graphs differ only at splitting level 4, the dendrograms differ from splitting levels 4 through 9. One might legitimately question whether the decision on the preferred clustering $(o_1o_2o_3)(o_4o_5)$ for Q as opposed to $(o_1)(o_2o_3o_4o_5)$ at splitting level 9 should be based on the simple rank order interchange of links o_1o_3 and o_2o_4 that occurred at a rank order so far removed from the threshold represented by splitting level 9.

This extreme sensitivity of the overall clustering provided by complete-linkage to a single transposition in the proximity data order suggests that a practitioner is well advised to avoid the complete-linkage method and utilize instead one of the k-linkage methods when a stronger interrelation than single-linkage is desired in the identified clusters.

In this section we have discussed over a half-dozen stratified clustering methods where the stratified clustering is composed of a sequence of level clusterings each associated with a splitting level in the link order. If the original proximity data exhibits only a few or perhaps

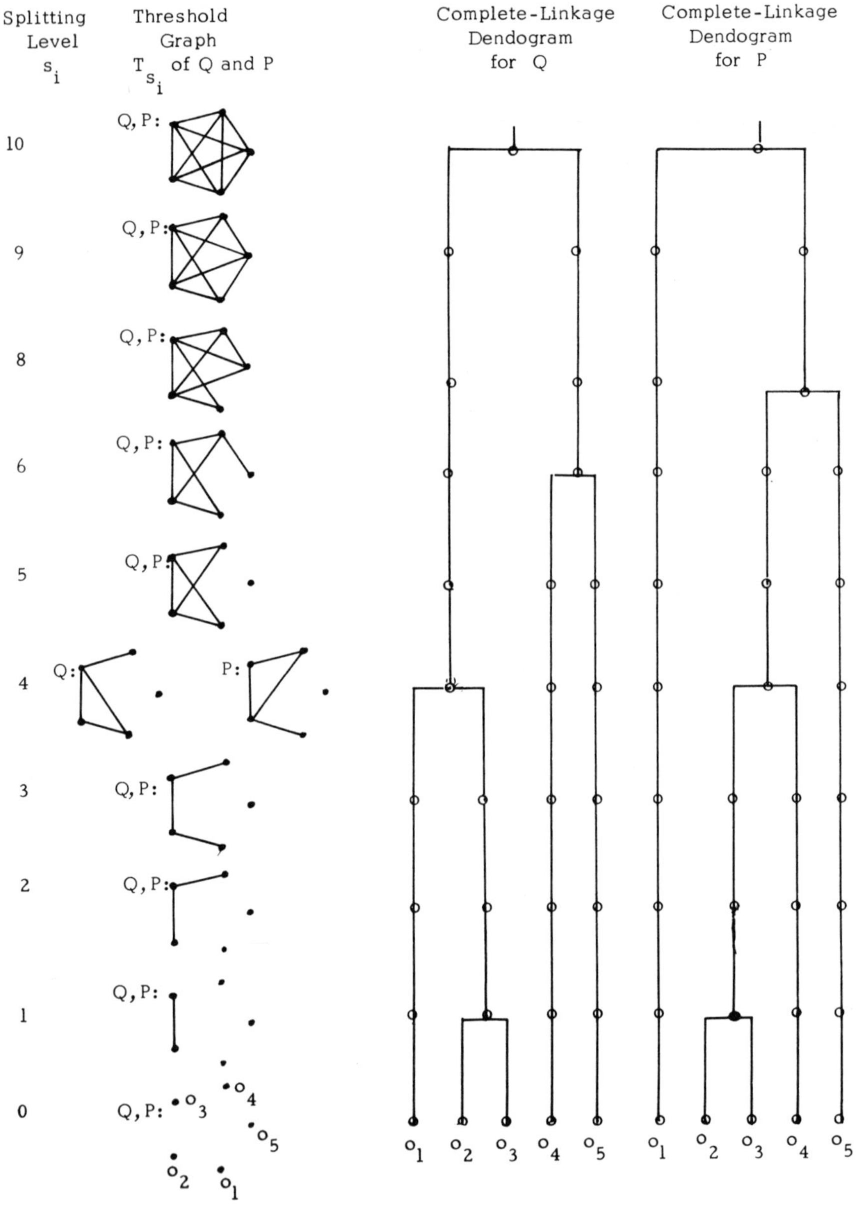
Splitting Level s_i
Threshold Graph T_{s_i} of Q and P
Complete-Linkage Dendogram for Q
Complete-Linkage Dendogram for P
10
Q, P:
9
Q, P:
8
Q, P:
6
Q, P:
5
Q, P:
4
Q:
P:
3
Q, P:
2
Q, P:
1
Q, P:
0
Q, P:
o_1 o_2 o_3 o_4 o_5

Figure 7

a single non-trivial splitting level, such as when the original proximity data is dichotomous, then the insight into cluster structure provided by these methods is necessarily limited. The next section presents an alternative criterion for realizing stratified clustering methods which is of particular value when the number of splitting levels is small.

IV. Connectivity Stratified Clustering Methods.

The level of connectivity of the induced proximity subgraphs of a proximity graph can be utilized parametrically to realize a stratified clustering of the object set of a proximity graph. For connectivity stratified clustering it is sufficient to ignore the order of the links of the proximity graph and refer simply to the underlying graph, thus a connectivity stratified clustering method may be given by a mapping $\sigma : \mathcal{G} \to \mathcal{S}$. Three distinct measures of the intensity of connectivity will be utilized here to form three different clustering methods, and these methods rely on identification of the following subgraphs parametrically in k,

k-bond: a maximal connected induced subgraph $\langle A \rangle$ where each object of A has degree at least k in the subgraph $\langle A \rangle$.

k-component: a maximal induced subgraph $\langle A \rangle$ with the property that for every partition A_1, A_2 of A, at least k links of $\langle A \rangle$ are each incident with an object of A_1 and of A_2,

k-block: a maximal induced subgraph $\langle A \rangle$ with $|A| \geq k+1$ where $\langle A' \rangle$ is connected for any $A' \subset A$ such that $|A'| \geq |A| - k+1$.

The nested sequences of k-bonds and of k-components and the partially overlapping sequence of k-blocks for $k = 1,2,3,4$ are shown for a sample graph G in Figure 8. The k-components and k-blocks are seen to provide better resolution into cohesive groupings than the k-bonds in Figure 8, but in general they are harder to compute. Note for the graph G in Figure 8 that for any $1 \leq k \leq 4$ and any pair of objects in the same k-component, there are at least k link-disjoint paths between those objects in the k-component. A fundamental theorem of

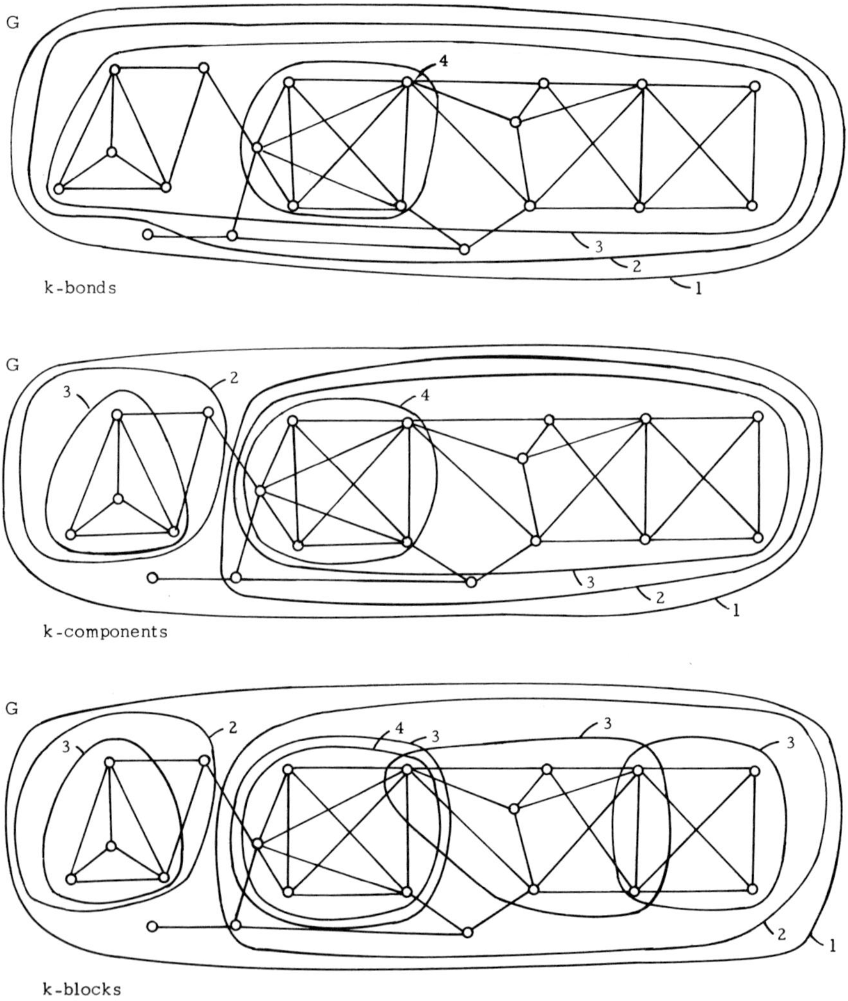

Figure 8. The k-bonds, k-components and k-blocks of a graph G for k = 1, 2, 3, 4.

graph theory due to Menger is now stated which allows alternative interpretations of the subgraphs termed k-components and k-blocks in terms of the number and type of non-overlapping paths between pairs of objects of the subgraph.

Theorem 3 [Menger - 8: p. 47-50, 22]

a) The minimum number of links whose removal separates any two particular objects in a graph equals the maximum number of link-disjoint paths between those two objects.

b) The minimum number of objects whose removal separates any two particular nonadjacent (unlinked) objects in a graph equals the maximum number of disjoint (except for endpoints) paths between those two objects.

Corollary 1: Every pair $u, v \in A$ of distinct objects of the k-component $\langle A \rangle$ of the graph G is joined by k link-disjoint paths of the subgraph $\langle A \rangle$, and $\langle A \rangle$ is maximal with this property.

Corollary 2: Every pair $u, v \in A$ of distinct objects of the k-block $\langle A \rangle$ of the graph G is joined by k disjoint (except for end points) paths of the subgraph $\langle A \rangle$, and $\langle A \rangle$ with $|A| \geq k+1$ is maximal with this property.

The equivalent alternative characterizations of k-components and k-blocks provided by these corollaries of Menger's theorem yield further credence to the significance of these subgraphs for cluster characterization and stratification.

For any $k \geq 1$, the k-bonds, k-components, and k-blocks can each be utilized to effect a level clustering of the objects of the graph G simply by appending as trivial clusters those objects not occurring in any k-bond, k-component or k-block, respectively. Specifically, recall definitions (7), (8) and (9) for $k \geq 1$ and $G \in \mathcal{G}$,

(7) $$\delta_k(G) = \{A \mid \langle A \rangle \text{ is a k-bond of } G\} \cup \{\{o\} \mid o \in V \text{ is in no k-bond of } G\},$$

(8) $$\lambda_k(G) = \{A \mid \langle A \rangle \text{ is a k-component of } G\} \cup \{\{o\} \mid o \in V \text{ is in no k-component of } G\},$$

(9) $\kappa_k(G) = \{A \mid \langle A \rangle$ is a k-block of $G\} \cup \{\{o\} \mid o \in V$ is in no k-block of $G\}$,

where then δ_k, λ_k and κ_k are each graphical level clustering methods mapping $\mathscr{G}$ into $\mathscr{L}$.

Theorem 4: For any $n \geq 1$ and any graph $G = (V,E) \in \mathscr{G}$ having $|V| = n$ objects, let

$$\bar{\delta}(G) = (\delta_n(G), \delta_{n-1}(G), \ldots, \delta_1(G)),$$

$$\bar{\lambda}(G) = (\lambda_n(G), \lambda_{n-1}(G), \ldots, \lambda_1(G)), \tag{12}$$

$$\bar{\kappa}(G) = (\kappa_n(G), \kappa_{n-1}(G), \ldots, \kappa_n(G)).$$

Then $\bar{\delta}(G)$, $\bar{\lambda}(G)$ and $\bar{\kappa}(G)$ are each stratified clusterings of V.

Proof: Since an n-bond, n-component or an n-block must have at least n+1 objects, $\delta_n(G) = \lambda_n(G) = \kappa_n(G) = \{\{o\} \mid o \in V\}$. A k-bond [k-component, k-block] is a subgraph of some (k-1)-bond [(k-1)-component, (k-1)-block, respectively] for any $k \geq 2$. Thus, by the defining equation (1), $\bar{\delta}(G)$, $\bar{\lambda}(G)$, and $\bar{\kappa}(G)$ are each stratified clusterings of V.

Given an arbitrary proximity graph $P = (V,E) \in \mathcal{P}$, the cluster analysist may utilize both the link order on E and the connectivity intensity in the threshold subgraphs T_{s_i} to determine possible clusters. Specifically, the practitioner may compute the level clustering $\kappa_k(T_{s_i})$ [or $\lambda_k(T_{s_i})$ or $\delta_k(T_{s_i})$] for all splitting levels $0 \leq s_i \leq |E|$ and all connectivity levels $1 \leq k \leq |V|$. An array of such level clusterings is shown schematically in Figure 9.

Note that the rows of the array correspond to splitting levels of P, and the row vector for row s_i is $\bar{\kappa}(T_{s_i})$ in reverse order. Also the columns of the array correspond to connectivity intensity levels, and the column vector for column k is $\kappa_k^*(P)$. For proximity graphs with a reasonably small number of objects and splitting levels, the array of level clusterings as shown in Figure 9 can provide the practitioner an output yielding more insight into the cluster structure than any one stratified clustering method could display.

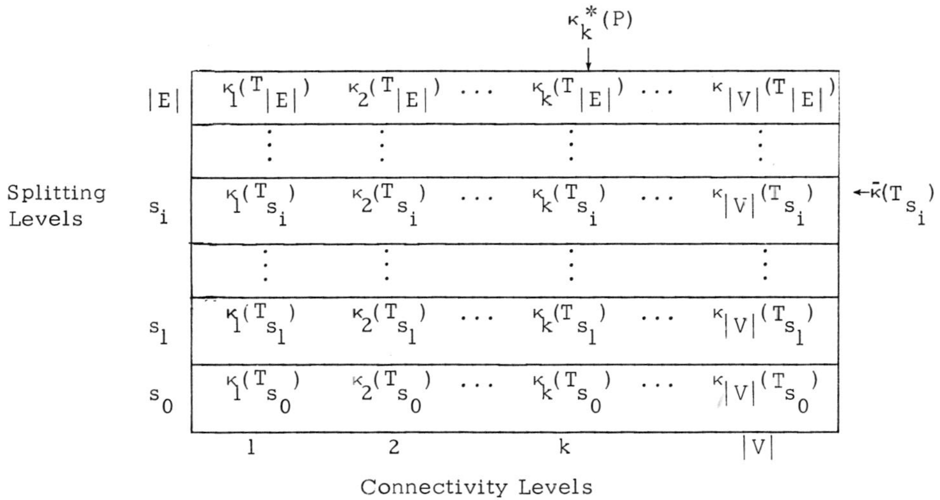

Figure 9: Array of level clusterings based on link order and connectivity for the proximity graph $P = (V, E)$.

V. Proximity Graph Data Structures and Clustering Algorithm Efficiency.

For the stratified clustering methods described in this paper it is generally not necessary to utilize all $n(n-1)/2$ proximity values to determine the stratified clustering. From a result of Erdos and Renyi [5] on random graphs, it follows that an average of order $n \log n$ links is sufficient to assure that a proximity graph on n objects is connected. Thus the single-linkage method can generally be determined from the proximity subgraph P_s with s of order $n \log n$ rather than $n(n-1)/2$. Sibson [13] and Anderberg [2:p. 149-150] describe single-linkage computer programs where the proximity data is assumed to be sequentially input in sorted form until the clustering is complete, therefore generally allowing the handling of much larger problems with time and space efficiency.

For clustering an n object set by any of the threshold stratified clustering methods α^*, δ_k^*, λ_k^*, κ_k^*, or any of the connectivity stratified clustering methods $\bar{\delta}$, $\bar{\lambda}$, and $\bar{\kappa}$, it is generally sufficient to process far less than $n(n-1)/2$ links to obtain the clustering. Appropriate graphical data structures and algorithms attuned to these structures are essential to realize the potential efficiencies in time and space requirements. Recent research in algorithmic complexity has provided much insight into data structures for graphs and their manipulation [1, 17, 26].

For a graph the list of objects linked to a given object is termed the <u>adjacency list</u> for that object, and the collection of such adjacency lists for all objects is termed the <u>adjacency structure</u> for the graph. For a <u>proximity graph adjacency structure</u>, it is required in addition that the order of each adjacency list be consistant with the ordering of the links. The adjacency structure for the proximity graph P of figure 1 is shown in Figure 10. In Figure 10a the adjacency lists are illustrated in "linked list" form, that is, each element of the list is coupled with a pointer to the location of the next element of the list. The initial segments of the adjacency lists up to the dividing line labeled P_6 constitute the adjacency structure for the proximity subgraph P_6 of P. In Figure 10b the

a)

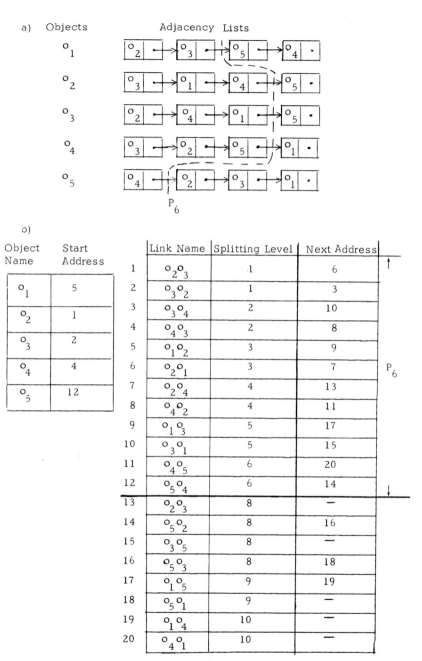

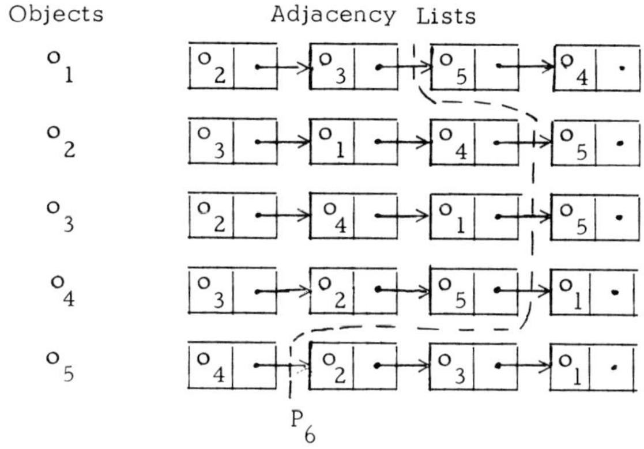

b)

Object Name	Start Address
o_1	5
o_2	1
o_3	2
o_4	4
o_5	12

	Link Name	Splitting Level	Next Address	
1	o_2o_3	1	6	↑
2	o_3o_2	1	3	
3	o_3o_4	2	10	
4	o_4o_3	2	8	
5	o_1o_2	3	9	
6	o_2o_1	3	7	P_6
7	o_2o_4	4	13	
8	o_4o_2	4	11	
9	o_1o_3	5	17	
10	o_3o_1	5	15	
11	o_4o_5	6	20	
12	o_5o_4	6	14	↓
13	o_2o_3	8	—	
14	o_5o_2	8	16	
15	o_3o_5	8	—	
16	o_5o_3	8	18	
17	o_1o_5	9	19	
18	o_5o_1	9	—	
19	o_1o_4	10	—	
20	o_4o_1	10	—	

Figure 10: The adjacency structure for the proximity graph P of Figure 1.

adjacency structure is shown as it may be realized in sequential storage fields in a computer, and it is evident that only the necessary initial segment of the link set need be input and processed to achieve the desired clustering for a proximity subgraph P_{s_i}.

A sufficient body of graph connectivity algorithms incorporating these data structures have appeared in the literature [7,16,17,20,26] to substantiate the data of table 1 on time and space complexity upper bounds for worst case performance of specific algorithmic realizations of the cited clustering methods.

Stratified Clustering Method	Time Complexity	Space Complexity
$\bar{\delta}(G)$	$O(\|E\|)$	$O(\|E\|)$
$\bar{\lambda}(G)$	$O(\min\{\|V\|^{8/3}\|E\|, \|V\|\ \|E\|^2\})$	$O(\|E\|)$
$\bar{\kappa}(G)$	$O(\|V\|^{3/2}\|E\|^2)$	$O(\|E\|)$
$\alpha^*(P_{s_i})$	$O(\|E\| \log \|E\|)$	$O(\|E\|)$
$\delta_k^*(P_{s_i})$	$O(\|E\| \log \|E\|)$	$O(\|E\|)$
$\lambda_k^*(P_{s_i})$	$O(\|V\|^2\ \|E\|)$	$O(\|E\|)$
$\kappa_k^*(P_{s_i})$	$O(\|V\|^{3/2}\|E\|^3)$	$O(\|E\|)$
$\omega_k^*(P)$	$O(2^{\|V\|}\|E\|)$	$O(\|V\|^2)$

Table 1: Algorithmic complexity upper bounds for specific procedures for certain clustering methods. Space requirement relates to processing space and not size of output.

In table 1, $O(|V|)$ denotes that the time or space requirement growth is proportional to the number of objects, and $O(|E|)$ denotes that the growth is proportional to the number of links. For the methods α^*, δ_k^*, κ_k^*, $|E|$ is assumed to be the number of links, s_i, in the proximity subgraph, P_{s_i}, sufficient to effect the clustering. For reasons previously described, it is likely in applications that $O(|E|) << O(|V|^2)$, and possible that $O(|E|)$ may be equivalent to $O(|V|)$. References [7,16,17,20] provide the basis for the algorithms and bounds for $\bar{\delta}$, $\bar{\lambda}$ and $\bar{\kappa}$. The methods for δ_k^* and λ_k^* are new. Specifically the method [21] for δ_k^* features a time complexity bound of $O(|E| \log |E|)$ essentially dictated by the time required to order the $|E|$ links. This represents a substantial improvement over the $O(|V|^3)$ bound in Ling's [14] original paper on the weak k-linkage method. Note that the connectivity stratified method $\bar{\delta}$ does not require the links to be ordered in the data structure, so there is no inconsistency in the $O(|E|)$ result for that case. The bound for $\kappa_k^*(P_{s_i})$ is quite poor, coming simply from application of the method $\bar{\kappa}(P_{s_j})$ for $0 \leq j \leq i$.

It is conjectured that $\kappa_k^*(P_{s_i})$ may also exhibit a time complexity bound no worse than $O(|V|^2|E|)$ by an algorithm similar to that for $\lambda_k^*(P_{s_i})$. Finally, the exponential growth cited for the ω_k^* clustering method is unavoidable, as evidenced by the possible occurrence of up to $3^{|V|/3}$ cliques in a graph [23]. The principal result for rapid practical computation indicated in table 1 is that the weak k-linkage method can be implemented by an algorithm with such good time and space efficiencies that it is a very desirable practical alternative when single-linkage is found unacceptable.

VI. Random Proximity Graphs and Cluster Significance.

Clustering methods generally force a determination of clusters for arbitrary input proximity data. In order to properly assess the significance of a given clustering it is relevant to compare the clustering with that expected from "random proximity data". The literature on random graphs provides some valuable insights into the significance of clusters determined by clustering methods applied to proximity graphs.

By a random graph $G_{n,p}$ we shall mean a graph on n objects where each of the $n(n-1)/2$ links is present independently with probability p. By a random proximity graph on n objects we shall mean a complete proximity graph with a strict link ordering determined uniformly over all $[n(n-1)/2]!$ such orderings. The properties of the random graph $G_{n,p}$ with $p = 2s/[n(n-1)]$ then essentially typify the properties of the threshold subgraph T_s of the random proximity graph on n objects.

The distribution of cliques in a random graph was investigated by this author in [15, 19], and these results are pertinent to both the k-clique and complete-linkage cluster methods. Of particular interest is the remarkably sharp value obtained for the largest complete subgraph size $Z_{n,p}$ of the random graph $G_{n,p}$.

Theorem 5 [19]: For $n \geq 1$, $0 < p < 1$, let $z = 2 \log n - 2 \log \log n + 2 \log \frac{e}{2} + 1$, where logarithms are to the base $1/p$. Then the maximum clique size $Z_{n,p}$ of the random graph $G_{n,p}$ satisfies

(12) $$\lim_{n \to \infty} \text{Prob}\,\{\lfloor z-\varepsilon \rfloor \leq Z_{n,p} \leq \lfloor z+\varepsilon \rfloor\} = 1,$$

where $\lfloor x \rfloor$ denotes the greatest integer less than or equal to x.

This result implies that the largest complete subgraph size for a random graph takes on one of at most two values depending on n and p with probability approaching unity as n grows to infinity. Computationally it can be shown [19] that the random graph on 1000 objects with link probability .5 has a largest clique size of 15 with probability greater than .8, and the random graph on 10^{10} objects with link probability .25 has a largest clique size of 30 with probability greater than .9997.

One must be cautious in drawing immediate results for cluster significance from this result on the largest clique size in a random graph. Substantive clustering problems may have an environment imposing some structure to the object proximity relation so that the concept of uniform independent probabilistic behavior for all links may not be representative of randomness in the application.

Nevertheless, this example shows that the behavior of a random variable (e.g. maximum clique size) of value in substantive clustering problems can be to exhibit nearly deterministic values when the underlying proximity data is quite random. Other results exhibiting surprisingly predictable behavior in probabilistic combinatorial problems have been described by Erdos and Spencer [6]. If an appropriate model of randomness for a class of substantive clustering problems yields a very narrow range of expected cluster characteristics, then a valuable tool for assessing the significance of clusters is obtained.

References

1. Aho, A. V., Hopcroft, J. E., Ullman, J. D., The Design and Analysis of Computer Algorithms, Addison-Wesley, Reading, 1974.

2. Anderberg, M. R., Cluster Analysis for Applications, Academic Press, New York, 1973.

3. Day, W. H. E., Flat Cluster Methods Based on Concepts of Connectedness, preprint.

4. Day, W. H. E., Flat Cluster Methods Based on Authentic Indicator Families, preprint.

5. Erdos, P. and Renyi, A., On Random Graphs, I, Publications Mathematicae (Debrecen) 6 (1957), 405-411.

6. Erdos, P. and Spencer, J., Probabilistic Methods in Combinatorics, Academic Press, New York, 1974.

7. Even, S. and Tarjan, R. E., Network Flow and Testing Graph Connectivity, SIAM J. Comp. 4 (1975), 506-518.

8. Harary, F., Graph Theory, Addison-Wesley, Reading, 1969.

9. Hubert, L., Data Analysis Implications of Some Concepts Related to the Cuts of a Graph, J. Math. Psych., to appear.

10. Janowitz, M. F., An Order Theoretic Foundation for Automatic Calssification, Computer J., to appear.

11. Janowitz, M. F., An Order Theoretic Model for Numerical Taxonomy, preprint.

12. Jardine, N. and Sibson, R., Mathematical Taxonomy, Wiley, London, 1971.

13. Sibson, R., SLINK: An Optimally Efficient Algorithm for the Single-Link Cluster Method, Computer Journal 16 (1973), 30-34.

14. Ling, R. F., On the Theory and Construction of k-Clusters, Computer J. 15 (1972), 326-332.

15. Matula, D. W., On the Complete Subgraphs of a Random Graph, Comb. Math. and Its Appl., (Chapel Hill, N.C., 1970), 356-369.

16. Matula, D. W., k-Components, Clusters, and Slicings in Graphs, SIAM J. Ap. Math. 22 (1972), 459-480.

17. Matula, D. W., Beck, L. L., and Feld, J. F., Breadth-Minimum Search and Clustering and Graph Coloring Algorithms, preprint.

18. Matula, D. W., k-Blocks and Ultrablocks in Graphs, J. C. T., to appear.

19. Matula, D. W., The Largest Clique Size in a Random Graph, SIAM J. Ap. Math., to appear.

20. Matula, D. W., Subgraph Connectivity Numbers of a Graph, in Theory and Applications of Graphs in America's Bicentennial Year, ed. Alavi, Y. and Lick, D. R., Springer-Verlag, Heidleberg, 1977.

21. Matula, D. W., Advances in k-Linkage Cluster Analysis (abstract) presented to The Classification Society meeting, Rochester, N. Y., May 24-25, 1976.

22. Menger, K., Zur allgemeinen Kurventheorie, Fund. Math. 10 (1927), 96-115.

23. Moon, J. and Moser, L., On Cliques in Graphs, Israel J. Math. 3 (1965), 23-28.

24. Sneath, P. H. A., A Comparison of Different Clustering Methods as Applied to Randomly Spaces Points, Classification Soc. Bull. 1 (2), (1966), 2-18.

25. Sneath, P. H. A. and Sokal, R. R., Numerical Taxonomy, Freeman, San Francisco, 1973.

26. Tarjan, R.E., Depth-First Search and Linear Graph Algorithms, SIAM J. Comp. 2 (1972), 146-160.

This research was supported in part by the National Science Foundation Grant DCR75-10930.

Department of Computer Science
School of Engineering and Applied Science
Southern Methodist University
Dallas, Texas 75275

An Empirical Comparison of Baseline Models for Goodness-of-Fit in r-Diameter Hierarchical Clustering

Lawrence J. Hubert
Frank B. Baker

1. Introduction.

The prior specification of a reasonable null model is a major difficulty in developing goodness-of-fit strategies for hierarchical clustering, and no generally accepted statistical method has been proposed in the literature for evaluating the adequacy of cluster analyses. Those methods that have been suggested for several specific techniques are defended mainly in terms of mathematical tractability (see the discussions in [8,14]). From an inference point of view, however, the availability of some type of reference distribution is an indispensable prerequisite if the techniques of cluster analysis are ever to be more than approximate descriptive tools. Without any baseline information whatsoever, it is always difficult to attribute the results of a cluster analysis to something other than the nonsystematic characteristics of a particular data set and the capriciousness of a particular technique.

Although the present paper will not solve the problem of defining a "best" baseline model that could aid statistical inference for all hierarchical clustering procedures, it does seek to give some further empirical information on a strategy followed by Ling [14, 15, 16], Hubert [8], Baker and Hubert [2,3], Baker [1], and others. The problem of defining some type of universal reference distribution for use in hierarchical clustering appears to be generally unsolvable due to the almost limitless variation that is possible in the choice of clustering strategy and proximity measure. Consequently, one of the main areas of interest of this paper is to illustrate a simple empirical fact: by restricting a proximity function in a manner that generally conforms to what could be found in actual data, a goodness-of-fit index has to be inflated over what would be expected from a general randomization model. A second interest is to show that clustering strategies defined by a compromise between the single-link and the complete-link criteria may prove more adequate in terms of average fit when restricted proximity functions are encountered. This compromise employs a class of r-diameter hierarchical clustering techniques (see [10, 17, 18]), which includes the prototypic complete-link and single-link methods as special cases. By using intermediate values of r, several common criticisms of the complete-link and the single-link criteria may be answered, i.e., that the complete-link method is too "strict," and conversely, that the single-link method is too "lenient." The class of r-diameter strategies provides a sequence of clustering alternatives that lie "between" the complete-link and the single-link extremes; at the same time, however, each such alternative maintains a sole dependence on the rank order of the proximity values assigned to the object pairs and used in constructing the partition hierarchy. The effect of restricted proximity measures and the r-diameter clustering criteria on an overall goodness-of-fit index will be assessed by comparing empirically obtained sampling distributions with published tables constructed with a randomization model. In addition, a fairly short reanalysis of a well-known Kahl-Davis [11] data set is performed, providing an example of how the r-diameter criteria may be applied in a traditional data analysis context.

2. r-Diameter Hierarchical Clustering.

The basic data to be used in a hierarchical clustering is in the form of a nonnegative real-valued symmetric function on $S \times S$, denoted by $s(\cdot,\cdot)$, where S is a set of n objects, $\{o_1, o_2, \ldots, o_n\}$. For technical convenience, the proximity function $s(\cdot,\cdot)$ is assumed to assign larger numerical values to the object pairs that are more dissimilar, and also, to assign zero if and only if the two arguments in $s(\cdot,\cdot)$ are the same.

The purpose of an r-diameter hierarchical clustering strategy, for any value of r from 1 to n-1, is to form a sequence of n partitions of S, $(\ell_0, \ldots, \ell_{n-1})$, that have the following three properties: ℓ_0 consists of n object classes each containing a single member of S ; ℓ_{n-1} consists of one object class containing all members of S ; ℓ_{k+1} is formed by uniting two and only two object classes in ℓ_k. In particular, r-diameter hierarchical clustering can be defined by the criterion used to obtain ℓ_{k+1} from ℓ_k. Suppose ℓ_k consists of the object classes $L_1, \ldots, L_{n-k}$, then L_i and L_j are united to form the new object class in ℓ_{k+1} if and only if

$$Q_r(L_i, L_j) = \min_{1 \leq t \neq s \leq n-k} Q_r(L_t, L_s) ,$$

where

$$Q_r(L_t, L_s) = \min\{s(o_i, o_j) \mid o_i, o_j \in S \text{ ; and for any two objects } o_{i_o} \text{ and } o_{i_r} \text{ in } L_t \cup L_s, \text{ there is a sequence of not necessarily distinct objects in } S \text{, say } o_{i_o}, o_{i_1}, \ldots, o_{i_{r-1}}, \text{ such that } s(o_{i_{j'}}, o_{i_{j'+1}}) \leq s(o_i, o_j),\ 0 \leq j' \leq r-1\} .$$

It should be observed that for any value of r only the rank order of the object pairs in terms of proximity is actually needed in the clustering process. Also, for convenience we assume that all proximity values are distinct, and thus, unique minimums for the Q_r measure are obtained; otherwise, some additional procedure would be necessary for resolving ties.

There are several ways of interpreting the r-diameter clustering criterion that may provide an additional clarification. First of all, for any subset $L \subseteq S$, the r-diameter of L may be defined as the minimum proximity value, say c , such that each pair of objects in L is connected by a sequence of at most r-1 intermediate objects, where each adjacent pair in the sequence has a proximity value less than or equal to c. It should be observed that the intermediate objects may be drawn from outside the set L . Thus, the 1-diameter of a subset is the proximity value that is maximum on all distinct object pairs in L ; alternatively, the r-diameter of L guarantees that L is "connected" at this proximity value with respect to the set S ; furthermore, each object in L can be reached from any other object in L by some sequence of at most r-1 intermediate objects in which adjacent pairs have proximities less than or equal to the r-diameter of L. Consequently, given this interpretation and using the clustering context discussed above, two subsets L_i and L_j in ℓ_k are united to form an object class in ℓ_{k+1} if and only if their union has the smallest r-diameter among all possible object classes that could have been formed from two of the sets $L_1, \ldots, L_{n-k}$.

As a second interpretation, it should be observed that an r-diameter of a subset $L \subseteq S$ using the proximity function $s(\cdot,\cdot)$ is really a 1-diameter of L but with a different proximity function $s^r(\cdot,\cdot)$ defined as follows:

$$s^1(\cdot,\cdot) \equiv s(\cdot,\cdot) \ ;$$

$$\{s^r(o_i,o_j)\} = \{s^{r-1}(o_i,o_j)\} \oplus \{s^1(o_i,o_j)\}, \ r \geq 2 \ ,$$

where the matrix operation $\oplus$ is specified for $s^r(o_i,o_j)$ by

$$s^r(o_i,o_j) = \min_k \{\max\{s^{r-1}(o_i,o_k), \ s(o_k,o_j)\}\} \ .$$

Since 1-diameter hierarchical clustering is equivalent to what is generally referred to as a complete-link clustering, any r-diameter partition hierarchy can be viewed as a complete-link partition hierarchy constructed from a modified proximity function. Moreover, since the (n-1)-diameter criterion is equivalent to what is usually called the single-link criterion, any r-diameter strategy can be considered as defining some point along a clustering continuum with the two extremes occupied by the complete-link procedure (1-diameter) and the single-link procedure ((n-1)-diameter). For related discussions of r-diameter hierarchical clustering the reader should consult [10].

Although the concept of an r-diameter clustering procedure as presented above is used in the later sections, an alternative definition should be pointed out that could have been considered. As mentioned previously, the r-diameter of a subset $L \subseteq S$ allows the use of indirect relationships between the objects in L obtained through objects in S - L. This notion may best be referred to as the indirect r-diameter of L. A second direct r-diameter of L could also be defined as the minimum proximity value, say c', such that each pair of objects in L is connected by a sequence of at most r-1 intermediate objects chosen from L, where each adjacent pair in the sequence has a proximity value less than or equal to c'. In the defining condition for the measure $Q_r(L_t, L_s)$, the phrase "objects in S" would be replaced by the phrase "objects in $L_t \cup L_s$." Again, the direct 1-diameter and (n-1)-diameter criteria would correspond to complete-link and single-link clustering, respectively, but the resulting r-diameter hierarchy cannot be obtained in general by using the modified proximity function $s^r(\cdot,\cdot)$ constructed from the matrix operation $\oplus$. Direct r-diameter clustering has been discussed in [9] but the use of the modified proximity functions $s^r(\cdot,\cdot)$ is incorrectly suggested in passing as a possible vehicle. Both the direct and indirect r-diameter criteria could be used and there appears to be arguments both for and against the selection of either. Computationally, however, the indirect notion is somewhat easier to deal with, and thus, the discussion here will be limited to the concept of an indirect r-diameter.

3. Goodness-of-Fit Indices.

There are a number of procedures that have been suggested in the literature for evaluating the adequacy of a partition hierarchy obtained either from the complete-link or the single-link method. The discussion below, however, concentrates on a single overall measure of fit between a partition hierarchy and the given proximity values. Since global comparisons of this type seem to be the primary concern for the practitioner (see [19]), other more detailed extensions will not be developed explicitly.

Nevertheless, the conclusions that we obtain for an overall index also have obvious and direct analogues for statistics that are appropriate for the individual partitions within the hierarchy [23] or that defined by related graph-theoretic ideas [15, 16].

Since only the rank order of all object pairs in terms of proximities is used in r-diameter hierarchical clustering, any sequence of obtained partitions can be viewed as a reranking of all object pairs. Specifically, an object pair is given a partition rank based on the level (i.e., 0 to n-1) at which that object pair is first placed within the same object class of a partition. Consequently, the degree to which the partition hierarchy matches the information provided by the proximity ranks can be partially assessed by comparing the partition ranks to the proximity ranks. In particular, the Goodman-Kruskal [7] γ index of rank correlation can be calculated and, at least for the 1-diameter and the n-1 diameter criteria, a hypothesis of random "agreement" between the partition and proximity ranks can be tested using the approximate tables for γ given in [8]. These reference distributions are based on the permutation model discussed by Ling [14] and assume that all possible assignments of proximity ranks to the object pairs are equally-likely. Most data sets that are subjected to a hierarchical clustering, however, use some type of proximity function that by its very nature is constrained to have certain properties. Consequently, the permutation model usually can be criticized as being inappropriate. For instance, since the triangle inequality holds for proximity functions with metric characteristics, and even though some type of randomness may be responsible for generating the observed proximities, all permutations of the proximity ranks are not equally-likely a priori. In other words, certain rankings of the object pairs should appear more often than others even though the proximities are being produced by some chance mechanism.

The section below is devoted to stating three possible ways of generating proximity functions from data that is originally obtained through a random process. For purposes of comparison, one of the suggestions produces the equally-likely permutation concept mentioned above in which all possible assignments of the proximity ranks to the object pairs are assumed to have the same probability of occurence. The two other alternatives are not intended to represent actual "null" models for use in statistical inference, but instead, merely define procedures that can induce a non-uniform distribution over the set of all possible permutations of the proximity ranks of a type expected to occur in actual data. Clearly, many alternative schemes for restricting proximities are possible but most appear to provide the same general type of nonuniform distribution and only the absolute degree of restriction distinguishes the various possibilities.

4. Proximity Function Generation.

One of the more direct ways of restricting the form of a proximity function for evaluating its effect on an overall index of fit is to first generate attribute data on the objects in S in some random fashion and then impose a further structure on a final proximity function by employing a summary distance or correlation measure. One of the most common proximity functions of this type is based on squared Euclidean distance in an m-dimensional attribute space and is labeled as the D^2 model below. The second approach, called the permutation model, is the one studied by Ling [14] and can be defined by the generation of uniform random numbers from the unit interval added to the same constant (e.g., zero) for all object pairs. The third model, called a shortest path model, is based on a metric measure constructed by means of a shortest path algorithm employed in operations research and graph theory. In particular, from an initial random proximity function, such as that used in the permutation model, the final proximity measure between two objects is the smallest sum of the randomly generated values over all paths between the given objects. Formally, these three final proximity measures can be characterized as follows:

D^2 model

$$s_1(o_i,o_j) = \sum_{k=1}^{m} (x_{ik} - x_{jk})^2 ,$$

where the variables x_{uv} are independent and $N(0,1)$, $1 \leq u \leq n$, $1 \leq v \leq m$.

Permutation model

$$s_2(o_i,o_j) = x_{ij} ,$$

where the variables x_{uv} are independent and uniform on $[0,1]$, $1 \leq u, v \leq n$.

Shortest path model

$$s_3(o_i,o_j) = s_2^{[n-1]}(o_i,o_j), \text{ where}$$

$$s_2^{[1]}(\cdot,\cdot) \equiv s_2(\cdot,\cdot) ;$$

$$\{s_2^{[k]}(o_i,o_j)\} = \{s_2^{[k-1]}(o_i,o_j)\} \otimes \{s_2^{[1]}(o_i,o_j)\} ,$$

for $k \geq 2$, and where the matrix operation $\otimes$ for the entry

$$s_2^{[k]}(o_i,o_j) = \min_h \{s_2^{[k-1]}(o_i,o_h) + s_2^{[1]}(o_h,o_j)\} .$$

Both the D^2 and the shortest path model generate proximity functions on $S \times S$ that have metric properties; specifically, the functions $(s_1(\cdot,\cdot))^{\frac{1}{2}}$ and $s_3(\cdot,\cdot)$ both satisfy the triangle inequality. The D^2 model was chosen due to its importance in the applied clustering literature, and the shortest path model was selected because of the simplicity of the metric measure actually constructed and its independence from any particular metric representation or constraints, e.g., variance conditions on the final proximities [4] or Euclidean realizability [13]. In particular, any proximity measure that is already metric would not be affected by an application of the shortest path algorithm.

It should be emphasized that the basic interest in using either the D^2 or the shortest path model results only from the restrictions placed upon the rank ordering of the final proximities. For instance, no constraints would be imposed on the final rank ordering of the proximities if the basic values used by the shortest path routine were defined by a large constant, say greater than 20, plus error uniform on the unit interval. In this case, the triangle inequality holds as a matter of course. What is crucial in our use of both the D^2 and the shortest path model is some adherence to the following property: two object pairs with a common object and small associated ranks (i.e., that are highly similar) imply that the rank of the third object pair defined in the triple is also relatively small.

5. Procedures and Results.

The empirical results to be reported below are limited to an n of 16, although exactly the same type of conclusions can be obtained for other values of n as well. Also, in the D^2 model, the number of attributes, m, was set at 4, but again, almost identical conclusions can be attained for other values of m.

Under the D^2 model, 1,000 attribute samples were obtained and for each of the 1,000 constructed proximity functions, the r-diameter clustering strategy was applied for values of r from 1 to 8 and for the extreme (single-link) value of 15. The sample means and standard deviations of the γ values are given in Table 1. Surprisingly, the mean γ index increases up to an r of 3 and thereafter decreases. Table 2 presents the same type of information for 1,000 proximity functions constructed under the permutation model. In this case there is a consistent decrease in the size of γ from an r of 1 to 15. The results for these two extreme criteria under the permutation model correspond directly to the tabled percentage points given in [8]. Finally, Table 3 provides the mean and variance of γ for the shortest path model using 100 randomly generated proximity functions. The sample size was reduced for this last simulation because of the increase in the computational cost involved in running the shortest path algorithm. Nevertheless, the same patterning of means and variances observed under the D^2 model is present here as well, including an increase in the mean γ up to an r of 3 and a decrease thereafter.

r	Mean γ	Standard deviation γ
1	.524	.104
2	.539	.101
3	.545	.096
4	.535	.095
5	.526	.092
6	.520	.096
7	.518	.097
8	.516	.097
15	.516	.098

Table 1. Sample Means and Standard Deviation for γ Under the D^2 Model-Sample Size is 1,000; n = 16, m = 4.

r	Mean γ	Standard deviation γ
1	.304	.058
2	.265	.077
3	.250	.070
4	.239	.067
5	.222	.066
6	.213	.066
7	.207	.064
8	.205	.064
15	.204	.064

Table 2. Sample Means and Standard Deviations for γ Under the Permutation Model - Sample Size is 1,000

r	Mean γ	Standard deviation γ
1	.664	.115
2	.673	.109
3	.687	.091
4	.667	.102
5	.661	.099
6	.648	.114
7	.648	.114
8	.648	.114
15	.648	.114

Table 3. Sample Means and Standard Deviations for γ Under the Shortest-Path Model - Sample Size is 100; n = 16

Tables 1, 2, and 3 provide a number of interesting empirical conjectures regarding the performance of the r-diameter clustering schemes and the use of the overall goodness-of-fit measure. For instance, given the permutation model results of Table 2, it is clear that the average value yielded by the single-link strategy ($\gamma = .204$) is much less than the average obtained for the complete-link ($\gamma = .304$). In comparison, both the D^2 and shortest path models generate similar values for these two extreme methods. Rather interestingly, the results for the D^2 model are at odds with what generally occurs when proximities are generated by imposing error on a perfect hierarchical structure, e.g., see [8], suggesting that the latter procedure obscures the underlying metric structure reflected by the D^2 model and possibly provides an unfair comparison for the single-link criterion. Secondly, the average γ values observed for both the D^2 and shortest path models are considerably greater than those found for the permutation model, roughly double. Clearly, constrained proximity functions impose a structure on the proximity ranks that could be identified as non-random under the permutation assumption. In addition, when some structure is present in the proximity ranks, the "best" value of r in terms of a mean γ value is not necessarily attained for the complete-link method, i.e., for an r of 1. Consequently, applied researchers in the behavioral scineces may wish to reconsider their strong reliance on the complete-link strategy, at least when the given proximity ranks are by definition constrained through the initial choice of a proximity function. In terms of an average index γ, some relaxing of the clustering criterion may be beneficial in producing a better "fit" between the proximity and the partition ranks.

Finally, in an initial screening of an obtained partition hierarchy for purposes of a further substantive interpretation, the percentage points available in [8] appear to be "minimum" acceptable values. For example, Table 4 compares the percentage points of the empirical γ distributions for the complete-link and the single-link strategies given in [8] to the percentage points obtained under the D^2 and shortest path models. The percentage points for the permutation model appear to be consistently much less than the values obtained with the other two models. Consequently, as a heuristic data analysis strategy, the approximate extreme percentage points under the permutation model should be exceeded easily whenever the proximity function is constrained to have certain properties, e.g., some form of the triangle inequality. Furthermore, given data of Table 5, the single-link percentage points under the permutation model appear to give minimal values for all of the r-diameter strategies under the same assumption. Thus, the single-link percentage points *per se* provide a very lenient baseline that may be of more general use when the proximity ranks are constrained.

Model	Cumulative proportions					
	.50	.70	.80	.90	.95	.99
			A. Complete-link method			
Permutation	.31	.34	.35	.38	.41	.45
Shortest path	.68	.73	.76	.80	.83	.85
D^2	.52	.58	.61	.66	.68	.73
			B. Single-link method			
Permutation	.21	.24	.26	.29	.32	.36
Shortest path	.67	.72	.74	.78	.82	.84
D^2	.51	.57	.59	.63	.66	.72

Table 4. Comparison of the Approximate Percentage Points for the Complete-Link and Single-Link Strategies Under the Permutation Model to the Approximate Percentage Points Under the D^2 and Shortest Path Models; $n = 16$

r	Cumulative proportions					
	.50	.70	.80	.90	.95	.99
1	.31	.34	.35	.38	.41	.45
2	.26	.30	.33.	.36	.38	.43
3	.25	.28	.31	.33	.36	.39
4	.23	.26	.28	.31	.34	.37
5	.22	.25	.27	.30	.32	.36
6	.21	.24	.26	.29	.32	.36
7	.21	.24	.26	.29	.32	.36
8	.21	.24	.26	.29	.32	.36
15	.21	.24	.26	.29	.32	.36

Table 5. Comparison of the Approximate Percentage Points for the r-Diameter Strategies Under the Permutation Model

6. Numerical Illustration.

Since intermediate values in r-diameter clustering appear to provide average fits in terms of γ that are slightly better than what is obtained with the complete-link procedure, at least when the proximities are not generated by a permutation model, it is of interest to demonstrate the effect of clustering for various values of r with an actual data set. As an illustration that has been discussed extensively in the literature [5, 12], we will use Kahl and Davis's [11] tetrachoric correlation matrix on the nineteen social status indices listed below:

1) Warner occupational category;
2) Occupation of friends, North-Hatt category;
3) Subject's education;
4) Census occupational category of subject;
5) North-Hatt occupational category of subject;
6) Wife's father's occupation, census category;
7) Interviewer's impressionistic rating of subject;
8) Self-identification of subject;
9) Subject's mother's education;
10) Source of income;
11) Census tract, mean monthly rent;
12) Interviewer's rating of residential area;
13) Subject's father's occupation, census category;
14) Interviewer's rating of house, Warner category;
15) Subject's father's education;
16) Wife's father's occupation, North-Hatt category;
17) Subject's wife's education;
18) Annual family income;
19) Subject's father's occupation.

The Kahl and Davis proximity matrix is given in Table 6, where an entry for any two of the indices is defined by one minus the tetrachoric correlation. All values of r from 1 to 18 were used but after an r of 5, the single-link hierarchy was consistently produced. The overall γ values were fairly similar, and for this value of n were much larger than needed to reject randomness in the assignment of the proximity ranks under the permutation model, e.g., at $\alpha = .05$. In particular, γ's of .50, .52, and .53 were found for r's of 1,2 and 3 respectively; for r's greater than 3, a γ value of .52 was consistently produced.

The partition hierarchies for the various values of r were very similar at the lower partition levels, but discrepancies appear in the partitions containing a smaller number of subsets. As representative examples, the results for the 11th, 13th, and 15th levels are given in Table 7 for r = 1, 2, 3, and 15; at least for these three specific levels, all values of r greater than 3 constructed the same partitions. As to be expected from the discussion in [10], as r increases, the subsets within a partition tend to become somewhat larger, and apparently, by decreasing r the researcher can exert some control over the chaining properties of procedures related to the single-link strategy.

The results listed in Table 7 exhibit a number of similarities as well as several discrepancies. For a convenience in interpreting the given partitions, several subsets of variables may be identified and provided with suggestive labels; these classes or their subclasses occur in some form or another within the various decompositions:

Subset	Heuristic interpretation
{1,3,4,5,7,8,10}	Indices relating directly to the subject;
{2,17}	Indices relating to the subject indirectly through wife and friends;
{1,2,3,4,5,7,8,10,17}	Indices relating either directly or indirectly to the subject;
{9,13,15,19}	Indices relating to the subject's family;
{6,16}	Indices relating to the wife's father;
{11,18}	Indices relating directly to the amount of money earned or spent;
{12,14}	Indices relating to the subject's house;
{11,12,14,18}	Indices relating to money and residence.

TABLE 6. KAHL AND DAVIS PROXIMITY MATRIX FOR NINETEEN STATUS INDICES

Index	1	2	3	4	5	6	7	8	9	10	11	12	13	14	15	16	17	18	19
1	X	.20	.23	.07	.19	.43	.25	.27	.52	.22	.47	.46	.46	.51	.63	.57	.51	.49	.71
2		X	.19	.30	.30	.37	.31	.31	.43	.41	.64	.52	.40	.54	.52	.61	.18	.66	.55
3			X	.30	.35	.48	.47	.25	.40	.35	.59	.55	.47	.57	.46	.55	.41	.64	.59
4				X	.50	.41	.26	.29	.61	.14	.53	.57	.47	.61	.64	.60	.50	.59	.78
5					X	.50	.37	.37	.45	.46	.54	.49	.52	.64	.54	.56	.57	.48	.71
6						X	.66	.41	.35	.38	.36	.52	.35	.50	.52	.32	.55	.66	.52
7							X	.40	.59	.38	.50	.50	.56	.62	.53	.57	.60	.44	.65
8								X	.47	.40	.57	.68	.59	.70	.60	.53	.60	.66	.68
9									X	.61	.62	.70	.43	.65	.14	.51	.61	.61	.52
10										X	.59	.66	.70	.55	.77	.65	.53	.66	.80
11											X	.24	.67	.38	.55	.57	.61	.37	.64
12												X	.72	.08	.63	.38	.80	.65	.73
13													X	.51	.39	.46	.75	.70	.48
14														X	.60	.54	.63	.78	.86
15															X	.71	.71	.79	.48
16																X	.84	.65	.56
17																	X	.77	.80
18																		X	.87

r	Level	Partition
1	11	{{1,3,4,5,10}{12,14}{9,13,15}{2,17}{11,18}{5,7}{6,16}{19}}
	13	{{1,3,4,5,7,8,10}{12,14}{9,13,15,19}{2,17}{11,18}{6,16}}
	15	{{1,2,3,4,5,7,8,10,17}{6,12,14,16}{9,13,15,19}{11,18}}
2	11	{{1,4,5,7,8,10}{11,12,14}{9,15}{2,3,17}{6,16}{13}{18}{19}}
	13	{{1,4,5,7,8,10}{11,12,14,18}{9,15}{2,3,17}{6,13,16}{19}}
	15	{{1,2,3,4,5,7,8,10,17}{11,12,14,18}{6,9,13,15,16}{19}}
3	11	{{1,2,3,4,5,7,8,10}{11,12,14}{9,15}{6}{13}{16}{18}{19}}
	13	{{1,2,3,4,5,7,8,10,17}{11,12,14}{9,13,15}{6,16}{18}{19}}
	15	{{1,2,3,4,5,7,8,10,17}{11,12,14,18}{6,9,13,15,16}{19}}
15	11	{{1,2,3,4,5,7,8,10,17}{11,12,14}{9,15}{6}{13}{16}{18}{19}}
	13	{{1,2,3,4,5,7,8,10,17}{11,12,14}{9,15}{6,13,16}{18}{19}}
	15	{{1,2,3,4,5,7,8,10,17}{6,9,11,12,13,14,15,16}{18}{19}}

Table 7. Representative Partitions for the Kahl and Davis Data Using r-Diameter Hierarchical Clustering

It may be reasonable to label in some meaningful manner several of the other subsets that appear in Table 7, but for our purposes, naming these additional classes appears to be more speculative than is really necessary. What is important to observe is that subtle differences in the various partition hierarchies do exist, and therefore, any reliance on a single value of r, or a single hierarchical clustering strategy in general, may promote a degree of certainty that is not justified by the data. As a further indication of the need for some caution, Figure 1 presents a two-dimensional configuration of the nineteen social status indices obtained with the well-known nonmetric multidimentional scaling algorithm developed by Guttman and Lingoes [6]. Referring to Figure 1, each of the partitions listed in Table 7 may be viewed as alternative techniques for subdividing the obtained placement in such a way that "close" objects are also placed together within the same subsets, and conversely. Interesting, both the complete-link and single-link partitions at, say, level 15 appear to "match" the two-dimensional configuration less well than either the 2 (or 3)-diameter partition at this same level. Moreover, given the placement of objects in this configuration, it is reasonable to expect that alternative clustering criteria will generate decompositions that organize the variables in slightly different patterns.

7. Summary.

Given the variety of data used in the behavioral and social sciences along with the enormous number of available clustering procedures, it is unrealistic to expect any general consensus on a single baseline model. Nevertheless, some reference guidelines are necessary since in their absence any clustering strategy is little more than an approximate descriptive tool. Unfortunately, by noting the simulation results reported earlier, it is apparent that the use of different models will also lead to rather different reference distributions, even for the same measure of goodness-of-fit.

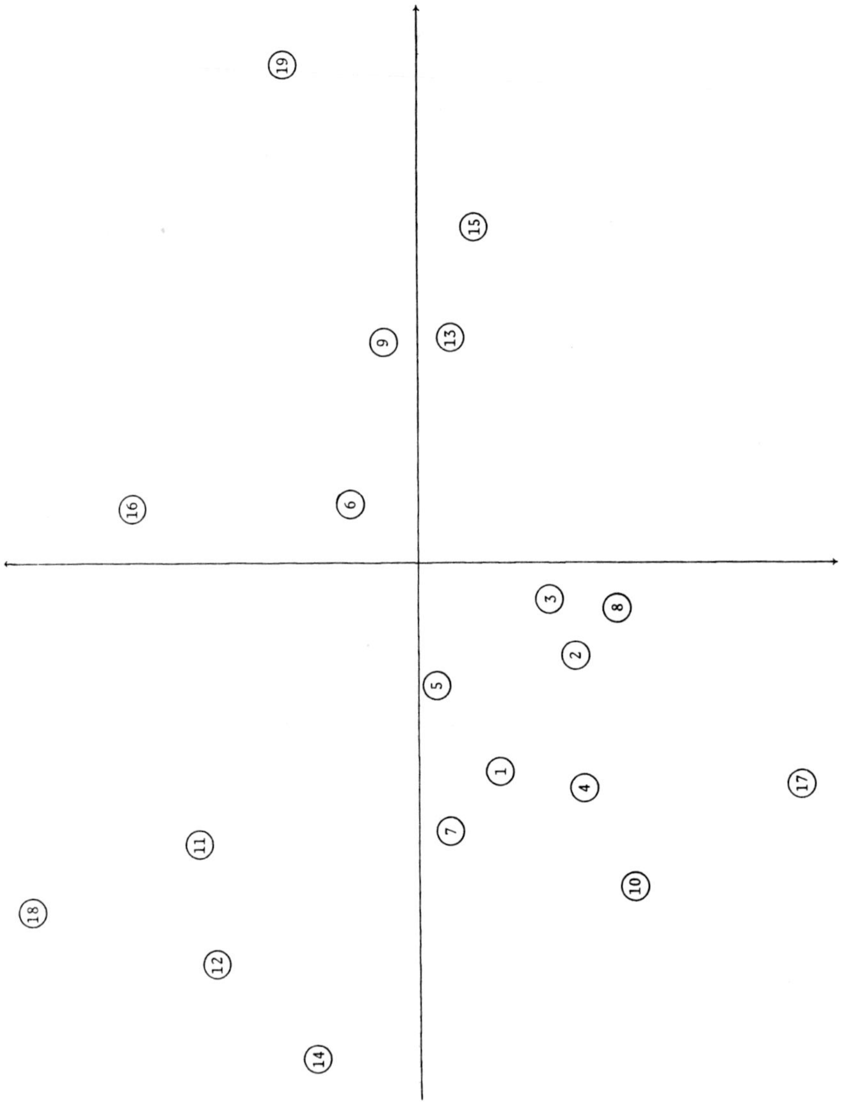
19
15
9
13
16
6
3
8
2
5
1
4
17
7
11
10
18
12
14

In conclusion, it is clear that the performance of any hierarchical clustering method needs to be evaluated in a number of alternative ways. For instance, even though the complete-link criterion has a definite superiority under the permutation model, the single-link and complete-link schemes are fairly close in terms of overall goodness-to-fit whenever the proximity functions are constrained in certain ways. Furthermore, considering proximity functions of the restricted form studied, neither the complete-link nor the single-link strategy may be the "best" clustering technique within the class of r-diameter methods; in fact, intermediate clustering criteria may prove more adequate with respect to an average overall goodness-of-fit index γ .

References

1. Baker, F. B., Stability of Two Hierarchical Grouping Techniques; Case I: Sensitivity to Data Errors, Journal of the American Statistical Association, 69 (June 1974), 440-445.

2. Baker, F. B. and Hubert, L. J., Measuring the Power of Hierarchical Cluster Analysis, Journal of the American Statistical Association, 70 (March 1975), 31-38.

3. Baker, F. B. and Hubert, L. J., A Graph-theoretic Approach to Goodness-of-Fit in Complete-link Hierarchical Clustering, Paper presented at the Psychometric Society Meeting, Iowa City, 1975.

4. Cunningham, J. P. and Shepard, R. N., Monotone Mapping of Similarities into a General Metric Space, Journal of Mathematical Psychology, 11 (November 1974), 335-363.

5. Fortier, J. J. and Solomon, H., Clustering Procedures, in P. R. Krishnaiah (Ed.), Multivariate Analysis, New York: Academic Press, 1966, 493-506.

6. Guttman, L., A General Nonmetric Technique for Finding the Smallest Coordinate Space for a Configuration of Points, Psychometrika, 33 (December 1968), 469-506.

7. Goodman, L. A. and Kruskal, W. H., Measures of Association for Cross Classification, Journal of the American Statistical Association, 49 (December 1954), 732-764.

8. Hubert, L., Approximate Evaluation Techniques for the Single-link and Complete-link Hierarchical Clustering Procedures, Journal of the American Statistical Association, 69 (September 1974), 698-704.

9. Hubert, L., Some Applications of Graph Theory to Clustering, Psychometrika, 39 (September 1974), 283-309.

10. Hubert, L. and Schultz, J. V., Hierarchical Clustering and the Concept of Space Distortion, The British Journal of Mathematical and Statistical Psychology, 28 (November 1975), 121-133.

11. Kahl, J. A. and Davis, J. A., A Comparison of Indexes of Socio-economic Status, American Sociological Review, 20 (June 1955), 317-325.

12. King, B., Step-wise Clustering Procedures, Journal of the American Statistical Association, 62 (March 1967), 86-101.

13. Kruskal, J. B., Multidimensional Scaling by Optimizing Goodness of Fit to a Nonmetric Hypothesis, Psychometrika, 29 (March 1964), 1-27.

14. Ling, R. F., A Probability Theory of Cluster Analysis, Journal of the American Statistical Association, 68 (March 1973), 159-169.

15. Ling, R. F., An Exact Probability Distribution on the Connectivity of Random Graphs, Journal of Mathematical Psychology, 12 (February 1975), 90-98.

16. Ling, R. F. and Killough, G. G., Probability Tables for Cluster Analysis Based on a Theory of Random Graphs, Journal of the American Statistical Association, 71 (June 1976), in press.

17. Luce, R. D., Connectivity and Generalized Cliques in Sociometric Group Structure, Psychometrika, 15 (June 1950), 169-190.

18. Peay, E., Nonmetric Grouping: Clusters and Cliques, Psychometrika, 40 (September 1975), 297-314.

19. Sokal, R. R. and Rohlf, F. J., The Comparison of Dendrograms by Objective Methods, Taxon 11 (February 1962), 33-39.

Professor Lawrence J. Hubert
Department of Educational Psychology,
University of Wisconsin-Madison
Madison, Wisconsin 53706

Professor Frank B. Baker
Department of Educational Psychology
University of Wisconsin-Madison
Madison, Wisconsin 53706

Data Dependent Clustering Techniques

Herbert Solomon

Introduction.

The major thrust in this study is data dependent clustering. In a large number of disciplines, large amounts of multivariate data are collected and the structure underlying the data is unknown or at best has received some initial and tentative exploration. We wish the data to tell us what is going on. It is not so important that an exact number of clusters with just the right elements in each cluster (an ideal not ever achieved) is determined. It is important that we partition the n multidimensional points in a p-dimensional space (measurements on p variables) in a valid, reliable, and parsimonious manner in an efficient way. This suggests that the consumer (i.e., supplier of the data), the statistician, and computing specialists state some satisfaction with a procedure.

The place of the statistician in this trinity is interesting. In a sense, modern statistical inference grew out of problems of classification and clustering, especially those problems posed in the latter half of the 19th century by the anthropometrists, biologists and those who followed Darwin, Galton, and the other empiricists of that day whose preoccupation was data collection. For example, Galton himself was interested in the classification scheme for fingerprints posed by Bertillon and essentially developed an unweighted factor analysis scheme as a result. Karl Pearson's 1901 paper [13] was essentially an attempt at classification and discrimination of n points in a p-dimensional space by construction of lines and planes of closest fit to these points.

However, any attempt at direct clustering or even classification of multivariate data had to be delayed until the advent of the computer as we knew it in the 1960's. In the interim period we find a number of mathematically elegant multivariate analysis models provided by the ingenious work of scholars like R. A. Fisher, Harold Hotelling, Sam Wilks, C. R. Rao, T. W. Anderson, and a number of others who directly or indirectly tried their hand at multivariate analysis. Fisher's multiple linear discriminant function technique (originally designed for two populations) is now a classic and a computer package. The iris data employed in his article [3] has been immortalized by usage in subsequent papers by other authors. This is true for any published data base and unfortunately these are rare occurrences. The scarcity of published data bases make it difficult for later investigators who wish to check their procedures on the same data base employed by earlier investigators. Fortunately the iris data appeared originally in an edition of the Bulletin of the Iris Society.

Clustering Techniques.

However, we are now at a juncture where a large number of clustering techniques have been developed and promulgated. The major activity along these lines has taken place in the last 15 years or so. In fact, we are now at the pretentious stage of thinking about the clustering of "clustering techniques" and we return shortly to this point. First a word about some specific clustering techniques. These appear here in connection with a study that will be discussed shortly. Here are some of the more popular varieties with brief comments about each.

1) Q-Factor Analysis: Factor analysis of elements rather than variables, number of clusters defined by factors and entry into cluster determined by highest factor loading [15].

2) Single Linkage (Nearest Neighbor): Groups initially consisting of single individuals are fused according to the distance between their nearest neighbors, groups with smallest distance being fused. Each fusion decreases by one the number of groups.

Distance between groups is defined as distance between their closest members. This leads to "serpentine" or "chained" clusters [6].

3) Complete Linkage (Furthest Neighbor): Distance between groups is now defined as distance between their most remote pair of individuals. Distance between merging clusters is the diameter of the smallest sphere which can enclose them. It yields tight, hyperspherical clusters that join others only with difficulty. Each fusion decreases by one the number of groups [6].

4) Average Linkage (King's Method): Distance between groups is judged by their centroids and closest centroids are fused. Each fusion decreases by one the number of groups [8].

5) k-means: Start with k-clusters (e.g., first k points); use minimum intracluster distance around the mean as criterion. As an element enters the cluster, the mean is updated and this continues until all points are placed [11].

6) ISODATA: Start with k-clusters and assign all elements by intracluster minimum criterion. After all elements have been assigned, update the means and do again until no gain occurs in intracluster minimum criterion [1].

7) Covariance Criterion Optimization: Place points in k-clusters and reassign according to a variance-covariance criterion: e.g., let T = matrix representing total scatter of n points in p-space, W_i = matrix of scatter of n_i points in i^{th} cluster, B = matrix of scatter of centroids of the k clusters; then $T = W + B$ where $W = \Sigma W_i$. Two criteria that can be developed are: minimize trace of W and maximize the determinantal ratio $|T|/|W|$. The latter is preferred for several reasons; one is that it is invariant under linear transformations while the former is invariant only under orthogonal transformations [5].

Note that clustering techniques (2), (3), (4) are hierarchical grouping procedures, i.e., one begins with n clusters each containing one element and each fusion reduces the number of clusters by one until only one cluster containing all points is achieved. In the k-means technique and the ISODATA technique, it appears that one deals only with k-clusters. This is not so, becuase the k-cluster configuration can be reduced or enlarged if the intracluster distances are either too small or too large for the finally selected k-clusters. This characteristic is true also for the covariance criterion optimization procedure.

There are some data representation and graphical techniques that are sometimes mistaken as clustering procedures. These techniques make visual clustering feasible by reducing the p-dimensional space to two dimensions or changing the multidimensional vector to a human face or some other analogue representation. The latter represents an interesting device developed by Chernoff [2], who translates the multidimensional data vector into a face and then judges are assigned to group the faces. Judges are also involved in Kruskal's multidimensional scaling technique [9, 10], which reduces the dimensionality of the p-dimensional space. Regular factor analysis can also be employed to reduce the p-dimensional space to two dimensions when measurement variables rather than elements are being clustered. Each variable is then a point in two dimensions, can be plotted, and the n points grouped by eye.

Evaluation of Clustering Techniques.

Since a number of clustering techniques are available, some evaluation of these techniques becomes a necessity. In a detailed and very productive study by Mezzich [12], the question of evaluation of clustering procedures is investigated. In order to accomplish this, he first constructs some evaluation indexes. They are: a measure of external criterion validity, a measure of internal criterion validity, and a measure of replicability. Mezzich measures external criterion validity by obtaining the percentage of concordance of expert predictions and the results of the clustering procedure. This can be accomplished by the use of a contingency table. Briefly, in this situation, the expert or the consumer decides how the actual clustering developed by the clustering procedure relates to the substance of the problem that produced the data in the first place.

The measure of internal criterion validity is the cophenetic correlation coefficient introduced by Sokal and Rohlf [14]. This is the ordinary product moment correlation coefficient between corresponding cell entries of the similarity matrix derived from the cluster configuration and the initial similarity matrix employed to initiate the clustering. The measure of replicability or stability is also essentially a correlation coefficient. In this situation the data base is divided at random into two equal data sets. For each of the two data sets, a clustering configuration is derived by the clustering procedure in question. From each of the two clustering configurations, a derived similarity matrix can be constructed. Accordingly, a correlation coefficient can then be computed over corresponding cells in each of the two similarity matrices, each of which is developed from the clustering configuration produced by the clustering technique. In this way, each of the clustering techniques previously described briefly can be evaluated; that is, in terms of each of the three measures: external criterion validity, internal criterion validity, and replicability or stability.

In order to accomplish this, some data bases have to be made available. We have previously mentioned that this is usually not an easy task, except for Fisher's iris data. Mezzich used four data bases, which are described below; one, of course, being Fisher's iris data. Let us view this first in the Mezzich study. In Fisher's data, there are 150 irises, 50 in each of three groups. For the Mezzich study, each iris group of 50 plants was divided into two groups of 25 plants at random. Thus, a clustering procedure is applied each time to each data base of 75 irises. The iris data is four-dimensional: sepal width and sepal length and petal width and petal length.

A second data base was developed in the following way. It has to do with treatment environments in connection with psychiatric impatient treatment. From a previous study, there were 72 psychiatric impatient environments (e.g., therapeutic community, etc.) which were placed in four groups, 18 members to a group. Each of the four groups was associated with systematic differences in such variables as institution affiliation, patient/staff ratios, etc. There was a random and equal division of each of the four 18-member subgroups resulting in four subgroups, each containing 9 treatment environments.

A third data base also related to psychiatric studies. In this case, 88 patients were fabricated to represent four diagnostic categories with 22 patients per category. Each of these 88 patients received scores on a 17-item inventory. Naturally, each diagnostic group of 22 patients was randomly divided into two groups of 11 each, thus permitting, once again, a chance at getting a measure of replicability or stability.

The fourth data base employed in the Mezzich study was an interesting one. In this case, there were ten human populations paired in five groups according to ethnicity. They are: American natives (Eskimo and Maya); Europeans (English and Italian); Africans (Pygmy and Bushman); Oriental (Chinese and Japanese); Oceanians (Australian aborigine and New Guinean). Thus, there were ten human populations on each of which measurements had been taken on 58 independent genetic units. To achieve a measure of replicability or stability in this case, the 58 independent genetic units were divided at random into two sets each with 29 variables. This is slightly different than the other three data bases because, here, the measurement variables were divided in half at random, rather than the elements being studied.

Mezzich has many results and many tables exhibiting the results, but overall it seems fair to say that the complete linkage technique [6] and the King centroid method technique [8] seem to do best over all data bases and over all three evaluation indexes. Let us recall that these are both hierarchical procedures. In a subsequent section, we will employ the King Stepwise Clustering Procedure on several additional data bases and give an account of the resulting clusters in each case. Before this is accomplished, we will look into Mezzich's results on how the clustering techniques themselves cluster. In order to examine the clustering of our clustering techniques, a similarity matrix has to be developed to serve as a basis for employing the clustering techniques. This was accomplished in the following way.

Suppose we consider any pair of clustering techniques to which we have already referred. Assume that A and B are the pair, and we have already accomplished the clustering by clustering procedure A and the clustering developed by clustering procedure B . Recall also that this is done only on half of the data base each time, as we have described it in the previous sections. For clustering method A and clustering method B , applied, say, to half of the first data base, we get a pair of clustering configurations. Each member of this will produce a derived similarity matrix. Thus we have two similarity matrices for tne same data base: one produced by clustering method A and the other by clustering method B . A correlation coefficient can be computed in a manner similar to the cophenetic correlation coefficient by taking the product moment correlation over all corresponding pairs of cell entries in the two derived similarity matrices. This can then be done for half of each of the other three data bases. This will give us four different correlation coefficients for the same cell if we construct a correlation matrix produced by the clustering procedures. Let us use the mean of the four correlations so developed so that we have only entry per cell in a correlation matrix where the rows and column categories are the different clustering procedures. A similar correlation matrix can be constructed from the other half of each of the four data bases.

Since complete linkage turned out to be the best clustering algorithm for the four data bases employed previously, let us now apply this to each of the two cluster correlation matrices and view the results. When this is accomplished, it develops that complete linkage and the covariance criterion optimization techniques fall in a cluster and together with ISODATA and the King clustering procedure, essentially form a cluster of four clustering procedures. Single linkage clustering and the k-means clustering technique each appear as isolates or clusters containing one element. This accounts for all the clustering techniques we described earlier, save for Q-factor analysis, but it is best to recall that Q-factor analysis is essentially a data representation rather than a clustering technique.

King Clustering Procedure Applied To Other Data Bases.

We now look at the King Stepwise Clustering Technique applied to several data bases. These had been done prior to the results provided by Mezzich simply because the technique was rather simple and efficient to use. For example, the covariance optimization technique is much less efficient on a computer than the King technique or the other two hierarchical techniques of complete linkage and single linkage. Two of the data bases we now consider concern the clustering of measurement variable, whereas four of the data bases refer to the clustering of individuals or elements.

In the first data base on measurement variables, there were 19 socio-economic variables employed on a personal inventory given to college students. This resulted in a 19×19 correlation matrix which had been computed in a paper by Kahl and Davis [7], who then employed the Tryon clustering technique. This technique led to 8 clusters. Table 1 shows these clusters. The King Stepwise Clustering Procedure applied to the same data provided a clustering configuration which is shown in the same table, and one can see a rather close relationship. Naturally, the King stepwise procedure is much, much faster than the method employed by the two authors. A factor analytic two-dimensional representation was also developed, and the visual clustering led to the clusters also shown in that table, along with a clustering produced by a total enumeration method developed by Fortier and Solomon [4]. The Fortier-

Clusters*	Tryon's method Variables	Fortier-Solomon C^* method	King step-wise procedure	Two-dimensional representation
1	12. Area rating 14. House rating	The same variables	The same variables	The same variables merged with 11
2	15. Subject's father's education 9. Subject's mother's education	The same variables	The same variables	The same variables merged with 13, 1
3	2. Friend's occupation 17. Wife's education	The same variables	The same variables	The same variables merged with 3, 8
4	4. Subject's occupation Census 1. Subject's occupation, Warner 10. Source of income	The same variables	The same variables merged with variables 3, 8, 5, and 7	The same variables
5	16. Wife's father's occupation, North-Hatt 6. Wife's father's occupation, Census	Not a cluster	Not a cluster	Not a cluster
6	11. Census tract 18. Income	Not a cluster	Not a cluster	Not a cluster
7	3. Subject's education 8. Subject's self-identification	The same variables	*See* cluster 4 above	*See* cluster 3 above
8	19. Subject's father's occupation, North-Hatt 13. Subject's father's occupation, Census	Not a cluster	Not a cluster	*See* cluster 2 above

* The clusters are ordered by decreasing values of Tryon's index.

Table 1. Clustering of the 17 of 19 variables in Kahl-Davis data (variables 5 and 7 are omitted) by 4 clustering procedures

Solomon method is a very tedious one and takes much time on the computer. The King stepwise procedure takes the least time. Yet as we see, the kind of clustering one accomplishes by any of the four methods is roughly the same to those who are interested in an initial look at the clustering of the 19 measurement variables.

In Table 2 we return to the Fisher iris data and give some indication of the kind of clustering one can achieve using the King method on the 150 irises, each of which is represented as a point in a four-dimensional space. Note that at the 137th pass the irises, which have been labeled 1, 2, and 3, because we know in which group each belongs, show 48 of the class 1 irises in one group, whereas the class 2 and class 3 irises are scattered essentially over four large clusters of 23 elements, 24 elements, 17 elements, and 24 elements. On the next pass, the 48 class 1 irises remain as before, but now we have three large clusters with 23 elements, 48 elements, and 17 elements. This is a situation where the class 2 irises and class 3 irises do overlap somewhat and what we are seeing here is how this is reflected in the King stepwise procedure.

Still another data base comes about in Table 3. Here we have the case of clinical vs. statistical prediction. Eighty-two sphasic children were tested and given a clinical diagnosis which placed them in one of seven categories. These categories and the number in each category follow.

(1) Mentally retarded

(2) Severe hearing impairment

(3) Neurologically handicapped/aphasic

(4) Oral apraxia

(5) Dysarthric

(6) Maturational lag

(7) Autistic .

The numbers appearing in the clusters in the table refer to the diagnosis already made. However, the clustering itself was accomplished by the King Stepwise Clustering Procedure on the 82 children where each was a

```
PASS=137          B=  1           M= 15
DISTANCE BETWEEN B,M = 0.318019E 01
  1      11111 11111 11111 11111 11110 11111 11111 11111 10111 11111 00000 00000 00000 00000 00000
         00000 00000 00000 00000 00000 00000 00000 00000 00000 00000 00000 00000 00000 00000 00000    48
 25      00000 00000 00000 00000 00001 00000 00000 00000 00000 00000 00000 00000 00000 00000 00000
         00000 00000 00000 00000 00000 00000 00000 00000 00000 00000 00000 00000 00000 00000 00000     1
 42      00000 00000 00000 00000 00000 00000 00000 00000 01000 00000 00000 00000 00000 00000 00000
         00000 00000 00000 00000 00000 00000 00000 00000 00000 00000 00000 00000 00000 00000 00000     1
 51      00000 00000 00000 00000 00000 00000 00000 00000 00000 00000 20202 00020 00000 20000 02002
         22200 00000 02000 00000 00200 00300 30330 03000 00030 00300 30003 30000 00000 00000 03000    23
 52      00000 00000 00000 00000 00000 00000 00000 00000 00000 00000 02000 02000 02002 00000 00000
         00000 00000 00000 00000 00000 00003 00003 30333 33000 03033 03330 00300 00033 30003 00300    24
 54      00000 00000 00000 00000 00000 00000 00000 00000 00000 00000 00020 00200 20200 00022 00200
         00002 22200 00202 00220 00020 00000 00000 00000 00003 00000 00000 00000 00000 00000 00000    17
 56      00000 00000 00000 00000 00000 00000 00000 00000 00000 00000 00000 20002 00020 02200 20020
         00020 00022 20020 22002 22002 03030 00000 00000 00000 00000 00000 00030 00300 00300 00003    24
101      00000 00000 00000 00000 00000 00000 00000 00000 00000 00000 00000 00000 00000 00000 00000
         00000 00000 00000 00000 00000 30000 00000 00000 00000 00000 00000 00000 03000 00000 00030     3
107      00000 00000 00000 00000 00000 00000 00000 00000 00000 00000 00000 00000 00000 00000 00000
         00000 00000 00000 00000 00000 00000 03000 00000 00000 00000 00000 00000 00000 00000 00000     1
118      00000 00000 00000 00000 00000 00000 00000 00000 00000 00000 00000 00000 00000 00000 00000
         00000 00000 00000 00000 00000 00000 00000 00000 00300 30000 00000 03000 00000 00030 00000     4
135      00000 00000 00000 00000 00000 00000 00000 00000 00000 00000 00000 00000 00000 00000 00000
         00000 00000 00000 00000 00000 00000 00000 00000 00000 00000 00000 00003 00000 00000 00000     1
136      00000 00000 00000 00000 00000 00000 00000 00000 00000 00000 00000 00000 00000 00000 00000
         00000 00000 00000 00000 00000 00000 00000 00000 00000 00000 00000 00000 30000 00000 00000     1
142      00000 00000 00000 00000 00000 00000 00000 00000 00000 00000 00000 00000 00000 00000 00000
         00000 00000 00000 00000 00000 00000 00000 00000 00000 00000 00000 00000 00000 03000 30000     2
TIME FOR THAT PASS HAS      0.37 SECONDS


PASS=138          B= 52           M= 56
DISTANCE BETWEEN B,M - 0.334658E 01
  1      11111 11111 11111 11111 11110 11111 11111 11111 10111 11111 00000 00000 00000 00000 00000
         00000 00000 00000 00000 00000 00000 00000 00000 00000 00000 00000 00000 00000 00000 00000    48
 25      00000 00000 00000 00000 00001 00000 00000 00000 00000 00000 00000 00000 00000 00000 00000
         00000 00000 00000 00000 00000 00000 00000 00000 00000 00000 00000 00000 00000 00000 00000     1
 42      00000 00000 00000 00000 00000 00000 00000 00000 01000 00000 00000 00000 00000 00000 00000
         00000 00000 00000 00000 00000 00000 00000 00000 00000 00000 00000 00000 00000 00000 00000     1
 51      00000 00000 00000 00000 00000 00000 00000 00000 00000 00000 20202 00020 00000 20000 02002
         22200 00000 02000 00000 00200 00300 30330 03000 00030 00300 30003 30000 00000 00000 03000    23
 52      00000 00000 00000 00000 00000 00000 00000 00000 00000 00000 02000 22002 02022 02200 20020
         00020 00022 20020 22002 22002 03033 00003 30333 33000 03033 03330 00330 00333 30303 00303    48
 54      00000 00000 00000 00000 00000 00000 00000 00000 00000 00000 00020 00200 20200 00022 00200
         00002 22200 00202 00220 00020 00000 00000 00000 00003 00000 00000 00000 00000 00000 00000    17
101      00000 00000 00000 00000 00000 00000 00000 00000 00000 00000 00000 00000 00000 00000 00000
         00000 00000 00000 00000 00000 30000 00000 00000 00000 00000 00000 00000 03000 00000 00030     3
107      00000 00000 00000 00000 00000 00000 00000 00000 00000 00000 00000 00000 00000 00000 00000
         00000 00000 00000 00000 00000 00000 03000 00000 00000 00000 00000 00000 00000 00000 00000     1
118      00000 00000 00000 00000 00000 00000 00000 00000 00000 00000 00000 00000 00000 00000 00000
         00000 00000 00000 00000 00000 00000 00000 00000 00300 30000 00000 03000 00000 00030 00000     4
135      00000 00000 00000 00000 00000 00000 00000 00000 00000 00000 00000 00000 00000 00000 00000
         00000 00000 00000 00000 00000 00000 00000 00000 00000 00000 00000 00003 00000 00000 00000     1
136      00000 00000 00000 00000 00000 00000 00000 00000 00000 00000 00000 00000 00000 00000 00000
         00000 00000 00000 00000 00000 00000 00000 00000 00000 00000 00000 00000 30000 00000 00000     1
142      00000 00000 00000 00000 00000 00000 00000 00000 00000 00000 00000 00000 00000 00000 00000
         00000 00000 00000 00000 00000 00000 00000 00000 00000 00000 00000 00000 00000 03000 30000     2
TIME FOR THAT PASS WAS      0.34 SECONDS
```

Table 2. Step-wise clustering for Fisher data on 150 Irises. Computer printout for 137th and 138th passes

```
P= 78           B= 3            M= 23
CORR. COEF. = 0.27540 00
          1     35004 72024 30330 40300 32005 03310 00320 02002 30300 30320 03060 06302 06030 00633 00303 26360 0000
          3     00720 00100 01001 05014 00110 10007 17001 30730 01017 04003 50003 70000 30106 30000 11030 00001 1000
         59     00000 00000 00000 00000 00000 00000 00000 00000 00000 00000 00000 00030 00000 00000 00000 00000 0000
         67     00000 00000 00000 00000 00000 00000 00000 00000 00000 00000 00000 00000 00000 06000 00000 00000 0000
         82     00000 00000 00000 00000 00000 00000 00000 00000 00000 00000 00000 00000 00000 00000 00000 00000 0300
         84     00000 00000 00000 00000 00000 00000 00000 00000 00000 00000 00000 00000 00000 00000 00000 00000 0006
P= 79           B= 1            M= 84
CORR. COEF. = 0.26190 00
          1     35004 72024 30330 40300 32005 03310 00320 02002 30300 30320 03060 06302 06030 00633 00303 26360 0006
          3     00720 00100 01001 05014 00110 10007 17001 30730 01017 04003 50003 70000 30106 30000 11030 00001 1000
         59     00000 00000 00000 00000 00000 00000 00000 00000 00000 00000 00000 00030 00000 00000 00000 00000 0000
         67     00000 00000 00000 00000 00000 00000 00000 00000 00000 00000 00000 00000 00000 06000 00000 00000 0000
         82     00000 00000 00000 00000 00000 00000 00000 00000 00000 00000 00000 00000 00000 00000 00000 00000 0300
P= 80           B= 1            M= 67
CORR. COEF. = 0.17820 00
          1     35004 72024 30330 40300 32005 03310 00320 02002 30300 30320 03060 06302 06030 06633 00303 26360 0006
          3     00720 00100 01001 05014 00110 10007 17001 30730 01017 04003 50003 70000 30106 30000 11030 00001 1000
         59     00000 00000 00000 00000 00000 00000 00000 00000 00000 00000 00000 00030 00000 00000 00000 00000 0000
         82     00000 00000 00000 00000 00000 00000 00000 00000 00000 00000 00000 00000 00000 00000 00000 00000 0300
P= 81           B= 1            M= 82
CORR. COEF. = 0.18140 00
          1     35004 72024 30330 40300 32005 03310 00320 02002 30300 30320 03060 06302 06030 06633 00303 26360 0306
          3     00720 00100 01001 05014 00110 10007 17001 30730 01017 04003 50003 70000 30106 30000 11030 00001 1000
         59     00000 00000 00000 00000 00000 00000 00000 00000 00000 00000 00000 00030 00000 00000 00000 00000 0000
P= 82           B= 1            M= 3
CORR. COEF. = 0.99660-01
          1     35724 72124 31331 45314 32115 13317 17321 32732 31317 34323 53063 76302 36136 36633 11333 26361 1306
         59     00000 00000 00000 00000 00000 00000 00000 00000 00000 00000 00000 00030 00000 00000 00000 00000 0000
P= 83           B= 1            M= 59
CORR. COEF. =-0.42690-01
          1     35724 72124 31331 45314 32115 13317 17321 32732 31317 34323 53063 76332 36136 36633 11333 26361 1306
```

Table 3. Step-wise clustering for 82 aphasic children. Computer printout

point in 25-dimensional space, since measurements were available on 25 variables. One can see in Table 3 that rather than seven nice clusters evolving, we essentially get to the point where there are only two clusters: one containing the class 1 and class 7 children and the other containing the class 2, class 3, class 4, class 5, and class 6 children. This was of interest to those in charge of the aphasia program, because it suggested that there were essentially only two categories, namely those who would not speak because of physiological handicaps, and those who did not speak becuase of non-physiological handicaps.

Still another interesting data base is one which has to do with the frequency usage by grammatical case of Russian nouns. A frequency count had been made for each of 560 Russian nouns on the basis of the frequency of usage in each of the six grammatical cases in Russian. Thus each noun was a point in a six-dimensional space. Some linguists were quite interested as to how these nouns would cluster as a function of their relative frequency of usage over the six grammatical cases. A King Stepwise Clustering Procedure was employed and Table 4 shows the results or partial results at the 400th pass. Naturally, the words have been translated into English. The linguists found much of interest in the way the clustering occurred at this pass and several other passes not published here, and once again, since the consumer was satisfied, the clustering was deemed to be a success.

In still yet another clustering situation, data had been collected on 238 individuals (males) convicted of first degree murder in California in the ten year period 1958-1968. In California in that period the jury bringing in the first degree verdict also served as the penalty jury and decided on life imprisonment or the death penalty. For each of the defendents involved, a number of measurements had been obtained on the defendant, on the defense attorney, on the prosecuting attorney, on the judge, and on the victim. What was at issue here, of course, was to see whether race or ethnic background or socio-economic status in some way was associated with the penalty decided upon by the jury. Thus

```
PASS=400          B=416              M=451   DISTANCE BETWEEN B,M = 0.806152E 00
  1     17     0   1   0   0   7   2   1   1   3   2    0.0   0.059 0.0    0.0   0.412 0.118 0.059 0.059 0.176 0.119
   7CC YEAR          9CC LIFE            800 WORK          419 SEA           543 VESSEL        401 HOUSE         403 CITY
   433 STATION       611 GRASS           109 STAFF         818 STUDIES       456 FOREST        414 SKY           962 RELATION
   4C5 EARTH         518 SHIP            837 PREPARATION
  2      7     0   0   1   1   C   0   1   0   1   3    0.0   0.0   0.143 0.143 0.0   0.0   0.143 0.0   0.143 0.429
   950 MATTER        305 HEART           954 PART          201 HORSE         963 CONDITION     815 STRUGGLE      674 TASTE

  3     15     0   0   0   2   0   4   0   5   2   2    0.0   0.0   0.0   0.133 0.0   0.267 0.0   0.333 0.133 0.133
   701 DAY           702 TIME            938 RANK          907 ANSWER        317 MOUTH         539 TICKET        554 WATCH
   72C HALF HOUR     325 HEAD            540 TRUNK         859 MEANING       810 TRANSMISSION  519 CHAIR         732 MOMENT
   709 WEEK
  4      5     0   0   0   2   0   0   0   0   2   1    0.0   0.0   0.0   0.400 0.0   0.0   0.0   0.0   0.400 0.200
   300 EYE           319 WING            939 ALARM         862 SPEED         872 ASSURANCE
  5      1     0   0   0   1   0   0   0   0   0   0    0.0   0.0   0.0   1.000 0.0   0.0   0.0   0.0   0.0   0.0
   3C1 HAND
  7      2     0   0   0   0   C   0   0   1   0   1    0.0   0.0   0.0   0.0   0.0   0.0   0.0   0.500 0.0   0.500
   951 TIME(EC.3TIM  716 PERIOD
  9     23    15   2   3   C   C   1   0   1   0   1    0.652 0.087 0.130 0.0   0.0   0.043 0.0   0.043 0.0   0.043
    33 PEOPLE         52 WORKER           63 WORKMAN        79 PHYSICIAN      89 ATHLETE        59 MEMBER OF KO   67 STUDENT
    E5 ARTIST         40 INHABITANT      200 BIRD          113 ARMY            2 PERSON        122 DETACHMENT     68 HERO
     7 CHILDREN       19 PARENT          992 COMMAND        16 CHILD         208 PIGEON        206 WILD ANIMAL   750 DOZEN
   551 FLAG           93 PEASANT
 11     40     1   0   1   1   0  10  10   4   6   7    0.025 0.0   0.025 0.025 0.0   0.250 0.250 0.100 0.150 0.175
   9C1 WORD          986 MOVEMENT        308 HAIR          725 ANNIVERSARY   501 MACHINE       533 TRACTOR       545 APPLE
   753 THOUSAND      527 MOTOR           607 TEAR          624 VODKA         568 PIPE          752 PAIR          534 LAMP
   604 MONEY         202 FISH            550 PLATE         618 FIREWOOD      724 DAY           601 BREED         621 MAIZE
    48 OPERATION     843 EQUIPMENT       615 GRAIN         869 BEAUTY        873 SKILL         821 FORMATION     999 SUPPER
   940 CONCERN       529 PEN             653 GLANCE        929 MATERIAL      827 LAUGHTER      672 SMELL         987 EXPERIENCE
   677 CLOUD         538 DISHES          914 FEELING       561 FLOWER        875 MERIT
 12     21    19   0   0   0   0   0   2   0   0   0    0.905 0.0   0.0   0.0   0.0   0.0   0.095 0.0   0.0   0.0
     1 CHILDREN       50 COMRADE           9 BROTHER        17 UNCLE           6 MOTHER        669 FROST           3 BOY
     4 GIRL            5 GIRL             10 WOMAN          46 GENERAL       655 WIND           82 AUTHOR         58 PUPIL
    7C CAPTAIN        71 COMMANDER        12 FATHER         78 WOMAN TEACHE   56 BRIGADEER      61 SECRETARY      23 MAN

 13      8     0   0   0   0   5   2   0   0   1   0    0.0   0.0   0.0   0.0   0.625 0.250 0.0   0.0   0.125 0.0
   400 PLACE         404 ROOM            417 APARTMENT     515 CLOTHES       524 ARMCHAIR      415 FIELD         847 FULFILLMENT
   429 KITCHEN
 14      4     0   0   0   1   0   1   1   0   0   1    0.0   0.0   0.0   0.250 0.0   0.250 0.250 0.0   0.0   0.250
   3C2 PERSON,FACE   608 GLASS           966 ROW           548 NOTEBOOK
 17      4     0   0   0   0   1   0   0   3   0   0    0.0   0.0   0.0   0.0   0.250 0.0   0.0   0.750 0.0   0.0
   721 MARCH         440 DISTRICT        719 FEBRUARY      729 AUGUST
 18      1     0   0   0   0   1   0   0   0   0   0    0.0   0.0   0.0   0.0   1.000 0.0   0.0   0.0   0.0   0.0
   402 SIDE
 19     12     0   0   0   0   7   0   0   0   3   2    0.0   0.0   0.0   0.0   0.583 0.0   0.0   0.0   0.250 0.167
   445 SCHOOL        811 CONSTRUCTION    446 FACTORY       424 SHOP          437 VILLAGE       968 LOT           842 COMPOSITION
   952 CLASS         412 WORLD           418 DISTRICT      448 COLL. FARM    822 CONFERENCE
 20      3     0   0   0   1   C   0   2   0   0   0    0.0   0.0   0.0   0.333 0.0   0.0   0.667 0.0   0.0   0.0
```

Table 4. Step-wise clustering for 560 Russian nouns on basis of frequency of usage in 6 grammatical cases. Computer printout

here we have 238 individuals we would like to cluster on the basis of measurements on a large set of variables which fall under the headings we have just mentioned. Twenty-five variables were employed for each defendant. In Table 5 we show the clustering at the 75th pass. Since we know which individuals received which penalty, we have labeled them already as class 2 and class 3, class 2 meaning life imprisonment and class 3 the death penalty. We note that in one cluster, there are 14 defendants, ten of whom have received life imprisonment and four the death penalty. It should be noted that overall the ratio of life imprisonment to death was essentially 55-45%. Thus, this cluster is quite meaningful since the percentage is now about 88% and it would be interesting to look inside the cluster to see what the characteristics of the defendants are. It turned out that in this cluster most of the defendants were Chicanos and yet they had received life imprisonment, a result somewhat at variance with what those who had conducted the study would expect.

In the last data base, we revert again to clustering of measurement variables. In this case there were 63 variables, each of which received responses from a large number of teenagers in connection with various aspects of the packaging and eating of chocolate candy. Obviously, one question here is whether all 63 variables are necessary. In Table 6 we show what happens at the 31st pass and we note the following. There seems to be a "chocolatey" cluster, for we see eight variables having to do with the "chocolatey" aspects of the candy. There seems to be a cluster that deals with the situational aspects of eating candy in which we find 20 variables that fall into that particular grouping. Then there is a third cluster which has five variables, all of which have to do with the "nuttiness" of the candy, or in effect, we have a "nuttiness" cluster.

Thus the King clustering procedure has in effect in these six data bases demonstrated how useful it can be to the individual who is exploring the data. In quick time, and at low cost, we get these exploratory clustering configurations. They are not a complete answer in themselves, but they are quite helpful to those who are exploring multivariate data

```
PASS=75          E= 16          P=153
DISTANCE BETWEEN E,M = C.131212E C2
  1     30CC0 CCCC0 0C000 C000C 30CC0 C0CCC CCCC0 0CCC0 00000 00CCC 00CC0 00000 C000C 00000 00CC0 0C0C0 C0000 00000 00000 00000
        CCCCC CCCCC CCCCC CCCCC CCCCC CCCCC C0CCC 00C0C 00000 00000 00000 00000 00000 C0CCC C0CC0 0CCCC C0C0C C0C00 C0000 0C000
        C0CC0 CCCC0 CC00C C0CCC CCCCC CCCCC CCCCC 00C                                          0.0    0   1   1
  2     C3CCC CCCCC CC0C0 C0C00 C0CC0 000C0 00CC0 CCC0C 00C00 C0C0C 00C0C CCCCC CCCCC C0000 000C0 0CCC0 CCCCC 00CC0 C0000 00000
        CCCC0 CCCC0 C00CC CCCCC CCCC0 CCCCC C0CCC CCCCC C0000 00000 00000 00000 00000 00000 00000 CC0CC C0C00 00C00 C0C0C 0C000
        CCCC0 CCCC0 C000C 00000 CCCCC CCCCC CCCCC C0C                                          0.0    0   1   1
  3     CC2CC CCCCC CCCCC C0CCC C0CC0 00C00 00000 00C00 00000 00000 00000 00000 0C00C C0CC0 00CCC C0CC0 00C00 00CC0 C0C0C 00000
        CCCC0 CCCC0 0C0C0 C0CCC CCCCC CCCCC CCCCC CCCCC CCC00 C0C0C 00C00 0C000 00000 00000 0000C 00000 C0C00 0C000 C0C0C CCC00
        000C0 CCCCC CC00C C000C CC0C0 0CCCC CCCC0 00C                                        100.0    1   0   1
  4     CC03C CCCCC CC00C CCCCC CCCCC CCCC0 000C0 0000C 00000 00C0C C0000 00000 0000C 00000 00000 0CCCC C0CC0 0CC00 C0C0C CCC00
        0CCC0 CCCCC CCCCC C0CC0 CCCCC CCCCC CCCCC CCCCC CC000 000CC C0CC0 C0000 000CC 00000 00000 00000 000C0 CCC00 0000C 00000
        CCCCC CCCCC CCCCC CCCCC CCCC0 C0C00 00000 00C                                          0.0    0   1   1
  5     CCCC3 2CCCC CCCCC C0CCC CCCCC CCCCC C0CCC 0CCC0 C0000 0000C 00000 00000 00000 00000 00000 00000 CCCCC CCCC0 C0C00 0CC00
        C0000 CCCC0 0C0CC C0000 CCCCC C0C0C C0CCC 00CCC CCCCC CCCCC C0CC0 CCC00 00000 00CCC 000C0 00000 00C00 00C0C 00000 00000
        CCCCC CCCCC CC0CC C000C 0CCCC CCCCC 000C0 000                                          0.C    0   2   2
  7     C00C0 C3CCC 0CCCC CCCCC CCCCC CCCC0 CCCCC CCCCC CCCC0 00000 00000 0CC00 0000C 00000 00000 00000 00C00 000C0 00C0C 0CCC0
        CCCC0 CCCC0 000CC CCCCC CCCCC CCCC0 00CCC C0C0C 0CCC0 3CCCC 00C00 C0CCC CCCCC 00000 0CCCC CC02C 00C00 000C0 C0C00 00000
        CCCCC CCCCC CCCCC CCCCC CCCCC CCCCC CCCCC 00C                                         50.C    1   1   2
  8     C0CC0 CC3CC CCC0C C000C CCCCC CCCCC C0CCC CCCCC CCC0C CCCCC 00C00 00000 0000C C0000 00000 00C00 CCCC0 00C00 00000 0CCCC
        CCCCC CCCCC CC0CC C0000 C0CC0 C0C0C CCCCC CCCCC CCCC0 CCCCC C0CC0 CCCCC CCCCC C0CCC 0000C 0CCCC CCCCC 0CC00 C0000 00000
        CCCCC CCCCC CCCCC CCCCC CCCC0 CCCCC CCCCC CCC.                                         0.0    0   1   1
  9     CCCC0 CC033 030CC C0000 CCCCC CCCCC CCCCC CCCCC CCC0C CCCCC C0CCC C0C00 0C00C 00000 0CCC0 C00C0 C0C00 00000 00000 00000
        CCCCC CCCCC CCCCC CCCCC CCCCC CCCCC C0CCC 00CC0 00000 00000 00000 0C000 0000C 00CC0 0CCCC CC0CC 00C0C 000CC C0C0C C0000
        C0CC0 CCCCC CC0CC CC0CC CCCCC CCCCC CCCCC 00C                                          0.0    0   2   3
 11     C000C C0CC0 3C0CC CCCCC 0CCCC CCCC0 00000 00CC0 00C00 0000C 00C00 0C0C0 0CCCC C0C00 000C0 00000 00000 00000 C0000 00000
        CCCCC CCCCC CCCCC CCCCC CCCCC CCCCC 00CCC 00C0C 00000 C000C 0CCCC CC000 CC00C 0000C 00CC0 0CCC0 C0C0C 00CC0 CC00C-CCCCC
        C0CCC CCCCC CCCCC CCCCC CCCC0 CCCCC CCCCC C0C                                          0.C    0   1   1
 13     CCCCC CCCC0 CC3CC CCCCC C0CC0 0C000 000C0 00CCC C0CCC CCCCC 00C00 00000 0C00C 00C00 00CC0 CCCCC C0CC0 00CC0 C0C00 00000
        CCCCC CCCCC CC0CC CCCCC CCCCC CCCCC CCCCC CCCC0 CCC00 00000 00000 00000 0000C 00000 00000 C00CC CCCCC C0000 C0000 00CC0
        CCCC0 CCCC0 C00CC CCCCC CCCCC CCCC0 CCCCC CCC                                          0.0    0   1   1
 14     CCCCC CCCCC 0C03C C0000 C00C0 C0CCC 00000 00000 00000 000C0 C0000 C0000 0C0CC C0000 00000 CCCCC CCCCC 00C00 00000 00000
        CCCCC CCCCC CCCCC CCCCC CCCCC CCCCC C0CCC CCCCC CCCCC C000C 00000 00000 0000C 00000 00000-00000 00C00 00C00 C0C0C CCC0C
        CCCCC CCCC0 C000C C000C 0CCC0 C0C00 C0000 00C                                          0.0    0   1   1
 15     CCCCC CCCCC CCCC2 CCCCC CCCCC C0CC0 C0CCC C0CC0 00000 000C0 00000 00000 00000 00000 00000 0000C CCCCC CCCCC CCC0C CCC00
        C00C0 CCCCC 0000C C0C0C CCCC0 CCC0C 000CC C0CCC.CCCCC CCCCC C0CC0 00000 00000 00000 00000 000CC 00C00 00000 C0000 00000
        CCCCC CCCCC CCCCC C0000 CCCC0 C0CCC CCCCC C0C                                        100.0    1   0   1
 16     CCCCC CCCC0 000CC 2CCCC CCC20 C0C00 00CCC CCCCC C0C00 00000 03030 00300 00000 00000 00CC0 0CCC0 00C20 00000 C0C2C 2CC00
        CCCC0 CCCCC 00000 02000 CCCC0 CCCCC 2CCCC 2CCCC CCCC0 0C0CC 02CC2 2C020 20020 2000C 030C2 00000 02000 00000 C2200 20002
        C222C C2C2C CC2CC 2CCCC CC0C2 CCCC2 C0CC0 00C                                         88.2   30   4  34
 17     CC0CC CCCCC 0000C C2CC0 CCCCC CCCCC CCCCC CCCCC C0CCC CCCCC 00C00 0C000 00000 00000 00000 00C00 00C00 00CC0 C0000 0CCC0
        C0CC0 CCCCC 000CC C00C0 CC0C0 C000C 00CCC 00CCC C0000 0000C 00CC0 CCCCC C00CC C000C 0000C 000C0 00C00 000C0 00000 00000
        CCCCC CCCCC CC0CC CCCCC CCCCC CCCCC C00C0 000                                          0.0    0   1   1
 18     C0CC0 CCCCC CCCCC CC3CC CCCCC CCCCC CCCCC 00CCC 0CC00 000CC C0CC0 C0000 00000 00000 00000 00000 C0C00 CCCCC C0C00 0CC00
        CCCCC CCCCC CCCCC CC0CC CCCC0 C0CC0 C00C0 0CCCC C0C0C CCCCC 0CCC0 CCCCC C0CCC C0CC0 0CCC0 CCCC0 CCC00 0CC00 C0C00 00000
        CCCCC CCCCC CCCCC CCCCC CCCCC CCCCC CCCCC CCC                                          0.0    0   1   1
 19     0000C CCCC0 0C00C C0C3C CCCC0 CCCCC 00CCC 00CCC CCCCC CCCCC C0000 000CC C000C 00C0C 00000 00000 00000 00C00 C0000 00000
        CCCCC CCCCC 000CC CCCCC CCCCC CCCC0 C0CCC 00C0C 00000 00000 00000 C000C 0C00C 00000 00000 CCCCC CCCCC 000CC CCCCC CCC00
        CCCC0 CCCCC CCCCC CCCCC CCCCC CCCCC CCCCC CCC                                          0.0    0   1   1
 2C     CCCCC CCCC0 00000 00002 CCCCC CCCCC C0000 C0C00 00000 C00CC 00C0C C0CC0 0C00C 00000 00CCC CCCC0 C0000 00000 00000 00000
        CCCCC CCCCC CCCCC CCCCC CCCCC CCCCC C0CCC 00CC0 00000 00000 00000 0000C CCCCC 00000 00000 000CC CCCCC CCC0C CCC0C C0CC0
        CCCC0 CCCCC 0C0CC CCCCC CCCC0 CCCC0 C0C00 00C                                        100.0    1   0   1
```

Table 5. Step-wise clustering for 238 individuals convicted of first degree murder in California in the period 1958-1968. Computer printout

1	1	1 OLD-FASHIONED			
2	1	2 SOFT			
3	1	3 SMOOTH TASTE			
4	1	4 CREAMY			
5	1	5 CHEWY			
6	1	6 FIRM			
7	1	7 COMB.FLAVORS			
8	2	8 CRUNCHY	9 CRISP		
10	1	10 MELTS IN MOUTH			
11	8	11 SWEET TASTE	12 CHOCOLATEY TASTE	58 MILK CHOCOLATE	18 HAS REAL CHOCOLATE
		42 PLAIN,SOLID CHOCOLATE BA	38 LIGHT CHOCOLATE	13 RICH TASTE	39 DARK CHOCOLATE
14	1	14 LIGHT TASTE			
15	1	15 FRESH TASTE			
16	1	16 HAS PEA.BUT.			
17	20	17 MODERN	24 COMES IN SEVERAL PIECES	29 MADE BY WELL-KNOWN RESPE	35 NICELY PACKAGED
		36 CONVENIENTLY PACKAGED	56 PIECES INDIVIDUALLY WRAP	57 FOIL WRAPPED	32 EASY TO CARRY AROUND
		33 COMES IN MANY SIZES	27 EASILY BROKEN INTO SMALL	40 EASY TO SHARE	25 FLAT SHAPE
		26 THICK SHAPE	31 CAN BE EATEN SECRETLY	22 DOESN'T STICK IN MOUTH	45 DOESN'T STICK TO TEETH
		30 NOT MESSY WHEN EATEN	34 GET A LOT FOR YOUR MONEY	19 FR,WHOL MLK	20 HI-QUAL INGREDIENTS
21	1	21 GOOD FOR ME			
23	1	23 NOT TOO FILLING			
28	1	28 BRAND FRIENDS EAT			
37	1	37 GOOD TASTING			
41	5	41 HAS WHOLE NUTS	49 HAS CHOPPED NUTS	51 HAS ALMONDS	52 HAS PEANUTS
		53 HAS NOUGAT			
43	1	43 SATISFYING			
44	1	44 DOESN'T TASTE TOO SWEET			
46	1	46 UNUSUAL TASTE			
47	1	47 GOOD W/ MILK OR OTHER DR			
48	1	48 TASTS GOOD FROZEN			
50	1	50 HAS COCONUT			
54	1	54 HAS CARAMEL			
55	1	55 HAS RAISINS			
59	1	59 TASTE NEVER CHANGES			
60	1	60 LASTS LONG TIME IN MOUTH			
61	1	61 HAS CRISP RICE			
62	1	62 HAS COMBINATIONS OF TEXT			
63	1	63 SEMI-SWEET CHOCOLATE			

Table 6. Step-wise clustering of 63 measurement variables related to characteristics of candy. Computer printout

available in a subject of interest to an investigator who would like to get some quick feeling as to what is going on.

References

1. Ball, G. H., and Hall, D. J. ISODATA, A Novel Method of Data Analysis and Pattern Classification. (AD 699616) California, Stanford Research Institute, 1965.
2. Chernoff, H. The use of faces to represent points in k-dimensional space graphically. Journal of the American Statistical Association, 1973, 68, 361-368.
3. Fisher, R. A. The use of multiple measurements in taxonomic problems. Annals of Eugenics, 1936, 7, 179-188.
4. Fortier, J. J., and Solomon, H. Clustering procedures. In Proceedings of the International Symposium on Multivariate Analysis, P. R. Krishnaiah (Ed.), Academic Press, New York, 1966.
5. Friedman, H. P., and Rubin, J. On some invariant criteria for grouping data. Journal of the American Statistical Association, 1967, 62, 1159-1178.
6. Johnson, S. C. Hierarchical clustering schemes. Psychometrika, 1967, 32, 241-254.
7. Kahl, J. A., and Davis, J. A. A comparison of indexes of socio-economic status. American Sociological Review, 1955, 20, 317-325.
8. King, B. F. Step-wise clustering procedures. Journal of the American Statistical Association, 1967, 62, 86-101.
9. Kruskal, J. B. Multidimensional scaling by optimizing goodness of fit to a nonmetric hypothesis. Psychometrika, 1964, 29, 1-27(a).
10. Kruskal, J. B. Non-metric multidimensional scaling: a numerical method. Psychometrika, 1964, 29, 115-129 (b).

11. MacQueen, J. B. Some methods for classification and analysis of multivariate observations. Proceedings of the Fifth Berkeley Symposium on Mathematical Statistics and Probability, 1967, 1, 281-297.

12. Mezzich, Juan E. An Evaluation of Quantitative Taxonomic Methods, Ph.D. dissertation, Ohio State University, 1975.

13. Pearson, Karl. On lines and planes of closest fit to systems of points in space. The London, Edinburgh, and Dublin Philosophical Magazine and Journal of Science, 2, 1901, 559-572.

14. Sokal, R. R., and Rohlf, F. J. The comparison of dendrograms by objective methods. Taxon, 1962, 11, 33-40.

15. Wherry, R. J., Sr., and Wherry, R. J., Jr. Wherry-Wherry Hierarchical Factor Analysis. In Computer Programs for Psychology, R. J. Wherry and J. Olivero, Department of Psychology, The Ohio State University, Columbus, Ohio, 1971.

Department of Statistics
Stanford University
Stanford, California 94305

Cluster Analysis Applied to a Study of Race Mixture in Human Populations

C. Radhakrishna Rao

Summary.

Two aspects of the technique of cluster analysis are considered. One is the choice of a dissimilarity coefficient (DC) between two populations and the other is the method of forming clusters which may be distinct or overlapping. A review is made of various DC's which are currently being used and their relative merits are discussed. Some methods of cluster analysis appropriate for studying race mixture are proposed and applied to the measurements collected on castes and tribes in two large scale anthropometric surveys in India.

1. Introduction.

The problems of cluster analysis as applied to human populations were considered by the author some years ago (Rao, 1948), and the methods developed in the paper were applied to three large scale anthropological studies (Mahalanobis, Majumdar and Rao, 1949; Majumdar and Rao, 1958 and Mukherji, Trevor and Rao, 1955). These methods and subsequent developments were reviewed in a recent paper (Rao, 1971a), where three types of taxonomial problems were mentioned - practical, pure and phylogenetic taxonomy. In the present paper, we shall discuss some aspects of these problems concerning the choice of an appropriate DC (dissimilarity coefficient) between two populations and methods of cluster analysis based on the matrix of DC's among populations.

A vast amount of literature has grown around the subject of cluster analysis. There are a number of books dealing with theoretical developments and applications to a wide variety of problems. Some of them are

Anderberg (1973), Everitt (1974), Hartigan (1975), Jardine and Sibson (1971) and Sneath and Sokhal (1973). A particularly interesting monograph is by Thompson (1975) on the construction of human evolutionary trees.

2. Dissimilarity Coefficient.

We consider a set S of N taxonomic units (TU) designated by $\pi_1,\ldots,\pi_N$. A TU may be a single individual or a population of individuals. Associated with each π_i is a distribution function F_i of some measurements taken on individuals of π_i. Let F be the set $(F_1,\ldots,F_N)$. We define a DC (dissimilarly coefficient) as a function d from $F \times F$ to the non-negative real numbers such that

$$
\begin{aligned}
d(F_i,F_j) &= d(F_j,F_i) \\
d(F_i,F_j) &= 0 \quad \text{if } F_i = F_j \\
&> 0 \quad \text{if } F_i \neq F_j
\end{aligned}
\tag{2.1}
$$

for all $F_i, F_j \in F$. Our problem is to describe the configuration of π_i in S on the basis of the DC's, $d(F_i,F_j)$, between all pairs of elements in F, with a view to study possible evolutionary relationships among the π_i. One method of describing the configuration is through what is called cluster analysis which may be broadly defined as forming subsets (possibly overlapping) of elements of S such that the elements within a subset are in some sense closer than those between subsets.

A DC may be defined in various ways, but the choice of a DC in a given problem naturally depends on the nature of the proposed investigation. If we are interested in evolutionary studies on a given set of TU's, the DC should reflect the inherent genetic differences between TU's. That is, the DC should measure the difference between distribution functions of relevant genetic variables. We call such a DC, a genetic dissimilarity coefficient GDC. But genetic variables are seldom directly observable. However, if the relationship between phenotypic and genetic variables is known, then GDC may be estimated from observable data.

A DC is less restrictive than a distance function in that it may not satisfy the triangular inequality. However, it may serve the purpose of cluster analysis, but a distance function is more appropriate for other analyses (besides cluster analysis) on the matrix of DC's such as representation of TU's in a coordinate space of suitable dimensions.

2.1. DC's based on qualitative data.

The most useful data for evolutionary studies are qualitative characters like blood groups, where the genetic mechanism of inheritance is known. From such data it may be possible to estimate for each TU, the proportions of different types of alleles at each of a number of loci on the chromosomes. Let p_{ijr} represent the proportion of the i-th type allele at the j-th locus for the r-th TU. (In theory the suffix j may stand for a set of loci in which case i would represent a combination of alleles at different loci). The problem reduces to the construction of a DC between the gene pools p_{ijr} and p_{ijs} of the r-th and s-th TU's. A variety of DC's have been suggested and their relative merits are discussed from time to time (see papers by Gower, Balakrishnan and Sanghvi, Edwards and others in Genetic Distance edited by J. F. Crow). A few of the proposed DC's are given below.

We suppose that there are k_j alleles at the j-th locus and there are m loci, in which case, in the following formulae, the summation over i is from 1 to k_j, and over j, from 1 to m.

(1) Minkowski distance:

Minkowski (1911) defined the ℓ^p norm

$$[\sum_j \sum_i |p_{ijr} - p_{ijs}|^p]^{1/p} \tag{2.2}$$

which is a distance function if $1 \leq p \leq \infty$. For $p < 1$, it can be used as a DC. When $p = 2$, we have the usual Euclidean distance. When $p = \infty$, we have the distance function

$$\sup_{i,j} |p_{ijr} - p_{ijs}| \tag{2.3}$$

(ii) Genetic distance of Sanghvi.

Following Sanghvi's earlier work, Balakrishnan and Sanghvi (1968) defined the DC between two populations as

$$(2.4) \qquad \sum_j \sum_i \frac{2(p_{ijr} - p_{ijs})^2}{(p_{ijr} + p_{ijs})}$$

which is, in fact, a distance function.

(iii) Hellinger distance:

Hellinger (1909) introduced a distance function between two probability measures, which when applied to the proportions

$$p_{1jr}, \; p_{2jr}, \; \ldots$$

$$p_{1js}, \; p_{2js}, \; \ldots$$

at the j-th locus is the form

$$(2.5) \qquad \frac{2}{\pi} \cos^{-1} \sum_i \sqrt{p_{ijr}\, p_{ijs}}$$

Bhattacharya (1946) advocated the use of the distance function (2.5) in studying differences between multinomial populations. A variant of (2.5) is the chord distance

$$(2.6) \qquad \frac{2\sqrt{2}}{\pi} [1 - \sum_i \sqrt{p_{ijr}\, p_{ijs}}\,]^{\frac{1}{2}}$$

used by Edwards and Cavilli-Sfroza (1967) who defined the distance over all loci by combining the individual distances (2.6) as

$$(2.7) \qquad \frac{2\sqrt{2}}{\pi} [\sum_j (1 - \sum_i \sqrt{p_{ijr}\, p_{ijs}})]^{\frac{1}{2}}$$

Edwards (1971) proposed the overall DC as

$$(2.8) \qquad \frac{4\sqrt{2}}{\pi} \left[\sum_j \frac{1 - \sum_i \sqrt{p_{ijr}\, p_{ijs}}}{(1+k_j^{-\frac{1}{2}} \sum_i \sqrt{p_{ijr}})(1+k_j^{-\frac{1}{2}} \sum_j \sqrt{p_{ijs}})} \right]^{\frac{1}{2}}$$

(iv) Genetic distance of Nei.

Nei (1975) defined two types of DC's and suggested their use in evolutionary studies. Let

(2.9) $$a^j_{rs} = \sum_i p_{ijr}\, p_{ijs}$$

and define

(2.10) $$J_{rs} = \sum_j a^j_{rs}, \quad J'_{rs} = \prod_j a^j_{rs} .$$

Then Nei gives two definitions of what he calls genetic distance,

(2.11) $$-\log_e \frac{J_{rs}}{\sqrt{J_{rr}\, J_{ss}}}, \quad -\log_e \frac{J'_{rs}}{\sqrt{J'_{rr}\, J'_{ss}}}$$

where J_{rr} and J'_{ss} are obtained by putting $r = s$ in (2.9) and (2.10).

A number of authors (see Chakravarthy (1974) for references) have found high correlations among various proposed DC's in empirical studies and conclude that cluster analyses based on different DC's are likely to give similar results. But most of such studies were conducted on local populations of a particular organism only where the true genetic differences are supposedly very small. To examine whether the high correlations are retained when organisms at different evolutionary ranks are compared, Chakravarthy (1974) made some computations using the data on fruit flies published by Ayala et al (1974). Distances were computed at the level of local populations, semispecies, and sibling and nonsibling species. The product moment correlations between the different distances were obtained as in Table 2.1.

Table 2.1: Correlation coefficients between different genetic distances at various evolutionary ranks (from Chakravarthy, 1974)

Pair of DC's	Evolutionary rank			
	local populations	semi-species	sibling species	non-sibling species
$D_N - D_{CE}$	.709	.812	.725	.457
$D_N - D_{BS}$	.899	.872	.739	.472
$D_{CE} - D_{BS}$	.913	.973	.990	.988

In Table 2.1 D_N stands for the first expression in (2.11) due to Nei, D_{CE}, for (2.7) due to Cavilli-Sfroza and Edwards and D_{BS} for (2.4) due to Balakrishnan and Sanghvi. The correlation coefficients between D_{CE} and D_{BS} is high at all evolutionary ranks, but that between D_N and D_{CE} or D_{BS} is high only at the level of local populations and decreases as more and more distant organisms are compared. Then the question arises as to which is a more appropriate DC. Nei (1975) has shown that his measure is linearly related with the evolutionary time of divergence of the two populations under the assumption of a constant rate of gene substitution. The relationships of the other measures with the time of divergence is not known or not well established. Thus Nei's distance appears to be suitable for phylogenetic studies and estimation of evolutionary trees, especially at higher evolutionary ranks.

2.2. DC's based on quantitative data.

The situation is somewhat complex when we are working with quantitative variables like stature, weight, head length, etc., for which the genetic basis is not so well known. However, we shall consider some DC's between distribution functions which are well known in mathematical statistics. For some theoretical considerations in the choice of a DC, the reader is referred to an interesting paper by Ali and Silvey (1966).

We consider two populations with density functions $p(x)$ and $q(x)$, where x is a vector measurement.

(i) Minkowski distance (1911).

$$[\int |p(x) - q(x)|^t \, dv]^{1/t}, \quad t \geq 1. \tag{2.12}$$

The case when $t = 1$

$$\int |p(x) - q(x)| \, dv \tag{2.13}$$

is of special interest and is known as Kolmogorov's variational distance. The function (2.13) is also connected with discrimination index between the two populations.

(ii) Discrimination index (overlap distance, Rao (1948)).

The best decision rule based on observed x for discriminating between two populations with prior probabilities in the ratio 1 : 1 is

select population 1 if $p(x) > q(x)$

select population 2 if $q(x) \geq p(x)$

where $p(x)$ and $q(x)$ are probability densities in the two populations. The probability of correct classification in such a case is

(2.14) $$\frac{1}{2} \int_{p>q} p(x)\, dv + \frac{1}{2} \int_{q>p} q(x)\, dv$$

which has the minimum value $\frac{1}{2}$ when $p = q$. A measure of separation between the two populations may be defined as

(2.15) $$\frac{1}{2} \int_{p>q} p(x)\, dv + \frac{1}{2} \int_{q>p} q(x)\, dx - \frac{1}{2}$$

which is equal to

(2.16) $$\frac{1}{4} \int |p(x) - q(x)|\, dv$$

i.e., a multiple of Kolmogorov's variational distance. The measure (2.15) is similar to what has been defined by the author (Rao, 1948) as the overlap distance.

(iii) Quadratic differential metric of Rao (1945).

We consider a family of probability measures $p(x,\theta)$, where $\theta = (\theta_1, \ldots, \theta_k) \in \Theta$, a k-vector parameter space. The Fisher information matrix at θ is $M = [m_{ij}(\theta)]$, where

(2.17) $$m_{ij}(\theta) = \int \frac{1}{p} \frac{dp}{d\theta_i} \frac{dp}{d\theta_j}\, dv .$$

We endow the space Θ with the quadratic differential metric

(2.18) $$\sum \sum m_{ij}(\theta)\, \delta\theta_i\, \delta\theta_j$$

which is the distance between two points θ and $\theta + \delta\theta$ as $\delta\theta \rightarrow 0$. Then the distance between any two points $\theta_1, \theta_2 \in \Theta$ is defined by Rao (1945) as the geodesic distance determined by the quadratic differential metric (2.18). The distance so defined may be useful in evolutionary studies where gradual changes take place in a population moving from one parameter point to another.

(iv) Divergence measures of Jefferys.

Jeffreys (1948) defined what are called invariants between two distributions

$$(2.19) \qquad I_m = \int |p^{1/m} - q^{1/m}|^m \, dv$$

$$(2.20) \qquad J = \int (q - p) \log \frac{q}{p} \, dv$$

where the second expression is the sum of Kullback and Leibler information numbers

$$(2.21) \qquad I(p,q) = \int p \log \frac{p}{q} \, dv, \quad I(q,p) = \int q \log \frac{q}{p} \, dv .$$

When $m = 1$

$$(2.22) \qquad I_1 = \int |p-q| \, dv$$

which is Kolmogorov's variational distance. When $m = 2$

$$(2.23) \qquad I_2 = \int (\sqrt{p} - \sqrt{q})^2 \, dv$$

$$= 2(1 - \int \sqrt{pq} \, dv)$$

which is a function of Hellinger's distance

$$(2.24) \qquad \cos^{-1} \int \sqrt{pq} \, dv .$$

We shall call

$$(2.25) \qquad H = -\log_e \int \sqrt{pq} \, dv$$

as Hellinger's dissimilarity coefficient (see Rao and Varadrajan, 1963).

(v) Information radius of Jardine and Sibson (1971).

Jardine and Sibson (1971) developed the concept of information radius which when applied to two populations with densities p and q reduces to

$$(2.26) \qquad \int [w_1 \, p \log_2 \frac{p}{w_1 p + w_2 q} + w_2 \, q \log \frac{q}{w_1 p + w_2 q}] \, dv$$

where w_1 and $w_2 \geq 0$ are weights such that $w_1 + w_2 = 1$. For the choice $w_1 = w_2$, (2.26) becomes

(2.27) $$\int \left[\frac{p}{2} \log_2 \frac{2p}{p+q} + \frac{q}{2} \log_2 \frac{2q}{p+q} \right] dv .$$

The formulae (2.26) and (2.27) refer to the joint density of the components $x_1, \ldots, x_k$ of a vector variable x.

Jardine and Sibson (1971) prefer to compute the information radius (2.27) for each x_i and take the sum over all i as a measure of dissimilarity between the populations, although the variables x_i are not independent. This was done to avoid the situation where singularity of distributions with respect to any one component implies singularity of distributions for the entire vector. However, they suggest a careful choice of measurements to avoid over representation of some attributes.

(vi) Mahalanobis D^2.

Let us consider two k-variate normal distributions $N_k(\mu_1, \Sigma_1)$ and $N_k(\mu_2, \Sigma_2)$. When $\Sigma_1 = \Sigma_2 = \Sigma$, both Kolmogorov's variational and Hellinger distances reduces to a monotone function of

(2.28) $$D^2 = (\mu_1 - \mu_2)' \Sigma^{-1} (\mu_1 - \mu_2)$$

which is Mahalanobis distance between two populations with mean vectors μ_1, μ_2 and common dispersion matrix Σ. When $\Sigma_1 \neq \Sigma_2$, Hellinger's dissimilarity coefficient defined in (2.25) reduces to

(2.29) $$H = \frac{1}{4} \rho + \frac{1}{8} D^2$$

(2.30) $$\rho = 2 \log |\Sigma| - \log |\Sigma_1| - \log |\Sigma_2|$$

(2.31) $$D^2 = (\mu_1 - \mu_2)' \Sigma^{-1} (\mu_1 - \mu_2)$$

where $\Sigma = (\Sigma_1 + \Sigma_2)/2$ as shown by Rao and Varadarajan (1963). It is seen that ρ measures the divergence in the dispersion matrices and D^2 between the mean values, and the total divergence is a linear combination of the two.

Thus Mahalanobis D^2 is an appropriate distance measure for measuring differences in mean values when we consider multivariate normal populations. However, its use can be recommended more generally when the measurements have a factor structure. D^2 has the following interesting properties.

(a) It is invariant under linear transformations.

(b) It is stable under addition or selection of measurements when a factor structure is assumed as shown by Rao (1954). Let g be the vector of a fixed number of (unobservable) factor variables which have mean values γ_1 and γ_2 in two populations with a common dispersion matrix Γ. Then Mahalanobis D^2 between the two populations based on factor variables is

$$D_f^2 = (\gamma_1 - \gamma_2)' \Gamma^{-1}(\gamma_1 - \gamma_2) . \tag{2.32}$$

Let X be a p-vector of observable variables with the structure

$$X = Ag + \varepsilon \tag{2.33}$$

where g and ε are uncorrelated and the dispersion matrix of ε is Λ. The variable ε may be viewed as random or environmental component of X and is assumed to have mean value zero. Then the mean values of X in the two populations are

$$A\gamma_1 \quad \text{and} \quad A\gamma_2 \tag{2.34}$$

and the dispersion matrix of X is

$$\Sigma = A \Gamma A' + \Lambda . \tag{2.35}$$

Mahalanobis D^2 based on the p-vector X is

$$\begin{aligned} D_p^2 &= (\gamma_1 - \gamma_2)' A' \Sigma^{-1} A(\gamma_1 - \gamma_2) \\ &\leq D_f^2 \end{aligned} \tag{2.36}$$

whatever p may be. But D_p^2 is an increasing function of p, and if p is sufficiently large D_p^2, being bounded by the fixed number D_f^2, reaches stability. Further the value of D_p^2 is not very much affected by the particular set of observable variables provided that they depend on the factor variable g, the matrix A of factor loadings in (2.33) is of full rank and the random component ε in (2.33) does not have large variance.

It is seen from (2.29) - (2.31) that when two populations differ in mean values as well as in dispersion matrices, the DC is composed of two elements

$$(2.37) \qquad H = \frac{1}{4}\rho + \frac{1}{8} D^2 .$$

In the context of the linear structure (2.33), D^2 reflects the differences in the distributions of factor variables and ρ in the dispersion matrices of random components for the two populations. In such a case cluster analysis based on D^2 values computed on mean values using an average dispersion matrix seems to be more meaningful from a biological view point. Of course, one could apply cluster analysis separately on ρ values to examine differences in dispersion matrices. Cluster analysis applied on the composite H values as recommended by some authors does not seem to be useful or easily interpretable.

3. Cluster Analysis.

A variety of clustering techniques are described in books listed in the reference. Sneath and Sokal (1973) mention most of these techniques. Jardine and Sibson (1971) provide a general theory of cluster analysis based on a set of axioms. Thompson (1975), following the work of Edwards and Cavilli-Sfroza , gives methods for constructing phylogenetic trees.

It is difficult to prescribe specific rules for cluster analysis which are applicable in all situations. This is partly due to the difficulty in laying down objective criteria for comparing clusters obtained by different procedures. Cluster analysis is a descriptive tool for studying configurations of objects with specified dissimilarity coefficients, or represented by points in a multidimensional space. Since a visual examination is not possible when the dimensions are more than three, we have to work with the matrix of mutual DC's and provide a description of the configuration of points to throw light on evolutionary aspects of the populations under study.

The description may not be simple as it depends on the complexity of the configuration of points. The study on the matrix of DC's should be such as to reveal inter-relationships between populations and suggest plausible hypotheses of their evolution rather than to fit particular models. For instance, fitting a tree structure estimating the evolutionary time of separation of populations under simplifying assumptions of isolation, genetic drift and constant rate of gene substitution may not be appropriate when we are considering classification of human populations living in a compact geographical region where questions of intermixture and social-ogical barriers between populations become important. I shall give two illustrations elaborating on the analyses employed in two large scale anthropometric studies carried out in India (Mahalanobis, Majumdar and Rao, 1949; Majumdar and Rao, 1958).

Before doing so, we shall consider some definitions of clusters, which seem to be appropriate in studying inter-relationships between different populations (which we shall call more generally as groups) and speculating on their origin. Let there be a set S of N populations with the matrix (d_{ij}) of dissimilarity coefficients (DC's). We give three definitions for a subset s of populations $i_1, \ldots, i_k$ to be called a cluster.

Definition (a). The set s is said to be a cluster at a threshold value h if it has the maximal number of elements such that

$$d_{ij} \leq h, \quad i, j \in s.$$

Definition (b). The set s is said to be a cluster at a threshold value h if it has the maximal number of elements such that

$$(k-1)^{-1} \sum_{j \in s} d_{ij} \leq h, \quad \text{for each } i \in s.$$

Definition (c). The set s is said to be a cluster at threshold values h and $r(>h)$ if it has the maximal number of elements such that

$$k(k-1)^{-1} \sum_{i} \sum_{j \in s} d_{ij} \leq h,$$

$$d_{ij} \leq r, \quad i, j \in s.$$

By cluster analysis is meant a method of obtaining a list of all possible clusters using a given definition of a cluster. Any two clusters may intersect in any number of elements. This method differs from the others which force the clusters to be disjoint such as the dendogram or place an upper limit on the number of elements common to any two clusters as in the B_k method of Jardine and Sibson.

The first illustration is from a statistical study of anthropometric measurements taken on castes and tribes (groups of individuals) in the United Provinces (now called Uttar Pradesh) in 1945. A sample of 100 to 200 individuals is chosen from each group and each individual was measured for nine characters. It may be mentioned that all the groups live in the same State but individuals belonging to two different groups do not generally marry each other.

Table 3.1 gives the values of D^2 based on 9 characters between a given group and each of the others arranged in an increasing order of magnitude. For details regarding the survey and the computation of D^2 values, the reader is referred to Mahalanobis, Majumdar and Rao (1949) or to Chapter 9 of the author's book (Rao, 1971b). The clusters according to definitions (a) and (b) are given in Table 3.2.

The clusters obtained by the two methods are nearly the same. The two Brahmin groups (B_1,B_2) form a close cluster, so also the Artisans (A_1,A_2,A_3,A_4) although A_4 is a bit removed from the others and specially from A_1, and Muslim and Chatri groups (M,Ch). At a higher threshold, Bhil and Dom (Bh,D) and the criminal tribes Bhatu and Habru (C_1,C_2) form distinct clusters. The Artisan cluster overlaps with the Brahmin cluster at one end (B_2 A_1) and the Muslim-Chatri cluster at the other end (A_4 M), showing the intermediate position occupied by the Artisans in the caste hierarchy.

The second illustration is from a statistical study of measurements made on individuals belonging to some castes and tribes (groups) of undivided Bengal. Details of the survey, description of the groups and the computation of D^2 values based on eleven characters can be found in

Table 3.1: Values of D^2 (based on 9 characters) arranged in increasing order of magnitude

U. P. Anthropometric Survey : 1941

Brahmin (Basti, B_1)	Brahmin (Other, B_2)	Bhatu (C_1)	Habru (C_2)	Dom (D)	Bhil (Bh)	Chattri (Ch)	Muslim (M)	Ahir (A_1)	Kurmi (A_2)	Other Artisan (A_3)	Kahar (A_4)
B_2 0.27	B_1 0.27	C_2 1.32	A_1 1.26	Bh 1.15	D 1.15	M 0.40	Ch 0.40	A_2 0.30	A_3 0.12	A_2 0.12	A_3 0.43
A_1 1.17	A_1 0.78	A_1 2.68	C_1 1.32	C_2 2.11	A_3 1.75	A_2 2.12	A_4 0.90	A_3 0.49	A_1 0.30	A_4 0.43	A_2 0.58
A_2 1.48	A_2 1.03	A_2 2.98	A_2 1.53	A_3 2.31	A_2 2.23	A_4 2.24	A_2 1.34	B_2 0.78	A_4 0.58	A_1 0.49	M 0.90
A_3 2.13	A_3 1.47	A_3 3.35	B_2 1.63	A_2 2.41	A_4 2.24	A_3 2.72	A_3 1.45	B_1 1.17	B_2 1.03	M 1.45	A_1 1.52
C_2 2.23	C_2 1.63	B_1 3.48	A_3 1.67	M 2.47	A_1 2.53	B_2 2.87	A_1 2.45	C_2 1.26	M 1.34	B_2 1.47	Bh 2.24
M 2.86	M 2.62	B_2 3.61	D 2.11	A_4 2.66	M 3.16	B_1 3.05	D 2.47	A_4 1.52	B_1 1.48	C_2 1.67	Ch 2.24
D 2.86	A_4 2.72	A_4 4.20	B_1 2.23	B_2 2.81	C_2 3.47	A_1 3.38	B_2 2.62	M 2.45	C_2 1.53	Bh 1.75	D 2.66
Ch 3.05	D 2.81	M 4.46	A_4 2.87	B_1 2.86	B_2 3.82	D 3.84	B_1 2.86	Bh 2.53	Ch 2.12	B_1 2.13	B_2 2.72
A_4 3.30	Ch 2.87	D 4.52	Bh 3.47	A_1 2.91	B_1 4.45	C_2 4.68	Bh 3.16	C_1 2.68	Bh 2.23	D 2.31	C_2 2.87
C_1 3.48	C_1 3.61	Bh 5.08	M 3.74	Ch 3.84	Ch 5.02	Bh 5.02	C_2 3.74	D 2.91	D 2.41	Ch 2.72	B_1 3.30
Bh 4.45	Bh 3.82	Ch 5.25	Ch 4.68	C_1 4.52	C_1 5.08	C_1 5.25	C_1 4.46	Ch 3.38	C_1 2.98	C_1 3.35	C_1 4.20

Table 3.2: Clusters of castes in the United Provinces at different threshold values.

Definition (a)			Definition (b)		
$h = 0.50$	$h = 1.0$	$h = 1.5$	$h = 0.50$	$h = 1.0$	$h = 1.5$
B_1 B_2	B_1 B_2	B_1 B_2 A_1 A_2	B_1 B_2	B_1 B_2 A_1 A_2	B_1 B_2 A_1 A_2 A_3
A_1 A_2 A_3	B_2 A_1	B_2 A_1 A_2 A_3	A_1 A_2 A_3	A_1 A_2 A_3 A_4	A_1 A_2 A_3 A_4
A_3 A_4	A_1 A_2 A_3	A_2 A_3 A_4 M	A_2 A_3 A_4	A_4 M	A_2 A_3 A_4 M
M Ch	A_2 A_3 A_4	M Ch	M Ch	M Ch	M Ch
Bh	A_4 M	Bh D	Bh	Bh	Bh D
D	M Ch	C_1 C_2	D	D	C_1 C_2
C_1	Bh	C_2 A_1	C_1	C_1	C_2 A_1
C_2	D		C_2	C_2	
	C_1				
	C_2				

Table 3.3: Values of D^2 (based on 11 characters) arranged in increasing order of magnitude

Bengal Anthropometric Survey : 1945

Baidya B_d	Brahmin (others) B^o	Kayastha (others) K^o	Namasudra (others) N^o	Brahmins Dacca B^{da}	Kayastha Dacca K^{da}	Muslim Dacca M^{da}	Namasudra Dacca N^{da}	Kayastha Barisal K^{ba}
B^o .24	B^d .24	B^d .44	M^{na} .24	K^{da} .39	M^{da} .21	K^{da} .21	M^{my} .41	M^{fa} .42
K^o .44	K^o .52	B^o .52	M^{bu} .36	M^{da} .59	B^{da} .39	N^{da} .43	M^{da} .43	M^{na} .62
M^{fa} .64	B^{da} .93	M^{fa} 1.13	M^{fa} .54	B^o .93	M^{my} .79	M^{my} .44	M^{na} .71	B_d .64
K^{ba} .64	K^{ba} .95	K^{ba} 1.26	K^{ba} .75	B_d .97	N^{da} .80	B^{da} .59	M^{bu} .71	M^{da} .67
M^{da} .80	K^{da} 1.09	M^{da} 1.35	M^{my} .80	K^{ba} .97	B_d .91	K^{ba} .67	K^{ba} .79	N^o .75
K^{da} .91	M^{fa} 1.14	B^{da} 1.39	N^{da} .92	M^{fa} 1.18	K^{ba} .92	M^{fa} .69	K^{da} .80	N^{da} .79
B^{da} .97	M^{da} 1.22	K^{da} 1.44	M^{da} .92	M^{my} 1.38	M^{fa} .99	B_d .80	M^{ma} .84	M^{my} .81
N° 1.30	N° 1.76	N° 2.14	K^{da} 1.08	K^o 1.39	N° 1.08	M^{na} .87	N^o .92	M^{bu} .88
M^{my} 1.51	M^{na} 2.04	M^{na} 2.34	M^{mu} 1.15	N^{da} 1.47	B^o 1.09	N^o .92	M^{fa} 1.00	K^{da} .92
M^{na} 1.56	M^{my} 2.14	M^{my} 2.41	B_d 1.30	N° 1.62	M^{na} 1.28	M^{bu} 1.01	M^{ba} 1.11	B^o .95
M^{bu} 1.63	M^{bu} 2.47	N^{da} 2.62	M^{ma} 1.57	M^{bu} 1.66	M^{bu} 1.33	B^o 1.22	M^{ra} 1.26	B^{da} .97
N^{da} 1.78	N^{da} 2.52	M^{bu} 2.73	B^{da} 1.62	M^{na} 1.69	K^o 1.44	M^{ba} 1.25	M^{mu} 1.35	K^o 1.26
M^{ba} 2.30	M^{ba} 2.97	M^{ba} 3.67	M^{ba} 1.63	M^{ba} 2.57	M^{ma} 1.71	K^o 1.35	B^{da} 1.48	M^{ma} 1.38
M^{my} 2.44	M^{ma} 3.13	M^{mu} 3.90	B^o 1.76	M^{ma} 2.58	M^{ba} 1.75	M^{ma} 1.38	B_d 1.78	M^{mu} 1.65
M^{mu} 2.54	M^{mu} 3.50	M^{ma} 3.95	M^{ra} 1.81	M^{mu} 3.60	M^{mu} 2.21	M^{mu} 1.92	B^o 2.52	M^{ba} 1.66
M^{ra} 3.22	M^{ra} 4.22	M^{ra} 4.47	K^o 2.14	M^{ra} 3.88	M^{ra} 2.76	M^{ra} 2.09	K^o 2.62	M^{ra} 1.74

Table 3.3: (continued)

Muslim Faridpur M^{fa}	Muslim Barisal M^{ba}	Muslim Burdwan M^{bu}	Muslim Murshidabad M^{mu}	Muslim Mymensingh M^{my}	Muslim Nadia M^{na}	Muslim Rangpur M^{ra}	Muslim Malda M^{ma}
K^{ba} .42	M^{my} .63	N^{o} .36	M^{ra} .62	N^{da} .41	N^{o} .24	M^{mu} .62	M^{my} .68
N^{o} .54	M^{ma} .90	M^{na} .48	M^{ma} .95	M^{da} .44	M^{bu} .48	M^{ma} .77	M^{ra} .77
M^{na} .57	N^{da} 1.11	N^{da} .71	M^{my} 1.10	M^{na} .49	M^{my} .49	M^{my} .81	N^{da} .84
M^{my} .63	M^{da} 1.25	M^{fa} .76	N° 1.15	M^{fa} .63	M^{fa} .57	M^{na} 1.19	M^{ba} .90
B_{d} .64	M^{na} 1.30	M^{my} .83	M^{na} 1.25	M^{ba} .63	K^{ba} .62	N^{da} 1.26	M^{mu} .95
M^{da} .69	M^{ra} 1.34	K^{ba} .88	N^{da} 1.35	M^{ma} .68	N^{da} .71	M^{ba} 1.34	M^{fa} 1.36
M^{bu} .76	M^{bu} 1.55	M^{da} 1.01	M^{bu} 1.49	K^{da} .79	M^{da} .87	M^{fa} 1.70	M^{na} 1.37
K^{da} .99	M^{fa} 1.62	K^{da} 1.33	M^{fa} 1.51	N^{o} .80	M^{ra} 1.19	K^{ba} 1.74	M^{da} 1.38
N^{da} 1.00	N° 1.63	M^{mu} 1.49	K^{ba} 1.65	M^{ra} .81	M^{mu} 1.25	N° 1.81	K^{ba} 1.38
K^{o} 1.13	K^{ba} 1.66	M^{ba} 1.55	M^{ba} 1.81	K^{ba} .81	K^{da} 1.28	M^{bu} 1.88	N° 1.57
B^{o} 1.14	K^{da} 1.75	M^{ma} 1.63	M^{da} 1.92	M^{bu} .83	M^{ba} 1.30	M^{da} 2.09	M^{bu} 1.63
B^{da} 1.18	M^{mu} 1.81	B_{d} 1.63	K^{da} 2.21	M^{mu} 1.10	M^{ma} 1.37	K^{da} 2.76	K^{da} 1.71
M^{ma} 1.36	B_{d} 2.30	B^{da} 1.66	B_{d} 2.54	B^{da} 1.38	B_{d} 1.56	B_{d} 3.22	B_{d} 2.44
M^{mu} 1.51	B^{da} 2.57	M^{ra} 1.88	B^{o} 3.50	B_{d} 1.51	B^{da} 1.69	B^{da} 3.88	B^{da} 2.58
M^{ba} 1.62	B^{o} 2.97	B^{o} 2.47	B^{da} 3.60	B^{o} 2.14	B^{o} 2.04	B^{o} 4.22	B^{o} 3.13
M^{ra} 1.70	K^{o} 3.67	K^{o} 2.73	K^{o} 3.90	K^{o} 2.41	K^{o} 2.34	K^{o} 4.47	K^{o} 3.95

Majumdar and Rao (1958). Table 3.3 gives the D^2-values between groups arranged as in Table 3.1. The clusters according to definition (a) are given in Table 3.4. Considering the groups as points and connecting the points with D^2 value not greater than 0.71 by an edge we obtain a graph as shown in Figure 1. A cluster is a maximal subgraph where every two points are connected.

Judging from the graph, the configuration of groups in undivided Bengal appears to be far more complicated in nature than in the United Provinces.

i) Overlapping of clusters occurs even at low threshold values, thus indicating paucity of distinct clusters of more than one group among the groups under study.

ii) Another striking feature is the clustering of groups more on regional basis rather than on caste or religion. For instance at the threshold value of 0.71, Brahmins, Kayasthas and Muslims of Dacca (B^{da}, K^{da}, M^{da}) form a cluster, so also Muslims and Namasudras of Dacca with Muslims in the neighbouring district of Mymensingh (M^{da}, N^{da}, M^{my}). The Kayasthas of Barisal are close to the Muslims in the contiguous districts of Faridpur and Nadia ($K^{ba}, M^{fa}, M^{na}, N^{o}$). Among the Muslim groups, affinities appear to be broadly related to the nearness of districts to which they belong. On the other hand the D^2 between Brahmins of Dacca and others is 0.93, that between Kayasthas of Barisal and Dacca is 0.92 and that between Namasudras of Dacca and others is 0.92, which are of a larger magnitude than the D^2 values between different caste groups within the same district. The regional affinities which appear to be stronger than caste affinities goes against accepted hypotheses about the caste system in India.

iii) An interesting feature is the close relationship of Namasudras (low caste Hindus) with the Muslims as indicated by the clusters ($N^{o}, M_{ma}, M_{fa}, K^{ba}$), ($N^{o}, M^{bu}, M^{na}$), ($N^{da}, M^{bu}, M_{na}$) and ($M^{da}, N^{da}, M^{my}$). This suggests the possibility that Namasudra provided a large proportion of converts to the Muslim religion.

Table 3.4: Clusters of castes in undivided Bengal at different threshold values according to definition (a)

h = 0.71*	h = 1.01*
K^o B^o B_d	K^o B^o B_d
B_d K^{ba} M^{fa}	M^{fa} B_d K^{ba} K^{da} M^{da}
K^{ba} M^{fa} M^{da}	B_d K^{ba} K^{da} M^{da} B^{da}
B^{da} K^{da} M^{da}	N^o N^{da} M^{my} M^{na} M^{fa} K^{ba} M^{da} M^{bu}
M^{da} N^{da} M^{my}	M^{ba} M^{my} M^{ma}
M^{my} M^{fa} M^{na}	M^{ma} M^{mu} M^{ra}
M^{na} M^{fa} K^{ba} N^o	
N^o M^{bu} M^{na}	
N^{da} M^{bu} M^{na}	
M^{my} M^{ba}	
M^{my} M^{ma}	
M^{mu} M^{ra}	

*The values 0.71 and 1.01 are chosen in such a way that for slightly higher threshold values the clusters remain the same and for slightly lower threshold values, clusters are broken up.

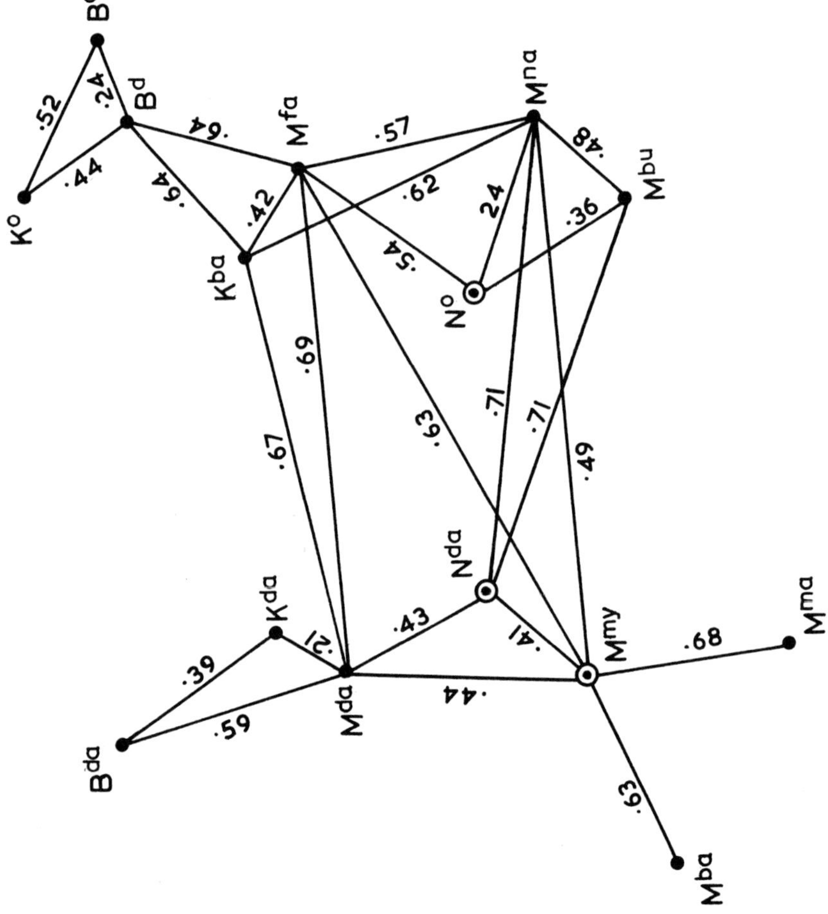

Figure 1. Graph of groups and maximal subgraphs

iv) The Muslims of Mymensingh have close affinities with Muslims of the other districts indicating the possibility of periodic migrations from and to Mymensingh and mixing with local people.

The simple type of cluster analysis used on the matrix of DC's has revealed a wealth of information. One might argue that the procedure becomes complicated when the number of groups under study is very large. In fact in the Bengal study, there were more groups than what have been chosen for illustration in the present paper. But the groups left out were quite distinct and would not cluster with others except at high threshold values. Thus, in problems involving large numbers of groups it may be possible to distinguish between broad and nearly distinct clusters to begin with and carry out detailed study (obtain sub-clusters) in each such cluster. Indeed, one could use different threshold values in different large clusters, and the proposed method has great flexibility.

References

Ali, S. M. and Silvey, S. D. (1966), A general class of coefficients of divergence of one distribution from another, J. Roy. Statist. Soc. B, 28, 131-142.

Anderberg, M. R. (1973), Cluster Analysis for Applications, Academic Press, New York.

Ayala, F. J., Tracey, M. L., Barr, L. G., McDonald, J. F. and Perez-Salas, S. (1974), Genetic variations in five Drosophila species and the hypothesis of the selective neutrality of protein polymorphism, Genetics, 77, 343-384.

Bhattacharya, A. (1946), On a measure of divergence between two multinomial populations, Sankhya 7, 401.

Balakrishnan, V. and Sanghvi, L. D. (1968), Distance between populations on the basis of attribute data, Biometrics, 24, 859-865.

Cavalli-Sfroza, L. L. and Edwards, A. W. F. (1967), Phylogenetic analysis: Models and estimation procedures, Amer. J. Hum. Genet. 19, 233-257.

Chakravarthy, R. (1974), Genetic distance measures and evolution: A review (Paper presented at the International Symposium, ISI, Calcutta, 1974).

Edwards, A. W. F. (1971), Distance between populations on the basis of gene frequencies, Biometrics, 27, 783-882.

Everitt, B. S. (1974), Cluster Analysis, Halstead Press, London.

Hartigan, J. (1975), Clustering Algorithms, Wiley, New York.

Hellinger, E. (1909), Neue bergrundung der theorie quadratisher formen von unendlichvielen veranderlichen, J. Fur reine and angew Mathematic, 136, 210 - .

Jardine, N. and Sibson, R. (1971), Mathematical Taxonomy, Wiley, New York.

Jeffreys, H. (1948), Theory of Probability, Second edition, Claredon Press, Oxford.

Mahalanobis, P. C. (1936), On the generalized distance in statistics, Proc. Nat. Inst. Sci. India, 2, 49-55.

Mahalanobis, P. E., Majumdar, D. N. and Rao, C. R. (1949), Anthropometric survey of the United Provinces, 1945: A statistical study, Sankhya, 9, 90-324.

Majumdar, D. N. and Rao, C. R. (1958), Bengal anthropometric survey, 1945: A statistical study, Sankhya, 19, 203-408.

Matusita, K. (1966), A distance and related statistics in multivariate analysis, Multivariate Analysis, P. R. Krishnaih, ed., Academic Press, New York, 187-202.

Minkowski, H. (1911), Gesammelte Abhandlungen Vol. II. Teubner, Berlin.

Mukherji, R. K., Trevor, J. C. and Rao, C. R. (1955), Ancient Inhabitants of Jebel Moya, Cambridge University Press, Cambridge.

Nei, M. (1975), Molecular Population Genetics and Evolution : A Statistical Study, North Holland and Elsevier.

Rao, C. Radhakrishna (1945), Information and the accuracy attainable in the estimation of statistical parameters, Bull. Cal. Math. Soc., 37, 81-91.

Rao, C. Radhakrishna (1948), The utilization of multiple measurements in problems of biological classification, J. Roy. Statist. Soc. B, 10, 159-193.

Rao, C. Radhakrishna (1954), On the use and interpretation of distance functions in statistics. Bull. Inst. Statist. Inst., 34, 90- .

Rao, C. Radhakrishna (1971a), Taxonomy in anthropology, Mathematics in the Archaeological and Historical Sciences, 19-29. Edin. Univ. Press.

Rao, C. Radhakrishna (1971b) Advanced Statistical Methods in Biometric Research, Haffner.

Rao, C. Radhakrishna and Varadarajan, V.S. (1963), Discrimination of Faussian Processes, Sankhya A, 25, 303-330.

Sneath, P.H.A. and Sokal, R.R. (1973), Numerical Taxonomy, Freeman, San Francisco.

Thompson, E.A. (1975), Human Evolutionary Trees, Cambridge University Press, Cambridge.

Indian Statistical Institute
7, S.J.S. Sansanwal Marg,
New Delhi - 110029
India

Linguistic Approach to Pattern Recognition

K. S. Fu

1. Linguistic (Structural) Approach to Pattern Recognition.

Most of the developments in pattern recognition research during the past decade deal with the decision-theoretic approach [1-11] and its applications. In some pattern recognition problems, the structural information which describes each pattern is important, and the recognition process includes not only the capability of assigning the pattern to a particular class (to classify it), but also the capacity to describe aspects of the pattern which make it ineligible for assignment to another class. A typical example of this class of recognition problem is picture recognition, or more generally speaking, scene analysis. In this class of recognition problems, the patterns under consideration are usually quite complex and the number of features required is often very large which make the idea of describing a complex pattern in terms of a (hierarchical) composition of simpler subpatterns very attractive. Also, when the patterns are complex and the number of possible descriptions is very large, it is impractical to regard each description as defining a class (for example, in fingerprint and face identification problems, recognition of continuous speech, Chinese characters, etc.). Consequently, the requirement of recognition can only be satisfied by a description for each pattern rather than the simple task of classification.

Example 1: The pictorial patterns shown in Figure 1(a) can be described in terms of the hierarchical structures shown in Figure 1(b).

In order to represent the hierarchical (tree-like) structural information of each pattern, that is, a pattern described in terms of simpler subpatterns and each simpler subpattern again be described in terms of even simpler subpatterns, etc., the linguistic (syntactic) or structural approach has been proposed [12 - 16]. This approach draws an analogy between the (hierarchical, tree-like) structure of patterns and the syntax of languages. Patterns are specified as building up out of subpatterns in various ways of composition just as phrases and sentences are built up by concatenating words and words are built up by cancatenating characters. Evidently, for this approach to be advantageous, the simplest subpatterns selected, called "pattern primitives", should be much easier to recognize than the patterns themselves. The "language" which provide the structural description of patterns in terms of a set of pattern primitives and their composition operations, is sometimes called "pattern description language". The rules governing the composition of primitives into patterns are usually specified by the so-called "grammer" of the pattern description language. After each primitive within the pattern is identified, the recognizing process is accomplished by performing a syntax analysis or parsing of the "sentence" describing the given pattern to determine whether or not it is syntactically (or grammatically) correct with respect to the specified grammer. In the meantime, the syntax analysis also produces a structural description of the sentence representing the given pattern (usually in the form of a tree structure).

The linguistic approach to pattern recognition provides a capability for describing a large set of complex patterns using small sets of simple pattern primitives and of grammatical rules. The various relations or composition operations defined among subpatterns can usually be expressed in terms of logical and/or mathematical operations. As can be seen later, one of the most attractive aspects of this capability is the use of recursive nature of a grammar. A grammar (rewriting) rule can be applied any number of times, so it is possible to express in a very compact way some basic structural characteristics of an infinite set of

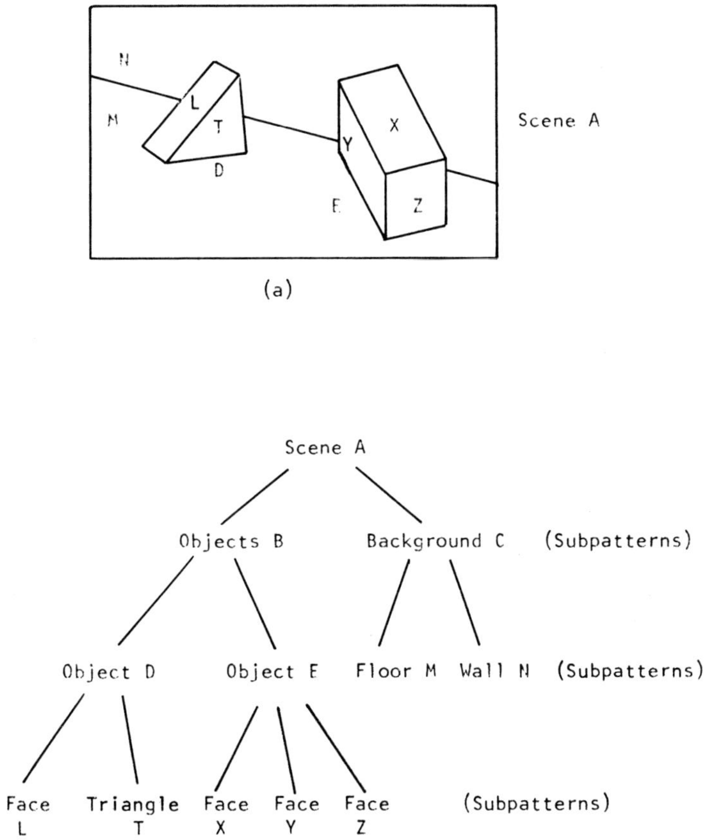

Fig. 1. The Pictorial Pattern A and Its Hierarchical Structural Descriptions

sentences. Of course, the practical utility of such an approach depends on our ability to recognize the simple pattern primitives and their relationships represented by the composition operations.

It should be noted that, for many practical applications, often both linguistic and decision-theoretic approaches are used [12, 88]. For example, decision-theoretic approaches are usually effective in the recognition of pattern primitives. This is primarily due to the fact that the structural information of primitives is considered not important and the (local) measurements taken from the primitives are sensitive to noise and distortion. On the other hand, in the recognition of subpatterns and the pattern itself which are rich in structural information, syntactic approaches are therefore required.

An alternative representation of the structural information of a pattern is to use a "relational graph." For example, a relational graph of Pattern A in Figure 1(a) is shown in Figure 2. Since there is a one-to-one corresponding relation between a linear graph and a matrix, a relational graph can certainly also be expressed as a "relational matrix." In using the relational graph for pattern description, we can broaden the class of allowed relations to include any relation that can be conveniently determined from the pattern. With this generalization, we may possibly express richer descriptions than we can with tree structures. However, the use of tree structures does provide us a direct channel to adapt the techniques of formal language theory to the problem of compactly representing and analyzing patterns containing a significant structural content.

We briefly introduce some important definitions and notations in this section.

Definition 1. A (phrase-structure) grammar G is a four-triple

$$G = (V_N, V_T, P, S)$$

where

V_N is a finite set of nonterminals,

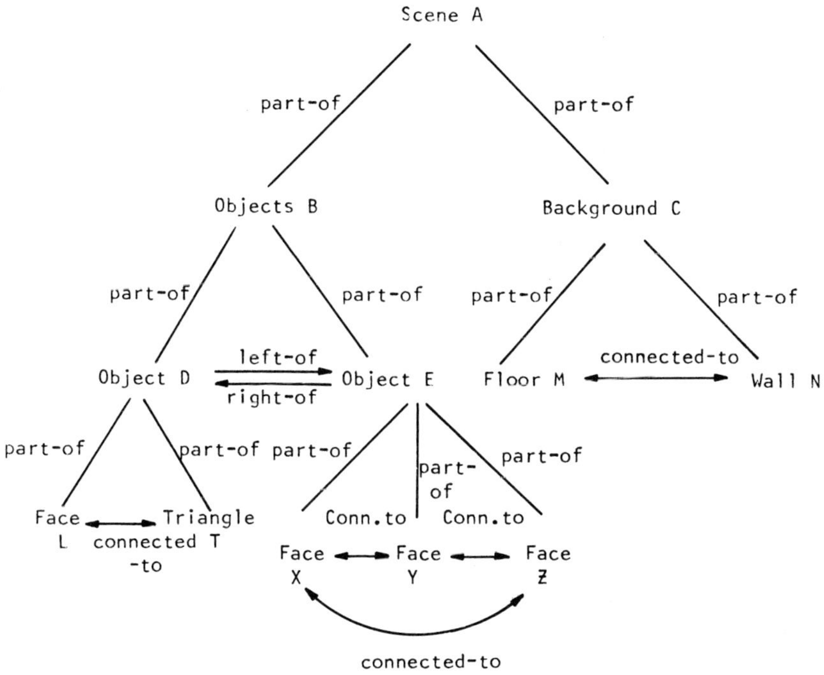

Fig. 2. A Relational Graph of Scene A

V_T is a finite set of terminals,

$S \in V_N$ is the start symbol,

and

P is a finite set of rewrite rules or productions denoted by

(1) $$\alpha \rightarrow \beta$$

α and β are strings over the union of V_N and V_T and with α involving at least one symbol of V_N.

The following notations are frequently used.

(1) V^* is the set of all strings of symbols in V, including λ, the string of length 0, $V^+ = V^* - \{\lambda\}$.

(2) If x is a string, x^n is x written n times.

(3) $|x|$ is the length of the string x, or the number of symbols in string x.

(4) $\eta \underset{G}{\Rightarrow} \gamma$, or a string η directly generates or derives a string γ if $\eta = \omega_1 \alpha \omega_2$, $\gamma = \omega_1 \beta \omega_2$, and $\alpha \rightarrow \beta$ is a production in P.

(5) $\eta \overset{*}{\underset{G}{\Rightarrow}} \gamma$, or a string η generates or derives a string γ if there exists a sequence of strings $\zeta_1, \zeta_2, \ldots, \zeta_n$ such that $\eta = \zeta_1$, $\gamma = \zeta_n$, $\zeta_i \Rightarrow \zeta_{i+1}$, $i = 1,2,\ldots,n-1$. The sequence of strings $\zeta_1, \zeta_2, \ldots, \zeta_n$ is called a derivation of γ from η.

<u>Definition 2.</u> The language generated by grammar G is

(2) $$L(G) = \{x | x \in V_T^* \text{ and } S \overset{*}{\underset{G}{\Rightarrow}} x\}.$$

That is, the language consists of all strings or sentences of terminals generated from the start symbol S.

<u>Definition 3.</u> In (1) if $|\alpha| \leq |\beta|$, the grammer is called a type 1 or context-sensitive grammar. If $\alpha = A \in V_N$, the grammar is called a type 2 or context-free grammar. If, in addition to $\alpha = A$, $\beta = aB$ or $\beta = a$, where $a \in V_T$ and $B \in V_N$, the grammar is called a type 3, or finite-state, or regular grammar.

The languages generated by context-sensitive, context-free, and finite-state (regular) grammars are called context-sensitive, context-free, and finite-state (regular) languages, resepctively.

Example 2: Consider the context-free grammar

$$G = (V_N, V_T, P, S)$$

where $V_N = \{S,A,B\}$, $V_T = \{a,b\}$, and $P^{\dagger}$:

(1) $S \rightarrow aB$ (5) $A \rightarrow a$

(2) $S \rightarrow bA$ (6) $B \rightarrow bS$

(3) $A \rightarrow aS$ (7) $B \rightarrow aBB$

(4) $A \rightarrow bAA$ (8) $B \rightarrow b$.

The language generated by G, L(G), is the set of all sentences or strings in V_T^+ consisting of an equal number of a's and b's. Typical generations or derivations of sentences include

$$S \overset{(1)}{\Longrightarrow} aB \overset{(8)}{\Longrightarrow} ab$$

$$S \overset{(1)}{\Longrightarrow} aB \overset{(6)}{\Longrightarrow} abS \overset{(2)}{\Longrightarrow} abbA \overset{(5)}{\Longrightarrow} abba$$

$$S \overset{(2)}{\Longrightarrow} bA \overset{(4)}{\Longrightarrow} bbAA \overset{(4)}{\Longrightarrow} bbbAAA \overset{(5)}{\Longrightarrow} bbbaAA$$

$$\overset{(5)}{\Longrightarrow} bbbaaS \overset{(5)}{\Longrightarrow} bbbaaa$$

where the parenthesized number indicates the production used.

An alternative method for describing any derivation in a context-free grammer is the use of derivation or parse trees. A derivation tree for a context-free grammar can be constructed according to the following procedure:

(1) Every node of the tree has a label, which is a symbol in V_N or V_T .

(2) The root of the tree has the label S .

(3) If a node has at least one descendant other than itself, and has the label A , then $A \in V_N$.

---

† For convenience, we can also use the shorthand notation $S \rightarrow aB|bA$ for representing productions (1) and (2). Similarly, we can use $A \rightarrow aS|bAA|a$ for productions (3), (4), and (5), and use $B \rightarrow bS|aBB|b$ for productions (6), (7), and (8) .

(4) If nodes $n_1, n_2, \ldots, n_k$ are the direct descendants of node n (with label A) in the order from left to right, with labels $A_1, A_2, \ldots, A_k$, respectively, then

$$A \rightarrow A_1 A_2 \ldots A_k$$

must be a production in P .

For example, the derivation $S \overset{*}{\Rightarrow} abba$ in Example 1.2 can be described by the following derivation tree:

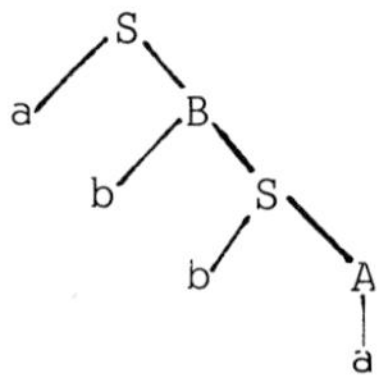

2. Linguistic Pattern Recognition System.

A linguistic pattern recognition system can be considered as consisting of three major parts; namely, preprocessing, pattern description or representation, and syntax analysis[†]. A simple block diagram of the system is shown in Figure 3. The functions of preprocessing include (i) pattern encoding and approximation, and (ii) filtering, restoration and enhancement. An input pattern is first coded or approximated by some convenient form for further processing. For example, a black-and-white picture can be coded in terms of a grid (or a matrix) of 0's and 1's, or a waveform can be approximated by its time samples or a truncated Fourier series expansion. In order to make the processing in the later stages of the system more efficient, some sort of "data compression" is often applied at this stage. Then, techniques of filtering, restoration and/or enhancement will be used to clean the noise, to restore the degredation, and/or to improve the quality of the coded (or approximated) patterns. At the output of the preprocessor, presumably, we have

[†] The division of three parts is for convenience rather than necessity. Usually, the term "linguistic pattern recognition" refers primarily to the pattern representation (or description) and the syntax analysis.

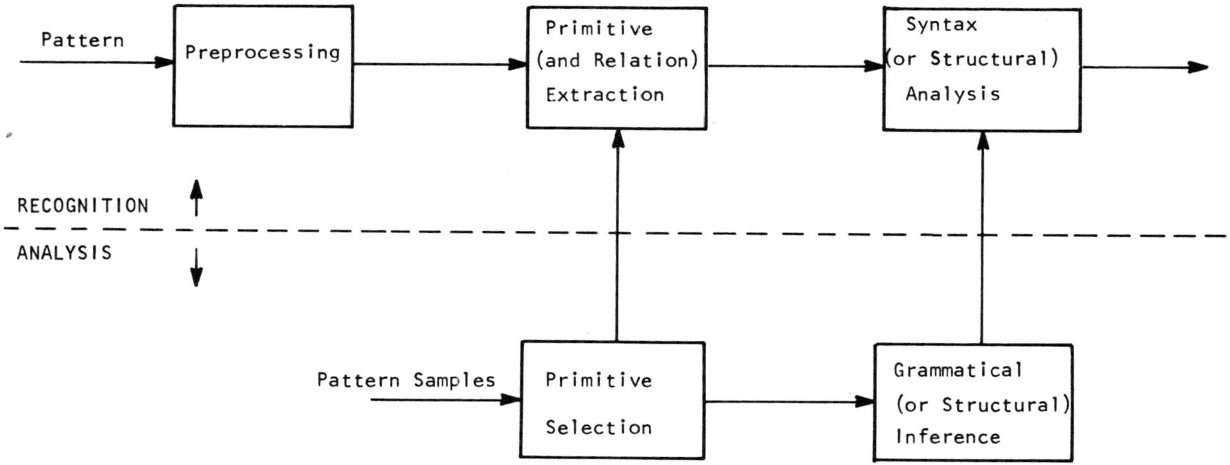

Fig. 3. Block Diagram of Linguistic Pattern Recognition System.

patterns with reasonably "good quality". Each preprocessed pattern is then represented by a language-like structure (for example, a string, a tree, or a graph). The operation of this pattern-representation process consists of (i) pattern segmentation, and (ii) primitive (feature) extraction. In order to represent a pattern in terms of its subpatterns, we must segmentize the pattern and, in the meantime, identify (or extract) the primitives and relations in it. In other words, each preprocessed pattern is segmentized into subpatterns and pattern primitives based on pre-specified syntactic or composition operations; and, in turn, each subpattern is identified with a given set of pattern primitives. Each pattern is now represented by a set of primitives with specified syntactic operations. For example, in terms of "concatenation" operation, each pattern is represented by a string of (concatenated) primitives. More sophisticated systems should also be able to detect various syntactic relations within the pattern. The decision on whether or not the representation (pattern) is syntactically correct (i.e., belongs to the class of patterns described by the given syntax or grammar) will be performed by the "syntax analyzer" or "parser". When performing the syntax analysis or parsing, the analyzer can usually produce a complete syntactic description, in terms of a parse or parsing-tree, of the pattern provided it is syntactically correct. Otherwise, the pattern is either rejected or analyzed on the basis of other given grammars, which presumably describe other possible classes of patterns under consideration.

Conceptually, the simplest form of recognition is probably "template-matching". The string of primitives representing an input pattern is matched against strings of primitives representing each prototype or reference pattern. Based on a selected "matching" or "similarity" criterion, the input pattern is classified in the same class as the prototype pattern which is the "best" to match the input. The hierarchical structure information is essentially ignored. A complete parsing of the string representing an input pattern, on the other hand, explores the complete hierarchical structural description of the pattern. In between, there are a number of intermediate approaches. For example, a series of tests can be designed to test the occurrences or non-occurrence of certain subpatterns (or primitives) or certain combinations of subpatterns or primitives. The result of the tests (for example, through a table look-up, a decision tree, or a logical operation) is used for a classification decision. Notice that each test may be a template-matching scheme or a parsing for a subtree representing a sub-pattern. The selection of an appropriate approach for recognition usually depends upon the problem requirement. If a complete pattern description is required for recognition, parsing is necessary. Otherwise, a complete parsing could be avoided by using other simpler approaches to improve the efficiency of the recognition process.

In order to have a grammar describing the structural information about the class of patterns under study, a grammatical inference machine is required which can infer a grammar from a given set of training patterns in language-like representations[†]. This is analogous to the "learning" process in a decision-theoretic pattern recognition system [1-11, 17-20]. The structural description of the class of patterns under study is learned from the actual sample patterns from that class. The learned description, in the form of a grammer, is then used for pattern description and syntax analysis (see Figure 3). A more general form of learning might include the capability of learning the best set of primitives and the corresponding structural description for the class of patterns concerned.

3. Selection of Pattern Primitives.

As we discussed in Section 1, the first step in formulating a linguistic model for pattern description is the determination of a set of primitives in terms of which patterns of interest may be described. This will be largely influenced by the nature of the data, the specific application in question, and the technology available for implementing the system. There is no general solution for the primitive selection problem at this time. The following requirements usually serve as a guideline for

[†]At present, this part is performed primarily by the designer.

selecting pattern primitives.

(i) The primitives should serve as basic pattern elements to provide a compact but adequate description of the data in terms of the specified structural relations (e.g., the concateration relation).

(ii) The primitives should be easily extracted or recognized by existing non-linguistic methods, since they are considered to be simple and compact patterns and their structural information not important.

For example, for speech patterns, phonemes are naturally considered as a "good" set of primitives with the concatenation relation†. Similarly, strokes have been suggested as primitives in describing handwriting. However, for general pictorial patterns, there is no such "universal picture element" analogous to phonemes in speech or strokes in handwriting‡. Sometimes, in order to provide an adequate description of the patterns, the primitives should contain the information which is important to the specific application in question. For example, if the size (or shape or location) is important in the recognition problem, then the primitives should contain information relating to size (or shape or location) so that patterns from different classes are distinguishable by whatever method is to be applied to analyze the descriptions. This requirement often results in a need for semantic information in describing primitives [12].

Requirement (ii) may sometimes conflict with requirement (i) due to the fact that the primitive selected according to requirement (i) may not be easy to recognize using existing techniques. On the other hand, requirement (ii) could allow the selection of quite complex primitives as long as they can be recognized. With more complex primitives, simpler structural descriptions (e.g., simple grammar) of the patterns could be used. This trade-off may become quite important in the implementation

†The view of continuous speech as composed of one sound segment for each successive phoneme is, of course, a simplification of facts.

‡It is also interesting to see that the extraction of phonemes in continuous speech and that of strokes is handwriting are not a very easy task with respect to the requirement (ii) specified above.

of the recognition system. An example is the recognition of two-dimensional mathematical expressions in which characters and mathematical notations are primitives. However, if we consider the characters as subpatterns and describe them in terms of simpler primitives (e.g., strokes or line segments), the structural descriptions of mathematical expressions would be more complex than the case of using characters directly as primitives.

Eden and Halle [22] have proposed a formal model for the abstract description of English cursive script. The primitives are four distinct line segments, called "bar" |, "hook" ◡, "arch" ∩, and "loop" J. These primitives can be transformed by rotation or by reflection about the horizontal or vertical axis. These transformations generate 28 strokes, but only nine of them are of interest in the English script commonly used. A word is completely specified by the stroke sequence comprising its letters.

No formal syntax was attempted for the description of handwriting. Interesting experimental results on the recognition of cursive writing were obtained by Earnest [23] and Mermelstein [24] using a dictonary and rather heuristic recognition criteria. In addition, the dynamics of the trajectory (in space and time) that the point of the pen traces out as it moves across the paper has also been studied [25]. The motion of the pen is assumed to be controlled by a pair of orthogonal forces, as if one pair of muscles controls the vertical displacement and another the horizontal.

More general methods for primitive selection may be grouped roughly into methods emphasizing boundaries and methods emphasizing regions. These methods are discussed in the following.

3.1 Primitive Selection Emphasizing Boundaries or Skeletons.

A set of primitives commonly used to describe boundaries or skeletons is the chain code given by Freeman [12, 26]. Under this scheme, a rectangular grid is overlaid on the two-dimensional pattern, and straight line segments are used to connect the grid points falling closest to the pattern. Each line segment is assigned an octal digit according to its slope. The pattern is thus represented by a chain (or string) or chains of octal digits. Figure 4 illustrates the primitives and the coded string describing a curve. This descriptive scheme has some useful properties. For example, patterns coded in this way can be rotated through multiples of 45° simply by adding an octal digit (modulo 8) to every digit in the string (however, only rotations by multiples of 90° can be accomplished without some distortion of the pattern). Other simple manipulations such as expansion, measurement of curve length, and determination of pattern self-intersections are easily carried out. Any desired degree of resolution can be obtained by adjusting the fineness of the grid imposed on the patterns. This method is, of course, not limited to simply-connected closed boundaries; it can be used for describing arbitrary two-dimensional figures composed of straight or curved lines and line segments.

Notable work using Freeman's chain code include efforts by Knoke and Wiley [28] and by Feder [29]. Knoke and Wiley attempted to demonstrate that linguistic approaches can usually be applied to describe structural relationships within patterns (hand-printed characters, in this case). Feder's work considers only patterns which can be encoded as strings of primitives. Several bases for developing pattern languages are discussed, including equations in two variables (straight lines, circles and circular arcs, etc.), pattern properties (self-intersections, convexity, etc.), and various measures of curve similarity. The computational power (automaton complexity) required to detect the elements of these pattern languages is studied. However, this problem is complicated considerably by the fact that (i) these languages are mostly context-sensitive and not context-free, (ii) the chain code yields only a piecewise linear approximation of the original pattern, and (iii) the coding of a typical curve is not unique, depending to a degree on its location and orientation with respect to the coding grid.

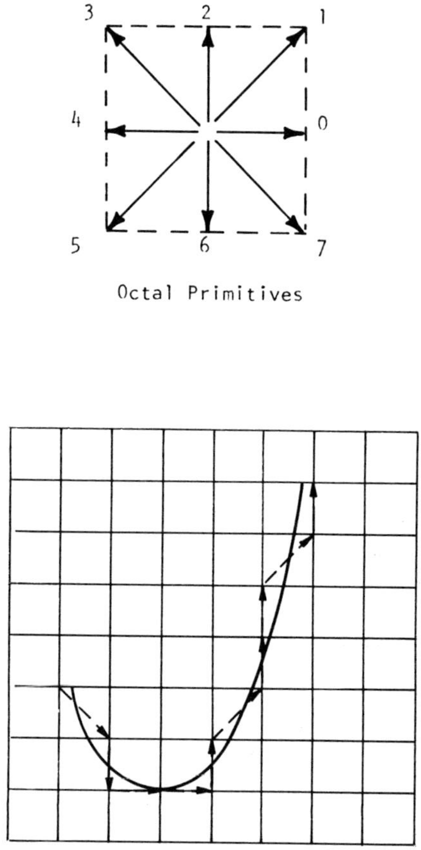

Coded String of the Curve = 7600212212

Fig. 4. Freeman's Chain Code

Other applications of the chain code include description of contour maps [30], "shape matching" [31], and identification of high energy particle tracks in bubble chamber photographs [32]. Contour lines can be encoded as chains. Contour map problems may involve finding the terrain to be flooded by a dam placed at a particular location, the water shed area for a river basin, the terrain visible from a particular mountain-top location, or the determination of optimum highway routes through mountainous terrain. In shape matching, two or more two-dimensional objects having irregular contours are to be matched for all or part of their exterior boundary. For some such problems the relative orientation and scale of the objects to be matched may be known and only translation is required. The problem of matching aerial photographs to each other as well as to terrain maps falls into this category. For other problems either orientation, or scale, or both may be unknown and may have to be determined as part of the problem. An example of problems in which relative orientation has to be determined is that of the computer assembly of potsherds and jigsaw puzzles [33].

Other syntactic pattern recognition systems using primitives with the emphasis on boundary, skeleton or contour information include systems for hand-printed character recognition [34-36], bubble chamber and spark chamber photograph classification [37-40], chromosome analysis [41-43], fingerprint identification [106-107], face recognition [44,45], and scene analysis [46-48].

3.2 Pattern Primitives in Terms of Regions.

A set of primitives for encoding geometric patterns in terms of regions has been proposed by Pavlidis [49]. In this case, the basic primitives are halfplanes in the pattern space† (or the field of observation). It can be shown that any figure (or arbitrary polygon) may be expressed as the union of a finite number of convex polygons. Each convex polygon can, in turn, be represented as the intersection of a finite

†This could be generalized to halfspace of the pattern space.

number of halfplanes. By defining a suitable ordering (a sequence) of the convex polygons composing the arbitrary polygon, it is possible to determine a unique minimal set of maximal (in an appropriate sense) polygons, called primary subsets, the union of which is the given polygon. In linguistic analogy, a figure can be thought of as a "sentence", the convex polygon composing it as "words" and the halfplanes as "letter". This process is summarized in this section.

Let A be a bounded polygon and let $s_1, s_2, \ldots, s_n$ be its sides. A point x in the plane will be said to be positive with respect to one side if it lies on the same side of the extension of a side as the polygon does with respect to the side itself. Otherwise, it will be said to be negative with respect to that side.

Example 3: For the polygon A given in Figure 5, the point x is positive with respect to the sides s_5 and s_6, but negative with respect to s_7. Similarly, y is positive with respect to s_4 and s_7, but negative with respect to s_5. Extending all the sides of A on both directions, A is intersected by some of these extensions, and it is subdivided into $A_1, A_2, \ldots, A_9$ convex polygons.

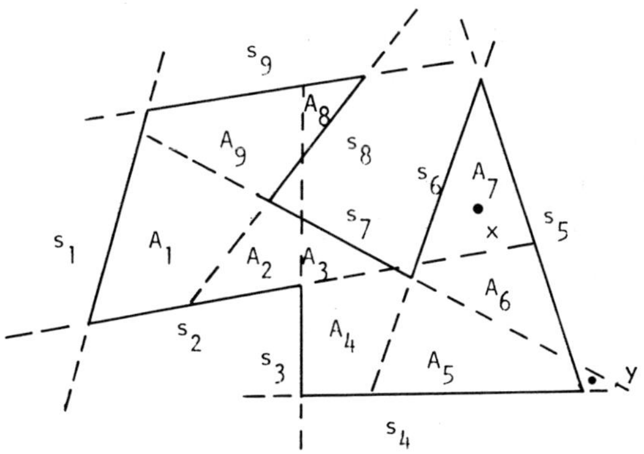

Fig. 5. Polygon A of Example 3

Obviously, the points which are positive with respect to a side form a halfplane whose boundary is the extension of the side. Let h_i denote the halfplane corresponding to the side s_i, and let Q denote the intersection of all the halfplanes $h_1, h_2, \ldots, h_n$ in A. If A is convex, then $A = Q$. If A is not convex, then Q may be empty or simply different from A. Let Q_I represent the intersection of all the halfplanes except $s_{i_1}, \ldots, s_{i_k}$ where $I = \{i_1, \ldots, i_k\}$, the index set. Then we can define a sequence of Q_I as follows:

$$
\begin{aligned}
Q &= \bigcap_{i=1}^{n} h_i \\
(4) \qquad Q_j &= \bigcap_{i=1}^{n} h_i \\
Q_{jk} &= \bigcap_{\substack{i=1 \\ i \neq j,\, i \neq k}}^{n} h_i \quad .
\end{aligned}
$$

This is an increasing sequence since $Q \subset Q_j \subset Q_{jk} \ldots$. The last element of the sequence will be the whole plane, and it is obtained for $I = \{1, \ldots, n\}$. If a sequence of the above form has a maximal element, then that set is called a primary (convex) subset of A. A nonempty member of such a Q-sequence which is also a subset of A is called a nucleus of A if all the previous elements of the sequence are empty. Consequently, it can be shown that the union of the primary subsets of A precisely equals A.

For a given polygon the primary subsets can be found by forming all the sequences $Q, Q_j, Q_{jk}, \ldots$ and searching for their maximal elements. This is a well-defined procedure and, hence, the primary subsets of A are unique.

It is noted that this approach provides a formalism for describing the syntax of polygonal figures and more general figures which can be approximated reasonably well by polygonal figures. The analysis or recongition procedure requires the definition of suitable measures of similarity between polygons. The similarity measures considered so far are quite sensitive to noise in the patterns and/or are difficult to implement practically on a digital computer. A somewhat more general selection procedure of pattern primitives based on regions has been recently proposed by Rosenfeld and Strong [50].

Another form of representing polygonal figures is the use of primary graphs [51, 52]. The primary graph of a polygon A is one whose nodes correspond to the nuclei and the primary subsets of A, and its branches connect each nucleus to all the primary subsets containing it. Another approach to the analysis of geometric patterns using regions is discussed primarily in the problem of scene analysis [9, 47]. Minsky and Papert [53] have considered the direct transformation of a gray scale picture to regions, bypassing the edge-finding, line-fitting procedures. Regions are constructed as the union of squares whose corners have the same or nearly the same gray scale. The method proposed by Guzman [54] assumes that a picture can be reduced by preprocessing to a list of vertices, lines and surfaces. Various heuristics, based on the analysis of types of intersections of lines and surfaces, are applied to this list to compose its elements into two- or three-dimensional regions. Some candidate pattern recognition schemes have been investigated, all of which involve methods for matching the reduced pattern descriptions against a prototype dictionary. The procedure studied by Brice and Fennema [55] decomposes a picture into atomic regions of uniform gray scale. A pair of heuristics is used to join these regions in such a way as to obtain regions whose boundaries are determined more by the natural lines of the scene than by the artificial ones introduced by quantization and noise. Then a simple line-fitting technique is used to approximate the region boundaries by straight lines and finally, the scene analyzer interprets the picture using some simple tests on object groups generated by a Guzman-like procedure.

4. Pattern Grammar

Assume that a satisfactory solution of the "primitive selection" problem is available for a given application. The next step is the construction of a grammar (or grammars) which will generate a language (or languages) to describe the patterns under study. Ideally, it would be nice to have a grammatical inference machine which would infer a grammar from a given set of strings describing the patterns under study. Unfortunately, such a machine has not been available except for some very special cases [56]. In most cases so far, the designer constructs the grammar based on the a priori knowledge available and his experience. It is known that increased descriptive power of a language is paid for in terms of increased complexity of the analysis system (recognizer or acceptor). Finite-state automata are capable of recognizing or accepting finite-state languages although the descriptive power of finite-state languages is also known to be weaker than that of context-free and context-sensitive languages. On the other hand, non-finite, nondeterministic devices are required, in general, to accept the languages generated by context-free and context-sensitive grammars. Except for the class of deterministic languages, nondeterministic parsing procedures are usually needed for the analysis of context-free languages. The trade-off between the descriptive power and the analysis efficiency of a grammer for a given application is, at present, almost completely justified by the designer. (For example, a precedence language may be used for pattern description in order to obtain good analysis efficiency; or, on the other hand, a context-free programmed grammar generating a context-sensitive language may be selected in order to describe the patterns effectively.) The effect of the theoretical difficulty may not be serious, in practice, as long as some care is exercised in developing the required grammars. This is especially true when the languages of interest are actually finite-state, even though the form of the grammars may be context-sensitive, or when the languages may be approximated by finite-state languages.

It should be remarked that a grammar is most appropriate for description when the pattern of interest is built up from a small set of primitives by recursive application of a small set of production rules. Also, the "primitive selection" and the "grammar construction" should probably be treated simultaneously rather than in two different stages. There is no doubt that a different selection of pattern primitives will result in a different grammar for the description of a given set of patterns. Sometimes, a compromise is necessary in order to develop a suitable grammar.

Although many classes of patterns appear to be intuitively context-sensitive, context-sensitive (but not context-free) grammars have rarely been used for pattern description simply because of their complexity. Context-free languages have been used to describe patterns such as English characters [57], chromosome images [41], spark chamber pictures [37], chemical structures [63], fingerprint patterns [106,107], plane projective geometry [58] and spoken digits [110].

Example 4: The following is a context-free grammar describing the chromosome images shown in Figure 6 [41].

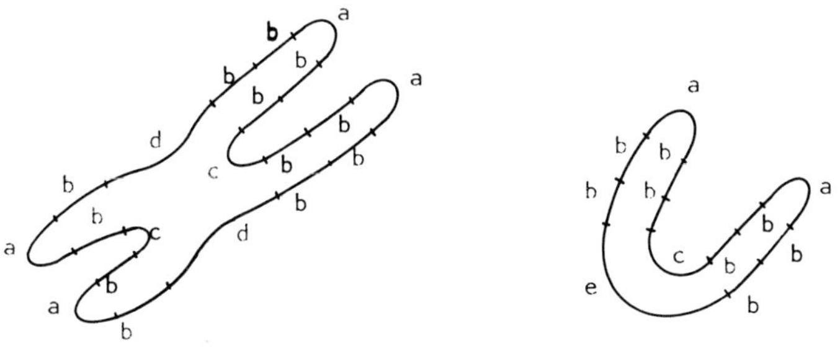

Fig. 6. (a) Submedian Chromosome and (b) Telocentric Chromosome

$G = (V_N, V_T, P, \{\langle\text{submedian chromosome}\rangle\ \langle\text{telocentric chromosome}\rangle\})$

where

V_N = {⟨submedian chromosome⟩, ⟨telocentric chromosome⟩
⟨arm pair⟩, ⟨left part⟩, ⟨right part⟩, ⟨arm⟩, ⟨side⟩, ⟨bottom⟩}

V_T = { 

}

and

P: ⟨submedian chromosome⟩ → ⟨arm pair⟩⟨arm pair⟩
⟨Telocentric chromosome⟩ → ⟨bottom⟩⟨arm pair⟩
⟨arm pair⟩ → ⟨side⟩⟨arm pair⟩
⟨arm pair⟩ → ⟨arm pair⟩⟨side⟩
⟨arm pair⟩ → ⟨arm⟩⟨right part⟩
⟨arm pair⟩ → ⟨left part⟩ ⟨arm⟩
⟨left part⟩ → ⟨arm⟩ c
⟨right part⟩ → c ⟨arm⟩
⟨bottom⟩ → b ⟨bottom⟩
⟨bottom⟩ → ⟨bottom⟩ b
⟨bottom⟩ → e
⟨side⟩ → b ⟨side⟩
⟨side⟩ → ⟨side⟩ b
⟨side⟩ → b
⟨side⟩ → d
⟨arm⟩ → b ⟨arm⟩
⟨arm⟩ → ⟨arm⟩ b
⟨arm⟩ → a

In addition to (i) the trade-off between the language descriptive power and the analysis efficiency, and (ii) the compromise sometimes necessary between the primitives selected and the grammar constructed, the designer should also be aware of the need to control the excessive strings generated by the constructed grammar. The number of pattern strings available in practice is always limited. However, in most cases, the grammer constructed would generate a large or infinite number of strings.[†] It is hoped that the excessive strings generated are similar to the available pattern strings. Unfortunately, this may not be true since the grammar, in many cases, is constructed heuristically. The problem may become very serious when the excessive strings include some pattern strings which should belong to other classes. In this case, adjustments should be made to exclude these strings from the language generated by the constructed grammar.

Recently, probably due to their relative effectiveness in describing natural language structures, transformational grammars have been proposed for pattern description [59-62]. Transformational grammars would allow the possibility of determining from the pattern generative mechanism a simple base grammar (deep structure) which generates a certain set of patterns and a problem-oriented set of transformations. Through the base grammar and the transformations, the original set of patterns can be described.

From the above discussion, it might be concluded that, before efficient grammatical inference procedures are available, a man-machine interactive system would be suitable for the problem of grammar construction. The basic grammar and the various trade-off's and compromises have to be determined by the designer. The results of any adjustment on the grammar constructed can be easily checked and displayed through a computer system.

[†]It may be argued that, in practice, a pattern grammar can always be finite-state since it is constructed from a finite number of pattern strings. However, the finite-state grammar so constructed may require a large number of productions. In such a case, a context-free or a context-free programmed pattern grammar may be constructed for the purpose of significantly reducing the number of productions.

5. High-Dimensional Pattern Grammars.

5.1 General Discussion.

In describing patterns using a string grammar, the only relation between subpatterns and/or primitives is the concatenation; that is, each subpattern or primitive can be connected only at the left or right. This one-dimensional relation has not been very effective in describing two- or three-dimensional patterns. A natural generalization is to use a more general formalism including other useful relations [57, 63-68]. Let R be a set of n-ary relations ($n \geq 1$). A relation $r \in R$ satisfied by the subpatterns and/or primitives $X_1, \ldots, X_n$ is denoted $r(X_1, \ldots, X_n)$. For example, TRIANGLE (a,b,c) means that the ternary relation TRIANGLE is satisfied by the line segments a, b, and c, and ABOVE (X,Y) means th that X is above Y . The following example illustrates pattern descirp-tions using this formalism of relations.

Example 5: The mathematical expression

$$\frac{a+b}{c}$$

can be described by

ABOVE (ABOVE (LEFT (a, LEFT(+,b)),——),c)

where LEFT(X,Y) means that X is to the left of Y .

A simple two-dimensional generalization of string grammars is to extend grammars for one-dimensional strings to two-dimensional arrays [69, 70]. The primitives are the array elements and the relation between primitives is the two-dimensional concatenation. Each production re-writes one subarray by another, rather than one substring by another. Relationships between array grammars and array automata (automata with two-dimensional tapes) have been studied recently [71].

Shaw, by attaching a "head" (hd) and a "tail" (tl) to each primi-tive, has used the four binary operators +, ×, - and * for defining binary concatenation relations between primitives [72, 73].

a + b

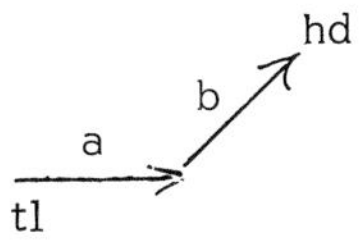

$$hd(a + b) = hd(b)$$
$$tl(a + b) = tl(a)$$

a × b

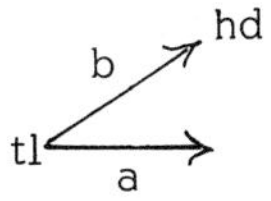

$$hd(a \times b) = hd(b)$$
$$tl(a \times b) = tl(b)$$

a - b

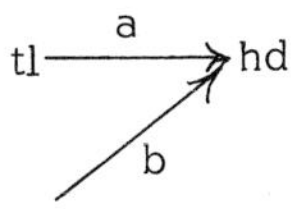

$$hd(a - b) = hd(a)$$
$$tl(a - b) = tl(a)$$

a * b

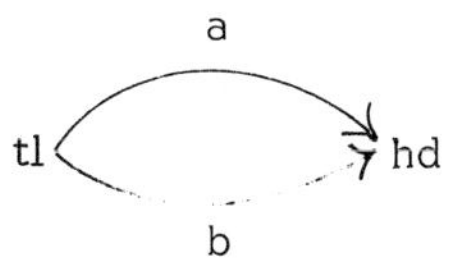

$$hd(a * b) = hd(a)$$
$$tl(a * b) = tl(a)$$

For string languages, only the operator + is used. In addition, the unary operator ~ acting as a tail/head reverser is also defined; i.e.,

~ a

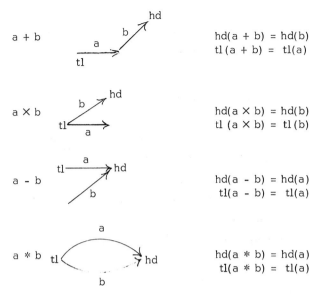

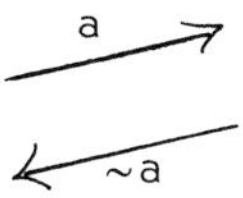

$$hd(\sim a) = tl(a)$$
$$tl(\sim a) = hd(a)$$

In the case of describing patterns consisting of disconnected subpatterns, the "blank" or "don't care" primitive is introduced. Each pictorial pattern is represented by a "labelled branch-oriented graph" where branches represent primitives.

The grammar which generates sentences (PDL expressions) in PDL (Picture Description Language) is a context-free grammar

$$G = (V_N, V_T, P, S)$$

where

$V_N = \{S, SL\}$

$V_T = \{b\} \cup \{+, \times, -, /, (,)\} \cup \{\ell\}$, b may be any primitive (including the "null point primitive" λ which has identical tail and head) and

$$S \to b,\ S \to (S\ \phi_b\ S),\ S \to (\sim S),\ S \to SL,\ S \to (/SL),$$
$$SL \to S^{\ell},\ SL \to (SL\ \phi_b\ SL),\ SL \to (\sim SL),\ SL\ \_\ (/SL),$$
$$\phi_b \to +,\ \phi_b \to S,\ \phi_b \to -,\ \phi_b \to * .$$

ℓ is a label designator which is used to allow cross reference to the expressions S within a description. The / operator is used to enable the tail and head of an expression to be arbitrarily located. A top-down parsing procedure (see Section 6) was used for the recognition of PDL expressions describing pictorial patterns [73].

Based on an idea in [34], Feder has formalized a "plex" grammar which generates languages with terminals having an arbitrary number of attaching points for connecting to other primitives or subpatterns [63]. The primitives of the plex grammar are called N-Attaching Point Entity (NAPE). Each production of the plex grammar is in context-free form in which the connectivity of primitives or subpatterns is described by using explicit lists of labelled concatenation points (called joint lists). While the sentences generated by a plex grammar are not directed graphs, they can be transformed by either assigning labelled nodes to both primitives and concatenation points as suggested by Pfaltz and Rosenfeld [74] or by transforming primitives to nodes and concatenations to labelled branches [75].

Pfaltz and Rosenfeld have extended the concept of string grammars to grammars for labelled graphs called webs. Labelled node-oriented graphs are explicitly used in productions. Each production described the rewriting of a graph α into another graph β and also contains an "embedding" rule E which specifies the connection of β to its surrounding graph (host web) when α is rewritten. A web grammar G is a 4-tuple

$$G = (V_N, V_T, P, S)$$

where V_N is a set of nonterminals; V_T is a set of terminals; S is a set of "initial" webs; and P is a set of web productions. A web production is defined as[†]

$$\alpha \rightarrow \beta \ , \ E$$

where α and β are webs, and E is an embedding of β. If we want to replace the subweb α of the web ω by another subweb β, it is necessary to specify how to "embed" β in ω in place of α. The definition of an embedding must not depend on the host web ω since we want to be able to replace α by β in any web containing α as a subweb. Usually E consists of a set of logical functions which specify whether or not each vertex of $\omega - \alpha$ is connected to each vertex of β.

Example 6: Consider the web grammar

$$G = (V_N, V_T, P, S)$$

where

$$V_N = \{A\}, \ V_T = \{a,b,c\}, \ S = \{A\}$$

and

P: (1) 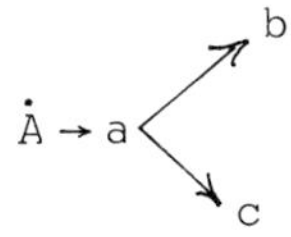

$E = \{(p,a) \mid (p,A) \text{ an edge in the host web}\}$

(2) 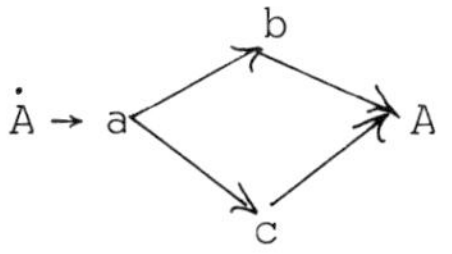

E is the same as in (1)

The language of this grammar is the set of all webs of the form

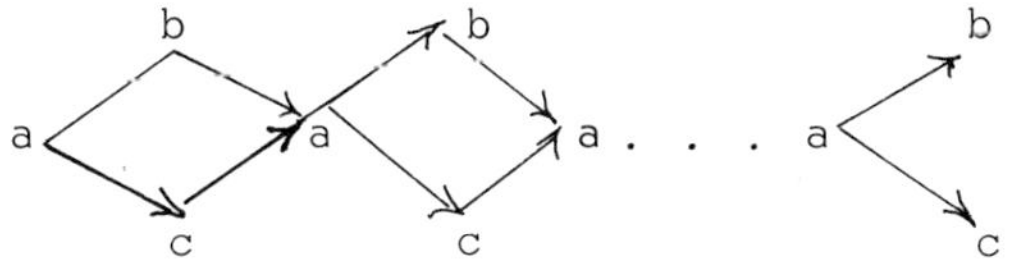

It is noted that web grammars are vertex or node-oriented compared with the branch - or edge- oriented grammars (e.g., PDL, Plex grammars, etc.). That is, terminals or primitives are represented as vertices in the graph rather than as branches. Some applications of web grammars

---

[†]In a most general formulation, the contextual condition of the production is added.

for picture description can be found in [76, 77].

An important special case of a web grammar is that in which the terminal set V_T consists of only a single symbol. In this case, every point of every web in the language has the same label, so that we can ignore the labels and identify the webs with their underlying graphs. This type of web grammar is called a "graph grammar", and its language is called graph language [78]. The formalism of web grammars has been rather extensively analyzed [74, 76]. The relations among PDL grammars, plex grammars and web grammars have been discussed by Shaw [75] and Rosenfeld [79].

Pavlidis [78, 80] has generalized string grammars to graph grammers by including nonterminal symbols which are not simple branches or nodes. An mth order nonterminal structure is defined as an entity which is connected to the rest of the graph by m nodes. In particular, a second-order structure is called a branch structure and a first-order structure a node structure. Then an mth-order context-free graph grammar G_g is a quadruple

$$G_g = (V_N, V_T, P, S)$$

where

V_N is a set of mth-order nonterminal structures: nodes, branches, triangles, ..., polygons with m vertices;

V_T is a set of terminals: nodes and branches;

P is a finite set of productions of the form $A \rightarrow \alpha$, where A is a nonterminal structure and α a graph containing possibly both terminals and nonterminals. α is connected to the rest of the graph through exactly the same nodes as A;

S is a set of initial graphs.

The expression $A * B$ denotes that the two graphs A and B are connected by a pair of nodes (Figure 7(a)), and $N(A + B + C)$ denotes that the graphs A, B and C are connected through a common node N (Figure 7(b)). Thus the production $A \rightarrow B * C$ where A, B and C are branch structures should be interpreted as: replace branch structure A by branch structures B and C connected to the graph through the same nodes as A . No other connection exists between B and C . Similarly, the production $N \rightarrow M(A + B)$ should be interpreted as: replace node structure N by a node structure M and two other structures A and B connected to the rest of the graph by the same node as N . When no ambiguity occurs, we can use simple concatenations, e.g., ANB to denote a nonterminal subgraph consisting of a branch structure A with nodes X and Y connected to the node structure N through Y and a branch structure B with nodes Y and Z connected to N through Y (Figure 7(c)). The subgraph is connected to the rest of the graph through the nodes X and Z.

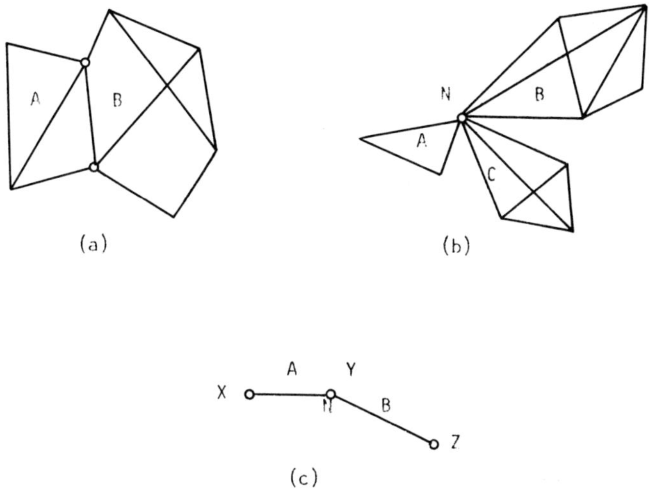

Fig. 7. Illustration of (a) A * B, (b) N(A + B + C) and (c) ANB

The following examples illustrate the use of graph grammars for pattern description:

Example 7: The following grammar describes graphs representing two terminal series-parallel networks (TTSPN)

$$G_g = (V_N, V_T, P, S)$$

where

$$V_N = \{S, B\}, \quad V_T = \{\overset{b}{\longrightarrow}, \overset{n}{\cdot}\}, \quad S = \{nBn\}$$

and

P: (1) $B \to B\dot{n}B$

(2) $B \to B * B$

(3) $B \to b$.

A typical generation would be

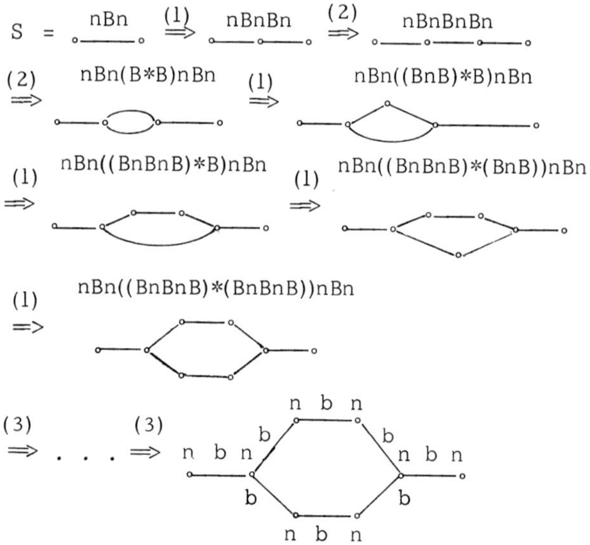

By extending one-dimensional concatenation to multi-dimensional concatenation, strings are generalized to trees. Tree grammars and the corresponding recognizers, tree automata, have been studied recently by a number of authors [81,82]. Naturally, if a pattern can be conveniently described by a tree, it will easily be generated by a tree grammar. For example, in Figure 8, patterns and their corresponding tree representations are listed in (a) and (b), respectively [83, 84]. Gips has recently used a formal syntactic technique in the analysis of line drawings of three-dimensional objects [85].

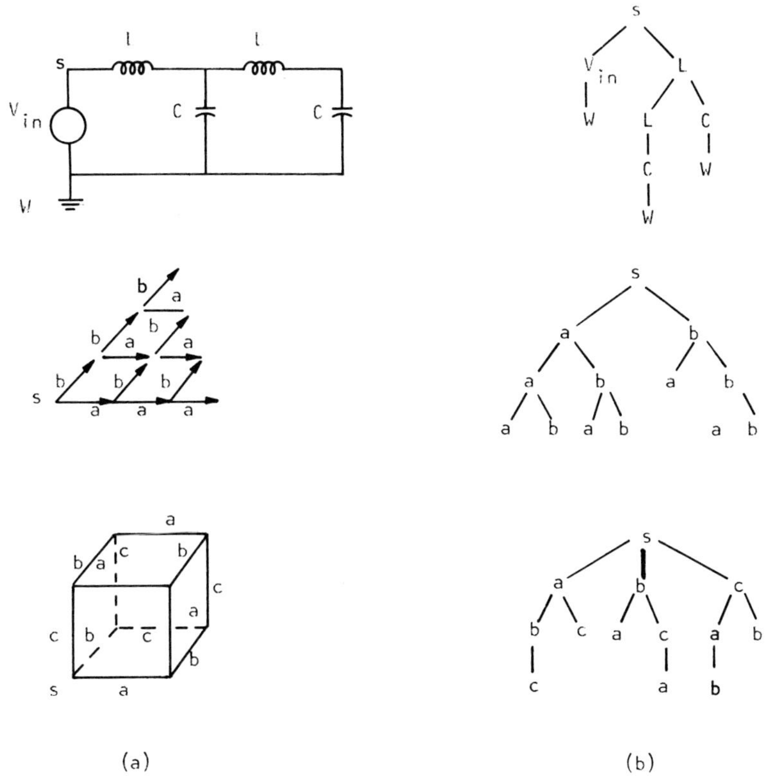

Fig. 8. (a) Patterns and (b) Corresponding Tree Representations

<u>Definition 4</u> A (regular) tree grammar G_t is a four-tuple

$$G_t = (V, r', P, S)$$

where $\langle V, r' \rangle$ is a finite ranked alphabet such that $V_T \subseteq V$ and $r'|V_T = r$. The elements of V_T and $V - V_T = V_N$ are terminals and nonterminals respectively. P is a finite set of productions of the form $\phi \rightarrow \psi$ where $\phi, \psi \in T_V$. $S \subseteq T_V$ is a finite set of axioms and T_V is a set of trees over $\langle V, r' \rangle$.

A generation or derivation $\alpha \Longrightarrow \beta$ is in G_t if and only if there is a production $\phi \rightarrow \psi$ in P such that $\alpha|a = \phi$ and $\beta = (a \rightarrow \psi)\alpha$.

That is, ϕ is a subtree of α at "a" and β is generated by replacing ϕ by ψ at "a." $\alpha \overset{*}{\Rightarrow} \beta$ is in G_t if and only if there exist trees t_0, $t_1, \ldots, t_m \in T_V$, $m \geq 0$ such that

$$\alpha = t_0 \Rightarrow t_1 \Rightarrow \ldots \Rightarrow t_m = \beta$$

in G_t. The sequence $t_0, t_1, \ldots, t_m$ is called a derivation of β from α.

Definition 5. The language generated by tree grammar G_t is

$$L(G_t) = \{t \in T_{V_T} \mid \text{there exists } \gamma \in S \text{ such that } \gamma \overset{*}{\Rightarrow} t \text{ is in } G_t\}.$$

Example 8: The tree grammar

$$G_t = (V, r, P, S)$$

where

$$V = \{S, a, b, \$, A, B\}$$

$$V_T = \{\overset{a}{\rightarrow}, \uparrow b, .\$\}$$

$$r(a) = \{2,1,0\},\ r(b) = \{2,1,0\},\ r(\$) = 2$$

and P:

```
S → $              A → a
   / \                / \
  A   B              A   B

B → b        A → a          B → b
   / \           |              |
  A   B          b              a
```

generates such patterns as

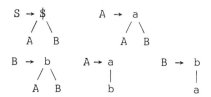

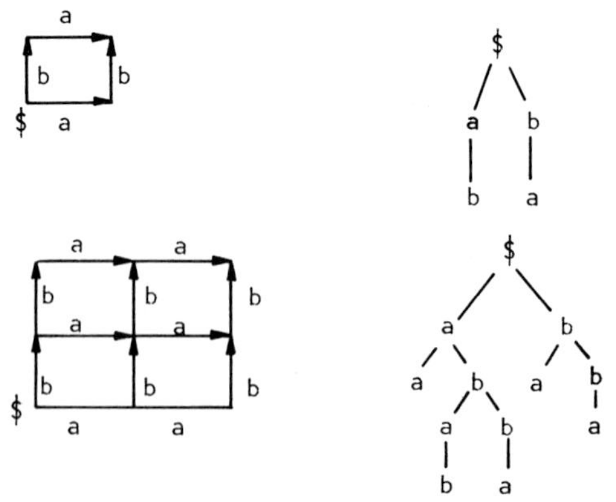

In specifying a selected primitive, a set of attributes is often required. A primitive (terminal) with different properties can be expressed in terms of its attribute values. This set of attributes may be considered as a semantic information of the primitive. Each attribute may be expressed by numerical functions or logical predicates. The semantic information of a subpattern (nonterminal) is, in general, evaluated either from the semantic information of the composed primitives according to the syntactic relations or operators and the semantic rules associated with each production of the grammar, or on the basis of a separate set of functions or rules which are not necessarily defined in conjunction with the productions of the grammar [12]. This class of grammars is sometimes called attribute grammars or "grammars with coordinates" [102-104].

6. Syntax Analysis as Recognition Procedure.

As it was pointed out in Section 2, a parsing or syntax analysis is necessary if a complete description of the input pattern is required for recognition. This requirement may be necessary due to the fact that the number of pattern classes is very large such as in a fingerprint recognition problem. It may also be necessary in the case that the complete description of each pattern will be stored for data retrieval purpose. In this section, syntax analysis for finite-state and context-free (string) languages will be briefly reviewed [86, 87]. Parsing of context-sensitive languages and web (and graph) languages is still an important topic for investigation. Regular tree languages are accepted by tree automata. The procedure of constructing a tree automatan to accept the language generated by a tree grammar is available [81 - 83].

6.1 Recognition of Finite-State Languages.

Finite-state automata are known to recognize or accept finite-state languages [87]. If a class of patterns can be described by a finite-state language, a finite-state automaton can then be constructed to recognize the strings or sentences describing this class of patterns.

Definition 6: A nondeterministic finite-state automaton is a quintuple $(\Sigma, Q, \delta, q_0, F)$, where Σ is a finite set of input symbols (alphabet), Q is a finite set of states, ψ is a mapping of $Q \times \Sigma$ into subsets of Q, $q_0 \in Q$ is the initial state, and $F \subseteq Q$ is the set of final states.

The interpretation of $\delta(q,a) = \{q_1, q_2, \ldots, q_\ell\}$ is that the automaton A, in state q, scanning a on its input tape, chooses any one of $q_1, \ldots, q_\ell$ as the next state and moves its input head one square to the right. The mapping δ can be extended from an input symbol to a string of input symbols by defining

$$\delta(q, \lambda) = \{q\}$$

$$\delta(q, xa) = \bigcup_{q_i \in \delta(q,x)} \delta(q_i,), \qquad a \in \Sigma^* \text{ and } a \in \Sigma .$$

Furthermore, we can define

$$\delta(\{q_1, q_2, \ldots, q_\ell\}, x) = \bigcup_{i=1}^{\ell} \delta(q_i, x) .$$

When $\delta(q,a)$ consists of only a single state, the automaton is called a deterministic finite-state automaton.

A string x is accepted by A if there is a state p such that

$$p \in \delta(q_0, x) \quad \text{and} \quad p \in F .$$

The set of all strings accepted by A is defined as

$$T(A) = \{x \mid p \in \delta(q_0, x) \quad \text{and} \quad p \in F\} .$$

Example 9: Given a nondeterministic finite-state automaton

$$A = (\Sigma, Q, \delta, q_0, F)$$

where

$$\Sigma = \{0,1\}$$

$$Q = \{q_0, q_1, q_2, q_3, q_4\}$$

$$F = \{q_2, q_4\} .$$

The state transition diagram of A is shown in Figure 9. A typical sentence accepted by A is 01011 since $\delta(q_0, 01011) = q_2 \in F$. In general, $T(A)$ is the set of all strings containing either two consecutive 0's or two consecutive 1's.

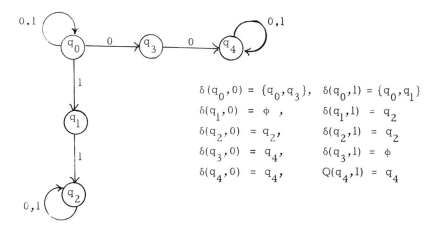

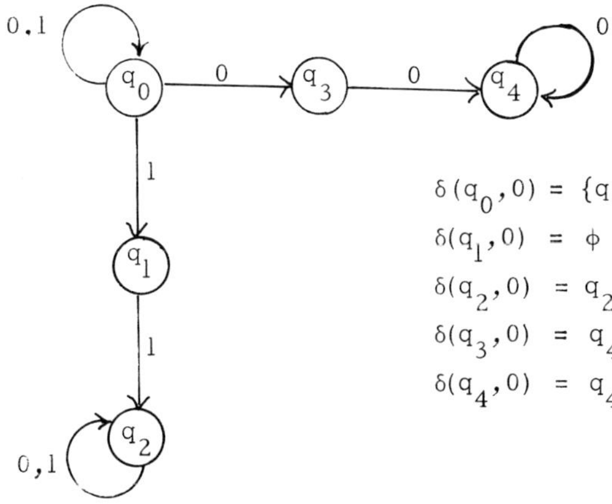

$$\delta(q_0,0) = \{q_0,q_3\}, \quad \delta(q_0,1) = \{q_0,q_1\}$$
$$\delta(q_1,0) = \phi\ , \quad \delta(q_1,1) = q_2$$
$$\delta(q_2,0) = q_2, \quad \delta(q_2,1) = q_2$$
$$\delta(q_3,0) = q_4, \quad \delta(q_3,1) = \phi$$
$$\delta(q_4,0) = q_4, \quad Q(q_4,1) = q_4$$

Figure 9. Transition Diagram for Automaton A

The relationship between deterministic finite-state automata and nondeterministic finite-state automata and the relationship between the finite-state languages and the sets of strings accepted by finite-state automata can be expressed by the following theorems.

Theorem 1: Let L be a set of strings accepted by a nondeterministic finite-state automaton $A = (\Sigma, Q, \delta, q_0, F)$. Then there exists a deterministic finite-state automaton $A' = (\Sigma', Q', \delta', q_0', F')$ that accepts L. The states of A' are all the subsets of Q, i.e., $Q' = 2^Q$, and $\Sigma' = \Sigma$. F' is the set of all states in Q' containing a state of F. A state of A' will be denoted by $[q_1, q_2, \ldots, q_i] \in Q'$ where $q_1, q_2, \ldots, q_i \in Q$. $q_0' = [q_0]$. $\delta'([q_1, \ldots, q_i], a) = [p_1, p_2, \ldots, p_j]$ if and only if

$$\delta(\{q_1, \ldots, q_i\}, a) = \bigcup_{k=1}^{i} \delta(q_k, a) = \{p_1, p_2, \ldots, p_j\} .$$

Since the deterministic and nondeterministic finite-state automata accept the same set of strings, we shall not distinguish between them unless it becomes necessary, but shall simply refer to both as finite-state automata.

Theorem 2: Let $G = (V_N, V_T, P, S)$ be a finite-state grammar. Then there exists a finite-state automaton $A = (\Sigma, Q, \delta, q_0, F)$ with $T(A) = L(G)$,

where

(i) $\Sigma = V_T$

(ii) $Q = V_N \cup \{T\}$

(iii) $q_0 = S$

(iv) if P contains the production $S \rightarrow \lambda$, then $F = \{S, T\}$, otherwise, $F = \{T\}$

(v) the state T is in $\delta(B, a)$ if $B \rightarrow a, b \in V_N$, $a \in V_T$ is in P;

and

(vi) $\delta(B, a)$ contains all $C \in V_N$ such that $B \rightarrow aC$ is in P and $\delta(T, a) = \phi$ for each $a \in V_T$.

Theorem 3: Given a finite-state automaton $A = (\Sigma, Q, \delta, q_0, F)$. Then there exists a finite-state grammar $G = (V_N, V_T, P, S)$ with $L(G) = T(A)$, where

(i) $V_N = Q$

(ii) $V_T = \Sigma$

(iii) $S = q_0$

(iv) $B \rightarrow aC$ is in P if $\delta(B, a) = C$, B, $C \in Q$, $a \in \Sigma$;

and

(v) $B \rightarrow a$ in in P if $\delta(B, a) = C$ and $C \in F$.

6.2 Syntax Analysis of Context-Free Languages.

When a context-free language is used to describe a class of patterns, the corresponding recognition device (or acceptor) is, in general, a nondeterministic pushdown automaton. Not every nondeterministic pushdown automaton can have an equivalent deterministic pushdown automaton. Therefore, the process or the algorithm of performing the recognition, called "syntax analysis" or "parsing" is in general a nondeterministic procedure. The output of the syntax analyzer usually includes not only the decision of accepting the string generated by the given grammar, but also the derivation tree of the string which, in turn, gives the complete structural description of the pattern. Alternatively speaking, given a sentence x and a context-free (or context-free programmed) grammar G, construct a self-consistent derivation tree to fill the interior of the following triangle [86].

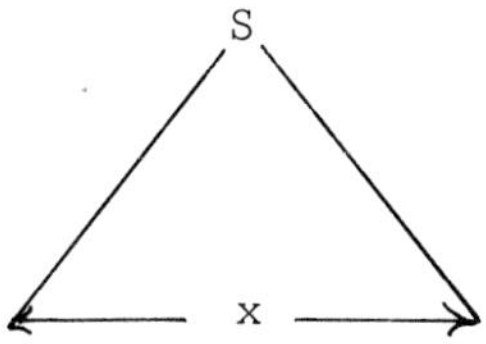

If the attempt is successful, $x \in L(G)$. Otherwise $x \notin L(G)$.

It is in principle unimportant how we attempt to fill the interior of the triangle. We may do it from the top (the root of the tree) towards the bottom (called "top-down" parsing), or from the bottom toward the top (called bottom-up parsing). A number of top-down and bottom-up parsing algorithms have been developed [86, 87]. Unless for special classes of grammars, backtrackings are often required because of the nondeterministic nature of the parsing processes. In the following, a parsing algorithm for general context-free languages proposed by Earley is briefly presented [87].

Let $G = (V_N, V_T, P, S)$ be a context-free grammar and let $x = a_1, a_2, \ldots, a_n$ be a string in V_T^*. An object of the form $[A \rightarrow X_1 X_2 \ldots X_k \cdot X_{k+1} \ldots X_m, i]$ is called an item for x if $A \rightarrow X_1 X_2 \ldots X_m$ is a production in P. For each integer j, $0 \leq j \leq n$, we can construct a list of items I_j such that $[A \rightarrow \alpha \cdot \beta, i]$ is in I_j for $0 \leq i \leq j$ if and only if for some γ and δ we have

$$S \stackrel{*}{\Longrightarrow} \gamma A \delta$$

where $\gamma \stackrel{*}{\Longrightarrow} a_1 \ldots a_i$, and $\alpha \stackrel{*}{\Longrightarrow} a_{i+1} \ldots a_j$. The sequence of lists $I_0, I_1, \ldots, I_n$ is called the parse lists for the string x. It is noted that $x \in L(G)$ if and only if there is some item of the form $[S \rightarrow \alpha \cdot, 0]$ in I_n.

The procedure of constructing the parse lists $I_0, \ldots, I_n$ for x consists of the following:

I. Construction of I_0

(1) If $S \rightarrow \alpha$ is a production in P, add $[S \rightarrow \cdot\, \alpha, 0]$ to I_0.

(2) Perform the following steps with $j = 0$ until no new items can be added to I_0:

(a) If $[A \rightarrow \alpha \cdot B\beta, i]$ is in I_j and $B \rightarrow \gamma$ is a production in P, add $[B \rightarrow \cdot\, \gamma, j]$ to I_j.

(b) If $[A \rightarrow \alpha \cdot, i]$ is in I_j, then for all items in I_1 of the form $[B \rightarrow \beta \cdot A\gamma, k]$ add $[B \rightarrow \beta A \cdot \gamma, k]$ to I_j.

II. Construction of I_j from I_{j-1}

(3) For all $[A \rightarrow \alpha \cdot a_j, i]$ in I_{j-1}, add $[A \rightarrow \alpha a_j \cdot \beta, i]$ to I_j.

(4) Perform (2) to I_j.

A parsing procedure for recognition is, in general, nondeterministic and hence, is regarded computationally inefficient. Efficient parsing could be obtained by using special classes of languages such as finite-state and deterministic languages for pattern description. Special parsers using sequential procedure or other heuristic means for efficiency improvement have recently been proposed [105, 106].

7. Stochastic Linguistic Pattern Recognition and Error-Correcting Parsing.

In some practical applications, a certain amount of uncertainty exists in the process under study. For example, due to the presence of noise and variations in the pattern measurements, ambiguities often occur in the languages describing real-data patterns. In order to describe and recognize noisy patterns under possible ambiguous situations the use of stochastic languages has been recently suggested [89-95].

A stochastic grammar is a four-tuple $G_s = (V_N, V_T, P_s, S)$ where P_s is a finite set of stochastic productions and all the other symbols are the same as defined in Definition 1. For a stochastic context-free grammar, a production in P_s is of the form

$$A_i \xrightarrow{P_{ij}} \alpha_j , \quad A_i \in V_N , \quad \alpha_j \in (V_N \cup V_T)^*$$

where P_{ij} is called the production probability. The probability of generating a string x, called the string probability $p(x)$, is the product of all production probabilities associated with the productions used in the generation of x. The language generated by a stochastic grammar consists of the strings generated by the grammar and their associated string probabilities.

By associating probabilities with the strings, we can impose a probabilistic structure on the language to describe noisy patterns. The probability distribution characterizing the patterns in a class can be interpreted as the probability distribution associated with the strings in a language. Thus, statistical decision rules can be applied to the classification of a pattern under ambiguous situations (for example, use the maximum-likelihood or Bayes decision rule). A block diagram of such a recognition system using maximum-likelihood decision rule is shown in Figure 10. Furthermore, because of the availability of the information about production probabilities, the speed of syntactic analysis can be improved through the use of this information [95, 96]. Of course, in practice, the production probabilities will have to be inferred from the observation of relatively large numbers of pattern samples [56, 90].

Other approaches for the recognition of distorted or noisy patterns using syntactic methods include the use of transformational grammar [59] and the application of error-correcting parsing techniques [97]. In the use of error-correcting parsing as a recognition procedure, different types of segmentation and primitive extraction error are introduced. The original pattern grammar is modified by taking these errors into consideration. The recognition process is then based on the parser designed according to the modified grammar. The error-correcting capability of this class of parsers can be achieved by using a minimum-distance decision criterion [98 - 101].

The most common errors on symbols of a noisy sentence or string can be classified into substitution error, deletion error and insertion error. We may describe these errors in terms of three transformations,

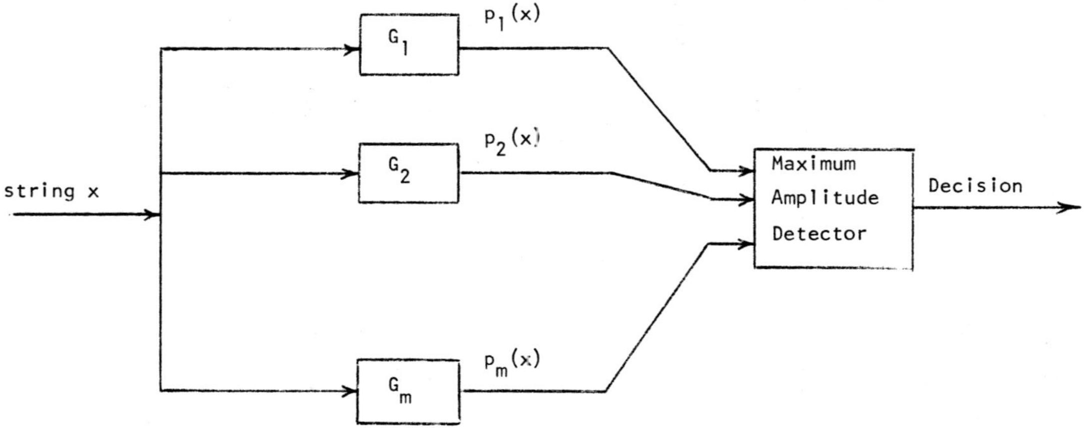

Figure 10. Maximum-Likelihood Syntactic Recognition System

T_S, T_D, T_I, respectively. For any $x, y \in \Sigma^*$

(1) substitution error: $xay \overset{T_S}{\vdash} xby$, for all $a \neq b$,

(2) deletion error: $xay \overset{T_D}{\vdash} xy$, for all a in Σ

(3) insertion error: $xy \overset{T_I}{\vdash} xay$, for all z in Σ.

Where the notation $\vdash$ represents error productions, $x \overset{T_i}{\vdash} y$ means $y \in T_i(x)$ for some $i \in \{S, D, I\}$.

The distance between two sentences x and y in Σ^* could be defined by letting $d(x,y)$ be the smallest integer k for which $x \overset{k}{\vdash} y$, i.e. the least number of transformation rules that are used to derive y from x. For example, given a sentence x = cbabdbb and a sentence y = cbbabbdb, then the distance is $d(x,y) = 3$. Since x = cbabsbb $\overset{T_S}{\vdash}$ cbabbbb $\overset{T_S}{\vdash}$ cbabbdb $\overset{T_I}{—}$ cbbabbdb $= y$, i.e. $x \overset{3}{\vdash} y$. The minimum number of transformations required is three.

Given a CFG $G = (V_N, V_T, P, S)$ an error-correcting parser is a parsing algorithm that takes any input string $y \in V_T^*$ will generate a parse for some string x in $L(G)$ such that $d(x,y)$ is as small as possible.

Based on the above three types of errors, we first construct a covering grammar G' by adding error productions to G such that $L(G') = L(G) \cup \{x | x \in T_\delta (L(G)),\ \delta \in \{S, D, I\}^+\}$, the following is an algorithm that constructs $G' = (V'_N, V'_T, P', S')$.

Step 1. For each production in P, substitute all terminal symbols $a \in V_T$ by a new non-terminal symbol E_a and add these productions to P'.

Step 2. Add to P' the productions

a. $S' \rightarrow S$

b. $S' \rightarrow SH$

c. $H \rightarrow HI$

d. $H \rightarrow I$

Step 3. For each $a \in V_T$, add to P' the productions

a. $E_a \rightarrow a$

b. $E_a \rightarrow b$ for all b in V_T, $b \neq a$

c. $E_a \rightarrow Ha$

d. $I \rightarrow a$

e. $E_a \rightarrow \lambda$ λ is the empty string.

Here $V'_N = V_N \cup \{S', H, I\} \cup \{E_a | a \in V_T\}$ and $V'_T = V_T$.

The production set P' obtained from step 1 and $S' \rightarrow S$, $E_a \rightarrow a$ for all a in V_T, will give $L(G)$; $S' \rightarrow SH$ will insert errors at the end of a sentence; $E_a \rightarrow HA$ will insert errors in front of a. Where productions of the form $E_a \rightarrow b$, $b \neq a$; $I \rightarrow a$; $E_a \rightarrow \lambda$ introduce substitution error, insertion error, and deletion error in a sentence respectively. We call them "terminal error productions".

The minimum-distance error-correcting parser is a modified Earley's algorithm [87]. It keeps counting the number of terminal error productions used and updates parse lists with the least number of terminal error productions.

When stochastic grammars are used for pattern description, the use of sequential decision procedure could result in reducing the parsing time by sacrificing slightly the probability of misrecognition [105]. Of course, when a sequential procedure is used, the parsing procedure stops most of the time before a sentence is completely scanned, and, consequently, in these cases, the complete structural information of the pattern cannot be recovered. When the probability distribution of various types of error is known or can be estimated, the concept of error-correcting parsing can be extended to include the probability information. The recognition process is based on the stochastic parsers designed according to the expanded pattern grammar with appropriate probability assignment. The error-correcting capability is then achieved by a maximum-likelihood or Bayes criterion [108].

8. Conclusions and Remarks.

The use of languages for the description of patterns is introduced in this paper. Sentences of a language are used to represent patterns in a class. Different classes of patterns are characterized by different languages or grammars. The classification of patterns into different classes then becomes a syntax analysis of the corresponding sentences with respect to the grammars characterizing each class. Stochastic languages are suggested for the description of noisy and distorted patterns. Maximum-liklihood and Bayes decision rules can be directly applied to the classification of noisy and distorted linguistic patterns as they are in decision-theoretic pattern recognition systems. Unfortunately, the similarity or distance measures between sentences, trees, and languages are not as easy to define as those in decision-theoretic pattern recognition. Only recently, the minimum-distance criterion is used for the classification of linguistic patterns [97]. Consequently, the clustering problem for linguistic patterns has not even been studied.

Finite-state and context-free languages are easier to analyze compared with context-sensitive languages; however, their descriptive power of complex patterns is less. The consistency condition for stochastic context-free languages can be determined through the theory of branching processes [12]. The consistency problem of stochastic context-sensitive languages is still a problem of investigation. This, of course, directly limits the use of stochastic context-sensitive languages for linguistic pattern recognition except in a few special cases (e.g., stochastic context-sensitive languages generated by stochastic context-free programmed grammars [12]). Only very limited results in grammatical inference are practically useful for linguistic pattern recognition [56]. Efficient inference algorithms are definitely needed before we can design a system to automatically infer a pattern grammar from sample pattern sentences of a language.

References

1. K. S. Fu: *Sequential Methods in Pattern Recognition and Machine Learning*, Academic Press, 1968.

2. G. S. Sebestyen: *Decision Processes in Pattern Recognition*, Macmillan, New York, 1962.

3. N. J. Nilsson: *Learning Machines-Foundations of Trainable Pattern-Classifying System*, McGraw-Hill, 1965.

4. J. M. Mendel and K. S. Fu: *Adaptive, Learning and Pattern Recognition Systems: Theory and Applications*, Academic Press, 1970.

5. W. Meisel: *Computer-Oriented Approaches to Pattern Recognition*, Academic Press, 1972.

6. K. Fukunaga: *Introduction to Statistical Pattern Recognition*, Academic Press, 1972.

7. E. A. Patrick: *Fundamentals of Pattern Recognition*, Prentice-Hall, 1972.

8. H. C. Andrews: *Introduction to Mathematical Techniques in Pattern Recognition*, Wiley, 1972.

9. R. O. Duda and P. E. Hart: *Pattern Classification and Scene Analysis*, Wiley, 1973.

10. C. H. Chen: *Statistical Pattern Recognition*, Hayden book Company, Washington, D. C., 1973.

11. T. Y. Young and T. W. Calvert: *Classification, Estimation, and Pattern Recognition*, American Elsevier, 1973.

12. K. S. Fu: *Syntactic Methods in Pattern Recognition*, Academic Press, 1974.

13. W. F. Miller and A. C. Shaw: Proc. AFIPS Fall Joint Computer Conference, 1968.

14. R. Narasimhan: Report 121, Digital Computer Laboratory, University of Illinois, Urbana, Illinois, 1962.

15. Special issues of PATTERN RECOGNITION on Syntactic Pattern Recognition, Vol. 3, No. 4, 1971 and Vol. 4, No. 1, 1972.

16. N. V. Zavalishin and I. B. Muchnik: Automatika i Telemekhanika, 86, 1969.

17. Ya. Z. Tsypkin: Foundations of the Theory of Learning System, Nauka, Moscow, 1970.

18. M. A. Aiserman, E. M. Braverman, L. I. Rozonoer: Potential Function Method in Theory of Learning Machines, Nauka, Moscow, 1970.

19. K. S. Fu: Pattern Recognition and Machine Learning, Plenum Press, 1971.

20. A. G. Arkadev and E. M. Braverman: Learning in Pattern Classification Machines, Nauka, Moscow, 1971.

21. R. L. Grimsdale, F. H. Summer, C. J. Tunis and T. Kilburn: Proc. IEEE, Vol. 106, Part B, No. 26, March, 1959, pp. 210-221; reprinted in Pattern Recognition, Wiley, 1966, ed. L. Uhr.

22. M. Eden and M. Halle: Proc. 4th London Symposium on Information Theory, Butterworth, London, 1961, pp. 287-299.

23. L. D. Earnest: Information Processing, ed. C. M. Poplewell, North Holland Publishing Co., Amsterdam, 1963, pp. 462-466.

24. P. Mermelstein and M. Eden: Information and Control 7, 255, 1964.

25. M. Eden and P. Mermelstein: Proc. 16th Annual Conference on Engineering in Medicine and Biology, 1963, pp. 12-13.

26. H. Freeman: IEEE Trans. Elec. Comp. EC-10, 260, 1961.

27. H. Freeman: Proc. National Electronics Conf. 18, 312, 1962.

28. P. J. Knoke and R. G. Wiley: Proc. IEEE Comp. Conf., 142, 1967.

29. J. Feder: Information and Control, 13, 230, 1968.

30. H. Freeman and S. P. Morse: J. of Franklin Inst. 284, 1, 1967.

31. J. Feder and H. Freeman: IEEE International Convention Record, Part 3, 1966, pp. 69-85.

32. C. T. Zahn: SLAC Rpt. 72, Stanford Linear Accelerator Center, Stanford, California 1966.

33. H. Freeman and J. Garder: IEEE Trans. on Elec. Comp. EC-13, 118, 1964 .

34. R. Narasimhan: Comm. ACM 9 , 166, 1966 .

35. R. J. Spinrad: Information and Control 8 , 124 (1965 .

36. J. F. O'Callaghan: Problems in On-Line Character Recognition, Picture Language Machines , ed. S. Kaneff, Academic Press, 1970.

37. A. C. Shaw: Rept. SLAC-84, Stanford Linear Accelerator Center Stanford, California, March, 1968 .

38. B. K. Bhargava and K. S. Fu: Application of Tree System Approach to Classification of Bubble Changer Photographs, Tech. Rept. TR-EE 72-30, School of Electrical Engineering, Purdue University West Lafayette, Indiana, November, 1972.

39. M. Nir: Recognition of General Line Patterns with Application to Bubble-Chamber Photographs and Handprinted Characters, Ph. D. thesis, Moore School of Electrical Engineering, University of Pennsylvania, December, 1967.

40. R. Narsimhan: Information and Control 7 , 151, 1964.

41. R. S. Ledley, L. S. Rotolo, T. J. Golab, J. D. Jacobsen, M. D. Ginsburg, and J. B. Wilson: Optical and Electro-Optical Information Processing, ed. J. T. Tippett, D. Beckowitz, L. Clapp, C. Koester, and A. Vanderburgh, Jr., MIT Press, Cambridge, Massachusetts, 591 , 1965 .

42. J. W. Butler, M. K. Butler, and A. Stroud: Automatic Calssification of Chromosomes, Proc. Conference on Data Acquisition and Processing in Biology and Medicine, New York, 1963.

43. H. C. Lee and K. S. Fu: A Syntactic Pattern Recognition System with Learning Capability, Proc. COINS-72 , December, 1972.

44. M. Nagao: Picture Recognition and Data Structure, in Graphic Languages, ed. F. Nake and A. Rosenfeld, North Holland Publishing Co., Amsterdam, London, 1972.

45. M. D. Kelley: Visual Identification of People by Computer, Ph. D. thesis, Dept. of Computer Science, Stanford University, Stanford, California, June, 1970.

46. L. G. Roberts: Optical and Electro-Optical Information Processing, ed. J. T. Tippett, D. Bechowitz, L. Clapp, C. Koester, and A. Vanderburgh, Jr., MIT Press, Cambridge, Mass. 159, 1965.

47. R. O. Duda and P. E. Hart: Experiments in Scene Analysis, Proc. First National Symposium on Industrial Robots, Chicago, April, 1970 .

48. J. A. Feldman, et al: The Stanford Hand-Eye Project, Proc. First International Joint Conference on Artificial Intelligence, Washington, D. C., May, 1969.

49. T. Pavlidis: Pattern Recognition 1 , 165, 1968 .

50. A. Rosenfeld and J. P. Strong: A Grammar for Maps, Software Engineering 2 ed. J. T. Tou, Academic Press, 1971.

51. T. Pavlidis: Pattern Recognition 4 , 5 (1972 .

52. T. Pavlidis: Structural Pattern Recognition: Primitives and Juxtaposition, Frontiers of Pattern Recognition ed. S. Watanabe, Academic Press, 1972.

53. M. L. Minsky and S. Papert: Project MAC Progress Rpt. IV, MIT Press, Cambridge, Mass., 1967.

54. A. Guzman: Proc. AFIPS FJCC 33 , Pt. 1, 291, 1968 .

55. C. R. Brice and C. L. Fennema: Artificial Intelligence 1 , 205, 1970.

56. K. S. Fu and T. L. Booth: Grammatical Inference - Introduction and Survey, IEEE Trnas. SMC, SMC-5 , 95 and 409, January and July, 1975 .

57. R. Narasimhan: On the Description Generation, and Recognition of Classes of Pictures, in Automatic Interpretation and Classification of Images, Academic Press, 1969, ed. A. Grasselli.

58. A. W. Laffan and R. C. Scott: A New Tool for Automatic Pattern Recognition: A Context-Free Grammar for Plane Projective Geometry, Proc. Second International Joint Conference on Pattern Recognition, Lyngby-Copenhagen, Denmark, August 13-15, 1974.

59. M. C. Clowes: Transformational Grammars and the Organization of Pictures, in Automatic Interpretation and Classification of Images, ed. A. Grasselli, Academic Press, 1969.

60. Laveen Kanal and B. Chandrasekaran: On the Linguistic, Statistical and Mixed Models for Pattern Recognition, in Frontiers of Pattern Recognition, ed. S. Watanabe, Academic Press, 1972.

61. R. Narasimhan: Picture Languages, in Picture Language Machines, ed. S. Kaneff, Academic Press, 1970 .

62. W. E. Underwood and L. N. Kanal: Structural Description, Transformational Rules and Pattern Analysis, Proc. First International Joint Conference on Pattern Recognition, Washington, D. C., October 30-November 1, 1973.

63. J. Feder: Information Sciences 3 , 225 , July, 1971 .

64. M. L. Minsky: Proc. IRE 49 , 8 , 1961 .

65. T. G. Evans: A Formalism for the Description of Complex Objects and its Implementation, Proc. Fifth International Congress on Cybernetics, Namur, Belgium, September, 1967 .

66. M. B. Clowes: Pictorial Relationships - A Syntactic Approach , in Machine Intelligence IV , ed. B. Meltzer and D. Michie, American Elsevier, New York, 1969 .

67. T. G. Evans: Descriptive Pattern Analysis Techniques, in Automatic Interpretation and Classification of Images , ed. A. Grasselli, Academic Press, 1969 .

68. H. G. Barrow and J. R. Popplestone: Machine Intelligence 6 , ed. B. Meltzer and D. Michie, Edinburgh University Press, 1971, pp. 377-396 .

69. R. A. Kirsch: IEEE Trans. on Elec. Comp. EC-13, 363, 1964.

70. M. F. Dacey: Pattern Recognition 2, 11, 1970.

71. D. M. Milgram and A. Rosenfeld: IFIP Congress 71, North-Holland, Amsterdam, August, 1971, Booklet TA-2, pp. 166-173.

72. A. C. Shaw: Information and Control 14, 9, January, 1969.

73. A. C. Shaw: J. ACM 17, 453, June, 1970.

74. J. L. Pfaltz and A. Rosenfeld: Proc. First International Joint Conference on Artificial Intelligence, May 1969, Washington, D. C., pp. 609 - 619.

75. A. C. Shaw: Picture Graphs, Grammars, and Parsing, in Frontiers of Pattern Recognition, ed. S. Watanabe, Academic Press, 1972.

76. J. L. Pfaltz: Web Grammars and Picture Description, Technical Report 70-138, Computer Science Center, University of Maryland, College Park, Maryland, 1970.

77. J. M. Brayer and K. S. Fu: Web Grammars and Their Application to Pattern Recognition, Technical Report 75-1, School of Electrical Engineering, Purdue University, W. Lafayette, Indiana 47907 1975.

78. T. Pavlidis: Jour. of ACM 19, No. 1, 11, 1972.

79. A. Rosenfeld: Picture Automata and Grammars: An Annotated Bibliography, Proc. Symposium on Computer Image Processing and Recognition, Vol. 2, August 24-26, 1972, Columbia, Mo.

80. T. Pavlidis: Graph Theoretic Analysis of Pictures, in Graphic Languages, ed. F. Nake and A. Rosenfeld, North-Holland, Amsterdam, 1972.

81. W. S. Brainerd: Information and Control 14, 217, 1969.

82. J. E. Donar: Tree Acceptors and Some of Their Applications, Jour. of Computer and System Sciences 4, 1970.

83. K. S. Fu and B. K. Bhargava: Tree Systems for Syntactic Pattern Recognition, IEEE Trans. on Comp., C-22, 1087, December 1973.

84. B. K. Bhargava and K. S. Fu: Transformation and Inference of Tree Grammars for Syntactic Pattern Recognition, Proc. IEEE International Conferences on Systems, Man and Cybernetics, October 2-4, Dallas, Texas.

85. J. Gips: Pattern Recognition 6 , 189 , 1974 .

86. J. M. Foster: Automatic Syntactic Analysis, American Elsevier, 1970.

87. A. V. Aho and J. D. Ullman: The Theory of Parsing, Translation, and Compiling, Vol. 1, Parsing , Prentice-Hall, 1972 .

88. F. W. Blackwell: Combining Mathematical and Structural Pattern Recognition, Proceeding Second International Joint Conference on Pattern Recognition, August 13-15, 1974, Copenhagen, Denmark.

89. V. Grenander: Syntax-Controlled Probabilities, Tech. Rept., Division of Applied Math., Brown University, Providence, Rhode Island, 1967.

90. K. S. Fu: Syntactic Pattern Recognition and Stochastic Languages, in Frontiers of Pattern Recognition, ed. S. Watanabe, Academic Press, 1972.

91. V. A. Kovalevsky: Sequential Optimization in Pattern Recognition and Pattern Description, Proc. IFIP Congress, Amsterdam, 1968.

92. K. S. Fu: Computer Graphics and Image Processing, 2 433, 1973.

93. L. W. Fung and K. S. Fu: Stochastic Syntactic Classification of Noisy Patterns, Proc. Second International Joint Conference on Pattern Recognition, August 13-15, 1974, Copenhagen Denmark.

94. V. Dimitrov: Multilayered Stochastic Languages for Pattern Recognition, Proc. First International Joint Conference on Pattern Recognition, Washington, D. C., October 30-November 1, 1973.

95. H. C. Lee and K. S. Fu: A Stochastic Syntax Analysis Procedure and Its Application to Pattern Classification, IEEE Trans. on Computers, C-21, 660 , 1972 .

96. T. Huang and K. S. Fu: Stochastic Syntactic Analysis for Programmed Grammars and Syntactic Pattern Recognition, Computer Graphics and Image Processing, 1, 257, November, 1972.

97. A. V. Aho and T. G. Peterson: A Minimum Distance Error-Correcting Parser for Context-Free Languages, SIAM J. Compt., 1 No. 4, 305, December, 1972.

98. L. W. Fung and K. S. Fu: IEEE Trans. on Computers, C-24, 662, 1975.

99. M. G. Thomason and R. C. Gonzalez: Classification of Imperfect Syntactic Pattern Structures, Proc. Second International Joint Conference on Pattern Recognition, August 13-15, 1974, Copenhagen, Denmark.

100. K. S. Fu: Error Correcting Parsing for Syntactic Pattern Recognition, in Data Structures, Computer Graphics and Pattern Recognition, ed. A. Klinger, K. S. Fu and T. Kunii, Academic Press, 1976.

101. M. G. Thomason: IEEE Trans. on Computers, C-24, 1211, 1975.

102. D. E. Knuth: Math. Syst. Theory, 2, 127, 1968.

103. W. T. Wilson: Formal Semantic Definition Using Synthesized and Inherited Attributes, in Formal Semantics of Programming Languages, ed. R. Rustin, Prentice-Hall, 1972.

104. D. L. Milgram and A. Rosenfeld: A Note on Grammars with Coordinates, in Graphic Languages, ed. F. Nake and A. Rosenfeld, North-Holland Publ., 1972.

105. E. Persoon and K. S. Fu: International Journal of Computer and Information Sciences, 4, 205, 1975.

106. B. Moayer and K. S. Fu: Pattern Recognition, 7, 1 1975.

107. B. Moayer and K. S. Fu: IEEE Trans. on Computers, C-25, 262, 1976.

108. S. Y. Lu and K. S. Fu: Efficient Error Correcting Syntax Analysis for Recognition of Noisy Patterns, Tech. Rept. 76-9, School of Electrical Engineering, Purdue University, West Lafayette, Indiana 47907, 1976.

109. K. S. Fu, ed: Digital Pattern Recognition, Springer-Verlag, Heidelberg, 1976.

110. K. S. Fu, ed: Applications of Syntactic Pattern Recognition, Springer-Verlag, Heidelberg, 1976.

School of Electrical Engineering
Purdue University
West Lafayette, Indiana 47907

Fuzzy Sets and Their Application to Pattern Classification and Clustering Analysis

L. A. Zadeh

1. Introduction.

The development of the theory of fuzzy sets in the early sixties drew much of its initial inspiration from problems relating to pattern classification -- especially the analysis of proximity relations and the separation of subsets of R^n by hyperplanes. In a more fundamental way, however, the intimate connection between the theory of fuzzy sets and pattern classification rests on the fact that most real-world classes are fuzzy in nature -- in the sense that the transition from membership to nonmembership in such classes is gradual rather than abrupt. Thus, given an object x and a class F, the real question in most cases is not whether x is or is not a member of F, but the degree to which x belongs to F or, equivalently, the grade of membership of x in F.

There is, however, still another and as yet little explored connection between the theory of fuzzy sets and pattern classification. What we have in mind is the possibility of applying fuzzy logic and the so-called linguistic approach [1]-[4] to the definition of the basic concepts in pattern analysis as well as to the formulation of fuzzy algorithms for pattern recognition. The principal motivation for this approach is that most of the practical problems in pattern classification do not lend themselves to a precise mathematical formulation, with the consequence that the less precise methods based on the linguistic approach may well prove to be better matched to the imprecision which is intrinsic in such problems.

Although the literature of the theory of fuzzy sets contains a substantial number of papers dealing with various aspects of pattern classification,[1] we do not, as yet, have a unified theory of pattern classification based on the theory of fuzzy sets. It is reasonable to assume that such a theory will eventually be developed, but its construction is likely to be a long-drawn task because it will require a complete reworking of the conceptual structure of the theory of pattern classification and radical changes in our formulation and implementation of pattern recognition algorithms.

In this perspective, the limited objective of the present paper is to outline a conceptual framework for pattern classification and cluster analysis based on the theory of fuzzy sets, and draw attention to some of the significant contributions by other investigators in which concrete pattern recognition and cluster analysis algorithms are described. For convenience of the reader, a brief exposition of the relevant aspects of the theory of fuzzy sets is presented in the Appendix.

2. Pattern Classification in a Fuzzy-Set-Theoretic Framework.

To place the application of the theory of fuzzy sets to pattern classification in a proper perspective, we shall begin with informal definitions of some of the basic terms which we shall employ in later analysis.

To begin with, it will be necessary for our purposes to differentiate between an object which is pointed to (or labeled) by a pointer (identifier) p, and a mathematical object, x, which may be characterized precisely by specifying the values of a finite (or, more generally, a countable) set of parameters. For example, in the proposition "Susan is very intelligent," Susan is a pointer to a person named Susan. The person in question, however, is not a mathematical object until a set of measurement procedures $\{M_1, \ldots, M_n\}$ is defined such that the application of $\{M_1, \ldots, M_n\}$ to the object p (or, more precisely, the object pointed to by p) yields an n-tuple of constants $(x_1, \ldots, x_n)$ which

[1] Some of the representative papers bearing on the application of fuzzy sets to pattern classification and cluster analysis are listed in the bibliography.

represent the feature-values (or attribute-values) of the object in question. The n-tuple $x \triangleq (x_1, \ldots, x_n)$, then, characterizes a mathematical object associated with p, expressed symbolically as[2]

$$x \triangleq M(p) \tag{2.1}$$

where $M = (M_1, \ldots, M_n)$. For example, M_1, M_2, M_3, M_4 could be, respectively, the procedures for measuring the height, weight and temperature, and determining the sex of the object in question. In this case, a 4-tuple of the form (5'7", 125, 98.6, F) would be a mathematical object associated with the person named Susan.

An important point that needs to be noted is that there are many -- indeed an infinity -- of mathematical objects that may be associated with p. In the first place, different combinations of attributes may be measured. And second, different mathematical objects result when the precision of measurement -- or, equivalently, the resolution level -- of an attribute is varied. Thus, to associate a mathematical object x with an object p it is necessary to specify, explicitly or implicitly, the resolution levels of the attributes of p. Usually this is done implicitly, rather than explicitly, which is the reason why the concept of a resolution level -- although important in principle -- does not play an overt role in pattern recognition.

Let U^o be a universe of objects, let U be the universe of associated mathematical objects, and let F be a fuzzy subset of U^o (or U). There are three distinct ways in which F may be characterized:

(a) Listing. If the support[3] of F is a finite set, then F may be defined by a listing of its elements together with their respective grades of membership in F. For example, if U^o is the set of persons pointed to by the labels John, Luise, Sarah and David, and F is the fuzzy subset labeled tall, then F may be characterized as the

[2] The symbol $\triangleq$ stands for "denotes" or "is equal to by definition."

[3] The support of a fuzzy set F is the set of elements of the universe of discourse whose grades of membership in F are positive.

collection of ordered pairs {(John, 0.9), (Luise, 0.8), (David, 0.7) and (Sarah, 0.8)}, which may be expressed more conveniently as the linear form (see A2)

(2.2) $$\underline{\text{tall}} = 0.9 \text{ John} + 0.8 \text{ Luise} + 0.7 \text{ David} + 0.8 \text{ Sarah}$$

where + denotes the union rather than the arithmetic sum.

(b) Recognition algorithm. Such an algorithm, when applied to an object p, yields the grade of membership of p in F. For example, if someone were to point to Luise and ask "What is the degree to which Luise is tall?" then a recognition algorithm applied to the object Luise would yield the answer 0.8.

(c) Generation algorithm. In this case, an algorithm generates those elements of U^o which belong to the support of F and associates with each such element its grade of membership in F. As a simple illustration, the recurrence relation

(2.3) $$x_n = x_{n-1} + x_{n-2}$$

with $x_0 = 0$, $x_1 = 1$ may be viewed as a nonfuzzy generation algorithm which defines the set of Fibonacci numbers $\{1, 2, 3, 5, 8, 13, \ldots\}$.[4] As an example of a generation algorithm which defines a fuzzy set, let U be the set of strings over a finite alphabet, say $\{a, b\}$, and let G be a fuzzy context-free grammer whose production system is given by

(2.4) $$\begin{array}{llll} S \overset{0.8}{\rightarrow} bA & & B \overset{0.4}{\rightarrow} b \\ S \overset{0.6}{\rightarrow} aB & & A \overset{0.3}{\rightarrow} bSA \\ A \overset{0.2}{\rightarrow} a & & B \overset{0.5}{\rightarrow} aSB \end{array}$$

in which S, A, B are nonterminals and the number above a production indicates its "strength." The fuzzy language, $L(G)$, generated by this grammar may be defined as follows. Let x be a terminal string derived from S by a sequence of substitutions in which the left-hand side of a production in G is replaced by its right-hand side member, e.g.,

---

[4] Many examples of nonfuzzy pattern generation algorithms may be found in the books by U. Grenander [5] and K. S. Fu [6].

(2.5) $S \overset{0.8}{\to} bA \overset{0.3}{\to} bbSa \overset{0.2}{\to} bbSa \overset{0.6}{\to} bbaBa \overset{0.4}{\to} bbaba$.

The strength of the derivation chain from S to x is defined to be the minimum of the strengths of constitutent productions in the chain, e.g., in the case of (2.5), the strength of the chain is $0.8 \wedge 0.3 \wedge 0.2 \wedge 0.6 \wedge 0.4 = 0.2$ (where $\wedge$ is the infix symbol for min). The grade of membership of x in L(G) is then defined as the strength of the strongest leftmost derivation[5] chain from S to x [7]. In the case of $x \triangleq bbaba$, there is just one leftmost derivation, namely,

(2.6) $S \overset{0.8}{\to} bA \overset{0.3}{\to} bbSA \overset{0.6}{\to} bbaBA \overset{0.4}{\to} bbabA \overset{0.2}{\to} bbaba$

whose strength is 0.2. Consequently, the grade of membership of the string $x \triangleq bbaba$ in the fuzzy set L(G) is 0.2. In this way, one can associate a grade of membership in L(G) with every string that may be generated by G , and thus the production system (2.4) together with the rule for computing the grade of membership of any string in U in L(G), constitutes a generation algorithm which characterizes the fuzzy subset, L(G), of U .

Opaque vs. Transparent Algorithms.

For the purposes of our analysis, it is necessary to differentiate between recognition algorithms which are opaque and those which are transparent. Informally, by an opaque recognition algorithm we mean an algorithm whose description is not known. For example, the user of a hand calculator may not know the algorithm which is employed in the calculator to perform exponentiation. Or, a person may not be able to articulate the algorithm which he/she uses to assign a grade of membership to a painting in the fuzzy set of beautiful paintings.

As its designation implies, a recognition algorithm is transparent if its description is known. For example, a parsing algorithm which

[5] In leftmost derivation, the leftmost nonterminal is replaced by the right-hand member of the corresponding production.

parses a string generated by a context-free grammar and thereby yields the grade of membership of the string in the fuzzy language generated by the grammar would be classified as a transparent algorithm.

Pattern Classification.

Within the framework of the theory of fuzzy sets, the problem of pattern classification may be viewed -- in its essential form -- as that of conversion of an opaque recognition algorithm into a transparent recognition algorithm. More specifically, let U^o be a universe of objects and let R_{op} be an opaque recognition algorithm which defines a fuzzy subset F of U^o. Then, pattern classification -- or, equivalently, pattern recognition -- may be defined as the process of converting an opaque recognition algorithm R_{op} into a transparent recognition algorithm R_{tr}.[6]

As an illustration of this formulation, consider the following typical problem. Suppose that U^o is the universe of handwritten letters and that when a letter, p, is presented to a person, P, that person -- by employing an opaque recongition algorithm R_{op} -- can specify the grade of membership, $\mu_F(p)$, of p in, say, the fuzzy set, F, of handwritten A's. Thus, in symbols,

$$\mu_F(p) = R_{op}(p), \quad \text{for } p \text{ in } U \tag{2.7}$$

Usually, P is presented with a finite set of sample letters $p_1, \ldots, p_m$, so that the result of application of R_{op} to $p_1, \ldots, p_m$ is a set of ordered pairs $\{(p_1, \mu_F(p_1)), \ldots, (p_m, \mu_F(p_m))\}$ which in the notation of fuzzy sets may be expressed as the linear form

$$S_F = \mu_F(p_1)p_1 + \ldots + \mu_F(p_m)p_m \tag{2.8}$$

where S_F stands for a fuzzy set of samples from F, and a term of the

---

[6] We assume for simplicity that only one fuzzy subset of U^o is defined by R_{op}. More generally, there may be a number of such subsets, say $F_1, \ldots, F_n$, with R_{op} yielding the grade of membership of p in each of these subsets.

form $\mu_F(p_i)p_i$, $i = 1,\ldots,m$, signifies that $\mu_F(p_i)$ is the grade of membership of p_i in F.

If, based on the knowledge of S_F, we could convert the opaque recognition algorithm R_{op} into a transparent recognition algorithm R_{tr}, then given any p we could deduce $\mu_F(p)$ by applying R_{tr} to p. Equivalently, we may view this as the process of interpolation of the membership function of F from the knowledge of the values which it takes at the points $p_1,\ldots,p_m$. It should be remarked that this is the way in which the problem of pattern classification was defined in [8], but the present formulation based on the conversion of R_{op} to R_{tr} appears to be more natural.

An important implicit assumption in pattern classification is that the recognition process must be automatic, in the sense that it must be performed by a machine rather than a human. This requires that the transparent recognition algorithm R_{tr} act on a mathematical object, $M(p)$, rather than on p itself, since an object must be well-defined in order to be capable of manipulation by a machine.

In more concrete terms, let U^o be a universe of objects and let M be a measurement procedure which associates with each object p in U^o a mathematical object $M(p)$ in U. Let F be a fuzzy subset of U^o which is defined by an opaque recognition algorithm R_{op} in the sense that

$$\mu_F(p) = R_{op}(p), \quad p \in U^o .$$

Denote by R_{tr} a transparent recognition algorithm which acting on the mathematical object $M(p)$ yields $\mu_F(p)$. Then, the problem of automatic (or machine) pattern recognition may be expressed in symbols as that of determining M and R_{tr} such that

$$\mu_F(p) = R_{op}(p) \tag{2.9}$$

$$R_{tr}(M(p)) = R_{op}(p), \quad p \in U^o. \tag{2.10}$$

Thus, the problem of automatic pattern recognition involves two distinct subproblems: (a) conversion of the object p into a mathematical object M(p); and (b) conversion of the opaque recognition algorithm R_{op} which acts on p's into a transparent recognition algorithm which acts on M(p)'s. Of these, problem (a) is by far the more difficult. In the conventional nonfuzzy approach to pattern classification, it is closely related to the problem of feature analysis -- a problem which falls into the least well-defined and least well-developed area in pattern recognition [35]-[46].

It is important to observe that, from a practical point of view, it is desirable that (i) M(p) be defined by a small number of attributes, and (ii) that the measurement of these attributes be relatively simple. With these added considerations, then, the problem of pattern classification may be reformulated in the following terms.

Given an opaque recognition algorithm R_{op} which defines a fuzzy subset of objects p in U^o.

Problem I. Specify a preferably small set of preferably simple measurement procedures which convert an object p in U^o into a mathematical object $M(p) = \{M_1(p), \ldots, M_n(p)\}$ in U .

Problem II. Convert R_{op} into a transparent recognition algorithm R_{tr} which acts on M(p) and yields the grade of membership of p in F as defined by R_{op} .

In the above formulation, the problem of pattern classification is not mathematically well-defined. In part, this is due to the fact that, as pointed out earlier, the notion of an object does not admit of precise definition and hence the functions $M_1, \ldots, M_n$ cannot be regarded as functions in the accepted mathematical sense. In addition, since the desired equality

$$R_{tr}(M(p)) = R_{op}(p) , \quad p \in U^o \tag{2.11}$$

cannot be realized precisely, the problem of pattern classification does not admit of exact solution. Furthermore, an added source of imprecision in pattern classification problems relates to the difficulty of assessing the goodness of a transparent recognition algorithm which may be offered as a solution to a given problem.

The main thrust of the above comments is that the problem of pattern classification is intrinsically incapable of precise mathematical formulation. For this reason, the conceptual structure of the theory of fuzzy sets may well provide a more natural setting for the formulation and approximate solution of problems in pattern classification than the more traditional approaches based on classical set theory, probability theory and two-valued logic [35]-[46].

3. The Linguistic Approach to Pattern Classification.

Most of the conventional approaches to pattern recognition are based on the tacit assumption that the mapping from the object space U^o to the feature space U has the property that if two mathematical objects $M(p)$ are "close" to one another in terms of some metric defined on U, then p and q are likely to be in the same class in U^o.[7] When F is a fuzzy set, this assumption may be expressed more concretely but not very precisely as the property of μ-continuity of M, namely: If p and q are objects in U^o and for almost all p and q $M(p)$ is close to $M(q)$ in terms of a metric defined on U, then the grade of membership of p in F, $\mu_F(p)$, is close to that of q, $\mu_F(q)$.

The importance of μ-continuity derives from the fact that it provides a basis for reducing Problem II to the interpolation of a "well-behaved" (i.e., smooth, slowly-varying) membership function. More significantly for our purposes, it makes it possible to employ the linquistic approach for describing the dependence of μ_F on the linguistic values of the attributes of an object.

[7] This assumption is implicit in perceptron-type approaches and is related to the notion of compactness in the potential function method of Aizerman, Braverman and Rozonoer [9]-[12].

More specifically, suppose that M(p) has n components $x_1 \triangleq M_1(p), \ldots, x_n \triangleq M_n(p)$, with x_i, $i = 1, \ldots, n$, taking values in U_i. Let $\mu_F(p)$ denote the grade of membership of p in F. We assume that the dependence $\mu_F(p)$ on $x_1, \ldots, x_n$ is expressible as an (n+1)-ary fuzzy relation R in $U_1 \times \ldots \times U_n \times V$, where $V \triangleq [0,1]$. In what follows, R will be referred to as the relational tableau defining $\mu_F(p)$.

An essential assumption which motivates the linguistic approach is that our perception of the dependence of $\mu_F(p)$ on $x_1, \ldots, x_n$ is generally not sufficiently precise or well-defined to enable us to tabulate $\mu_F(p)$ as a function of the numerical values of $x_1, \ldots, x_n$. As a coarser and hence less precise characterization of this dependence, we allow the tabulated values of $x_1, \ldots, x_n$ and $\mu_F(p)$ to be linguistic rather than numerical, employing the techniques of the linguistic approach to enable us to interpolate R for the untabulated values of $x_1, \ldots, x_n$.

To be more specific, it is helpful to assume, as in [86], that a linguistic value of x_i, $i = 1, \ldots, n$, is an answer to the question Q_i: "What is the value of x_i?" and that the corresponding linguistic value of $\mu_F(p)$ is the answer to the question Q: "If the answers to $Q_1, \ldots, Q_n$ are $r_1, \ldots, r_n$, respectively, then what is the value of $\mu_F(p)$?" A purpose of this interpretation of the values of $x_1, \ldots, x_n, \mu_F(p)$ is to express the recognition algorithm R_{tr} as a branching questionnaire, that is, a questionnaire in which the questions are asked sequentially, with the question asked at stage j depending on the answers to the previous questions. The conversion of a relational tableau to a branching questionnaire is discussed in greater detail in [86].

Typically, the entries in a relational tableau are of the form shown in Table 1, in which the rows correspond to different objects, with the entry under Q_i representing a linguistic value of x_i for a particular object. (For simplicity, we shall speak interchangeably of the values of x_i and Q_i.) The questions $Q_1, \ldots, Q_n$ will be referred to as the constituent questions of R (or Q).

Q_1	Q_2	Q_3	Q
true	small	wide	high
very true	very small	not wide	very high
not very true	medium	NA	not very high
borderline	very large	not wide	low
not true	not very small	not very wide	more or less low
true or not very true	small	not very wide	very low

Table 1. A relational tableau defining the dependence of Q on Q_1, Q_2, Q_3.

In this table, the entries in the column labeled Q_i constitute a subset of the term-set of Q_i (see A66), that is, the possible linguistic values that may be assigned to Q_i. For example, the term-set of Q_1 might be: {true, very true, not very true, borderline, very (not true), not true, not borderline, very very true, ... }. The elements of the term-set of Q_i are assumed to be generated by a context-free grammar. For instance, the elements of the term-set of Q_1 can be generated by the grammar

$$
\begin{array}{llll}
 & S \rightarrow A & & C \rightarrow D \\
 & S \rightarrow S \text{ or } A & & C \rightarrow E \\
(3.1) & A \rightarrow B & & D \rightarrow \text{very } D \\
 & A \rightarrow A \text{ and } B & & E \rightarrow \text{very } E \\
 & B \rightarrow C & & D \rightarrow \text{true} \\
 & B \rightarrow \text{not } C & & E \rightarrow \text{borderline}
\end{array}
$$

in which S, A, B, C, D, E are nonterminals and "or," "and," "not," "very," "true" and "borderline" are terminals. Using the production system of this grammar, the linguistic value "true or not very true" may be derived from S by the chain of substitutions

$$
\begin{array}{ll}
 & S \rightarrow S \text{ or } A \rightarrow A \text{ or } A \rightarrow B \text{ or } A \rightarrow C \text{ or } A \rightarrow D \text{ or } A \rightarrow \text{true or} \\
(3.2) & A \rightarrow \text{true or } B \rightarrow \text{true or not } C \rightarrow \text{true or not } D \rightarrow \text{true or not} \\
 & \text{very } D \rightarrow \text{true or not very true}
\end{array}
$$

The linguistic values of Q_i play the role of labels of fuzzy subsets of a universe of discourse which is associated with Q_i. For example, in the case of Q_1 the universe U_1 is the unit interval $[0,1]$, and "true" is a fuzzy subset of U_1 whose membership function might be defined in terms of the S-function (see A17) by

$$\mu_{true}(v) = S(v;0.6,0.75,0.9), \qquad v \in [0,1] \tag{3.3}$$

where $S(v;\alpha,\beta,\gamma)$ is an S-shaped function which vanishes to the left of α, is unity to the right of γ and takes the value 0.5 at $\beta = \frac{\alpha+\gamma}{2}$. Similarly, the membership function of the fuzzy subset labeled "borderline" may be defined in terms of the π-function (see A18) by

$$\mu_{borderline}(v) = \pi(v;0.3,0.5) \tag{3.4}$$

where $\pi(v;\beta,\gamma)$ is a bell-shaped function whose bandwidth is β and which achieves the value 1 at γ.

By the use of a semantic technique which is described in [2], it is possible to compute in a relatively straightforward fashion the membership function of the fuzzy set which plays the role of the meaning of a linguistic value in the term-set of Q_i. For example, the membership functions of "not true," "very true," "not very true" and "true or not very true" are related to that of "true" by the equations (in which the argument v is suppressed for simplicity)

$$\mu_{not\ true} = 1 - \mu_{true} \tag{3.5}$$

$$\mu_{very\ true} = (\mu_{true})^2 \tag{3.6}$$

$$\mu_{not\ very\ true} = 1 - (\mu_{true})^2 \tag{3.7}$$

$$\mu_{true\ or\ not\ very\ true} = \mu_{true} \vee (1 - (\mu_{true})^2) \tag{3.8}$$

where $(\mu_{true})^2$ denotes the square of the membership function of true and $\vee$ stands for the infix form of max.

A fuzzy set (or fuzzy sets) in terms of which the meaning of all other linguistic values in the term-set of Q_i may be computed is termed a primary fuzzy set (or sets). Thus, in the case of Q_1 the primary fuzzy set is labeled "true;" in the case of Q_2 the primary fuzzy sets are "small," "medium" and "large;" and in the case of Q the primary fuzzy sets are "high" and "medium," with "low" defined in terms of "high" by

$$\mu_{low}(v) = \mu_{high}(1-v), \qquad v \in [0,1] . \tag{3.9}$$

In effect, a primary fuzzy set plays a role akin to that of a unit whose meaning is context-dependent and hence must be defined a priori. The important point is that once the meaning of the primary terms is specified, the meaning of non-primary terms in the term-set of each Q_i may be computed by the application of the semantic rule which is associated with that Q_i.

The entry NA in Q_3 stands for "not applicable." What this means is that if the answer to Q_1 is, say, "not very true" and the answer to Q_2 is "medium," then Q_3 is not applicable to the object corresponding to the third row in the table. As a simple illustration of non-applicability, if the answer to the question "Is p a prime number?" is "true," then the question "What is the largest divisor of p other than 1?" is not applicable to p .

In the representation of R in the form of a relational tableau, it is helpful to divide the constituent questions into two categories: attributional and classificational. As its name implies, an attributional question is one which asks for the value of an attribute of p , e.g., Q_2 and Q_3 in Table 1 are attributional questions. A classificational question, on the other hand, relates to the degree to which a specified property is possessed by the object in question. Thus, the answer to a classificational question is either a truth-value, as in Q_1, or the grade of membership, as in Q . In both cases, the universe of discourse associated with a classificational question is assumed to be the interval [0,1]. Generally, we shall assume that "high" is equivalent to "true;" "medium" to "borderline;" and "low" to "false," where, by analogy with (3.9), "false" is defined by

(3.10) $$\mu_{false}(v) = \mu_{true}(1-v), \qquad v \in [0,1].$$

As an illustration of the above approach, assume that we wish to characterize the concept of an oval[8] contour, with U being the space of curved, smooth, simply-connected and non-self-intersecting contours in a plane.[9] To simplify the example, we assume that the constituent questions are limited to the following.

Classificational: $Q_1 \overset{\Delta}{=}$ Does p have an axis of symmetry?

Classificational: $Q_2 \overset{\Delta}{=}$ Does p have a second axis of symmetry?

Classificational: $Q_3 \overset{\Delta}{=}$ Are the two axes of symmetry orthogonal?

Classificational: $Q_4 \overset{\Delta}{=}$ Does p have more than two axes of symmetry?

Attributional: $Q_5 \overset{\Delta}{=}$ What is the ratio of the lengths of the major and minor axes?

Calssificational: $Q_6 \overset{\Delta}{=}$ Is p convex?

For simplicity, the answers to the classificational questions are allowed to be only true, borderline and false, abbreviated to t, b and f, respectively, with the membership functions of t, b and f expressed in terms of the S and π functions by (3.3), (3.4) and[10]

(3.11) $$\mu_f(v) = \mu_t(1-v)$$
$$= 1 - S(v;0.1,0.25,0.4).$$

Similarly, the term-set for Q_5 is assumed to be

$$T(Q_5) = \{\text{about } 1, \text{ about } 1.5, \text{ about } 2, \text{ about } 2.5, \text{ about } 3, \text{ about } 4, \text{ about } 5, > \text{about } 5\}$$

[8] For purposes of this example, by oval we mean a shape resembling that of an egg.

[9] Note that the point of departure in this example is U rather than U^o because we assume that a contour is a mathematical object.

[10] It should be understood that true and false in the present context do not have the same meaning as they do in classical logic. Rather, as in fuzzy logic [3], true in the sense of (3.3) means "approximately true," and likewise for "false".

where about α , $\alpha = 2,\ldots,5$, is defined by (with the arguments of π and S suppressed for simplicity)

$$\text{about } \alpha \;=\; \pi(0.4,\alpha) \tag{3.12}$$

and

$$\text{about } 1 = 1 - S(1,0.2,0.4)\ . \tag{3.13}$$

The answer to Q_6 is assumed to be provided by a subquestionnaire with an unspecified number of classificational constituent questions $Q_{61}, Q_{62}, \ldots$ which are intended to check on whether the slope of the tangent to the contour is a monotone function of the distance traversed along the contour by an observer. Thus, if an observer begins to traverse the contour in, say, the counterclockwise direction starting at a point a_0, and $a_1, \ldots, a_m$ are regularly spaced points on the contour, with $a_{m+1} = a_0$, then Q_{6i} would be the question

$Q_{6i} \overset{\Delta}{=}$ Is the slope of the tangent at a_i greater than that at a_{i-1}, $i = 1, 2, \ldots, m+1$?

The answer to Q_6 is assumed to be true if and only if the answers to all of the constituent questions $Q_{61}, Q_{62}, \ldots$ are true.

In terms of the constituent questions defined above, the relational tableau characterizing an oval object may be expressed in a form such as shown in Table 2. For simplicity, only a few of the possible combinations of answers to these questions are exhibited in the table (NA stands for not applicable).

Q_1	Q_2	Q_3	Q_4	Q_5	Q_6	Q
t	t	t	f	about 1	t	b
t	t	t	f	about 1.5	t	t
f	f	t	f	about 1	t	f
t	f	NA	f	about 1	t	f
t	b	NA	f	about 1	t	b
t	b	NA	f	about 1.5	t	b

Table 2. Relational tableau characterizing an oval object

The first row in this table signifies that if the answer to Q_1 is t (i.e., p has one axis of symmetry); the answer to Q_2 is t (i.e., p has a second axis of symmetry; the answer to Q_3 is t (i.e., the two axes of symmetry are orthogonal); the answer to Q_4 is f (i.e., p has two and only two axes of symmetry); the answer to Q_5 is about 1 (i.e., the major and minor axes are about equal in length); and the answer to Q_6 is t (i.e., Q_6 is convex), with the answer to Q_6 provided by the subquestionnaire; then the answer to Q is b (i.e., p is an oval object to a degree which is approximately equal to 0.5, with "approximately equal to 0.5" defined by (3.4)).

Similarly, the fifth row in the table signifies that if the answer to Q_1 is t; the answer to Q_2 is b; the answer to Q_3 is NA; the answer to Q_4 is f; the answer to Q_5 is about 1 and the answer to Q_6 is t; then the answer to Q is b. Comparing the entries in row 5 with those of row 6, we note the answer to Q remains b when we change the answer to Q_5 from about 1 to about 1.5.

4. Translation Rules and the Interpolation of a Relational Tableau

Assuming that we have a characterization of M(p) in the form of a relational tableau R, the question that arises is: How can we deduce from R the grade of membership of an object p in F?

As a preliminary to arriving at an approximate answer to this question, we have to develop a way of converting R into an (n+1)-ary fuzzy relation in $U_1 \times \ldots \times U_n \times V$. To this end, we shall employ the translation rules of fuzzy logic -- rules which provide a basis for translating a composite fuzzy proposition into a system of so-called relational assignment equations [14].

More specifically, let p be a pointer to an object and let q be a proposition of the form

(4.1) $$q \triangleq p \text{ is } F$$

where F is a fuzzy subset of U. For example, q may be

(4.2) $$q \triangleq \text{Pamela is tall.}$$

Translation rule of Type I asserts that q translates into

(4.3) $$p \text{ is } F \rightarrow R(A(p)) = F$$

where A(p) is an implied attribute of p and R(A(p)) is a fuzzy restriction[11] on the variable A(p). Thus, (4.3) constitutes a relational assignment equation in the sense that the fuzzy set F -- viewed as a unary fuzzy relation in U -- is assigned to the restriction on A(p). For example, in the case of (4.2), the rule in question yields

$$\text{Pamela is tall} \rightarrow R(\text{Height(Pamela)}) = \text{tall}$$

where R(Height(Pamela)) is a fuzzy restriction on the values that may be assigned to the variable Height(Pamela).

Now let us consider two propositions, say

(4.4) $$q_1 \triangleq p_1 \text{ is } F_1$$

and

(4.5) $$q_2 \triangleq p_2 \text{ is } F_2$$

where p_1 and p_2 are possibly distinct objects, and F_1 and F_2 are fuzzy subsets of U_1 and U_2, respectively. For example, q_1 and q_2 might be $q_1 \triangleq X$ is large and $q_2 \triangleq Y$ is small.

By (4.3), the translations of q_1 and q_2 are given by

(4.6) $$p_1 \text{ is } F_1 \rightarrow R(A_1(p_1)) = F_1$$

(4.7) $$p_2 \text{ is } F_2 \rightarrow R(A_2(p_2)) = F_2$$

where $A_1(p_1)$ and $A_2(p_2)$ are implied attributes of p_1 and p_2.

By the rule of conjunctive composition [4], the translation of the composite proposition q_1 and q_2 is given by

(4.8) $$q_1 \text{ and } q_2 \rightarrow R(A_1(p_1), A_2(p_2)) = F_1 \times F_2$$

where $F_1 \times F_2$ denotes the cartesian product of F_1 and F_2 (see A56)

[11]A fuzzy restriction is a fuzzy relation which acts as an elastic constraint on the values that may be assigned to a variable [2], [14].

which is assigned to the restriction on $A_1(p_1)$ and $A_2(p_2)$. Dually, by the rule of disjunctive composition, the translation of the composite composition q_1 or q_2 is given by

$$q_1 \text{ or } q_2 \rightarrow R(A_1(p_1), A_2(p_2)) = \bar{F}_1 + \bar{F}_2 \tag{4.9}$$

where $\bar{F}_1$ and $\bar{F}_2$ are the cylindrical extensions of F_1 and F_2 (see A59) and + denotes the union.

As we shall see presently, these two rules provide a basis for constructing a translation rule for relational tableaus. More specifically, consider a tableau of the form shown in Table 3

A_1	A_2	...	A_n
r_{11}	r_{12}		r_{1n}
r_{21}	r_{22}	.	r_{2n}
⋮	⋮	⋮	⋮
r_{m1}	r_{m2}		r_{mn}

Table 3. A relational tableau

in which $A_1, \ldots, A_n$ are variables taking values in $U_1, \ldots, U_n$, and the r_{ij} are linguistic labels of fuzzy subsets of U_j. (In relation to Table 1, the A_j play the roles of Q_j and Q.)

Expressed in words, the meaning of the tableau in question may be stated as:

(4.10) A_1 is r_{11} and A_2 is r_{12} and ... and A_n is r_{1n}

or

A_1 is r_{21} and A_2 is r_{22} and ... and A_n is r_{2n}

or

⋮ .....

or A_1 is r_{m1} and A_2 is r_{m2} and ... and A_n is r_{mn}

Regarding (4.9) as a composite proposition and applying (4.8) and (4.9) to (4.10), we arrive at the tableau translation rule which is expressed by

(4.11)

A_1	...	A_n
r_{11}	...	r_{1n}
.	...	.
r_{m1}	...	r_{mn}

$$\rightarrow R(A_1,\ldots,A_n) = r_{11} \times \ldots \times r_{1n} + \ldots + r_{m1} \times \ldots \times r_{mn}$$

where $r_{11} \times \ldots \times r_{1n} + \ldots + r_{m1} \times \ldots \times r_{mn}$ is an n-ary fuzzy relation in $U_1 \times \ldots \times U_n$ which is assigned to the restriction $R(A_1,\ldots,A_n)$ on the values of the variables $A_1,\ldots,A_n$.

As a very simple illustration of the tableau translation rule, assume that the tableau of R is given by [86]

(4.12)

Q_1	Q_2	Q
t	t	vf
f	f	t

where t, f and vf are abbreviations for true, false and very false, respectively, and

$$U_1 = U_2 = V = 0+0.2+0.4+0.6+0.8+1 \quad (4.13)$$

$$t = 0.6/0.8 + 1/1 \quad (4.14)$$

$$f = 1/0 + 0.6/0.2 \quad (4.15)$$

and by (3.6)

$$vf = 1/0 + 0.36/0.2 . \quad (4.16)$$

Applying the translation rule (4.11) to the table in question, we obtain the ternary fuzzy relation in $V \times V \times V$:

$$(4.17)\quad R(Q_1,Q_2,Q) = t\times t\times vf + f\times f\times t$$
$$= (0.6/0.8+1/1)\times(0.6/0.8+1/1)\times(1/0+0.36/0.2)$$
$$+(1/0+0.6/0.2)\times(1/0+0.6/0.2)\times(0.6/0.8+1/1)$$
$$= 0.36/((0.8,0.8,0.2)+(0.8,1,0.2)+(1,0.8,0.2)$$
$$+(1,1,0.2))+0.6((0,0,0.8)+(0,0.2,0.8)+(0.2,0,0.8)$$
$$+(0.2,0.2,0.8) + (0,0.2,1)+(0.2,0.2,1))$$
$$+ 1/((0,0,1)+(1,1,0))$$

as the expression for the meaning of the relational tableau (4.12).

The Mapping Rule

The translation rule expressed by (4.11) provides a basis for an interpolation of a relational tableau, yielding an approximate value for the answer to Q given the answers to $Q_1,\ldots,Q_n$ which do not appear in R.

Specifically, let $(r_{i1},\ldots,r_{in},r_i)$ denote the ith (n+1)-tuple in R and let $\tilde{R}$ denote the (n+1)-ary fuzzy relation in $U_1\times\ldots\times U_n\times V$, $V \triangleq [0,1]$, expressed by

$$(4.13)\qquad \tilde{R} = r_{11}\times\ldots\times r_{1n}\times r_1 + \ldots + r_{m1}\times\ldots\times r_{mn}\times r_m$$

where, as in (4.11), $\times$ and $+$ denote the cartesian product and union, respectively.

Now suppose that $g_1,\ldots,g_n$ are given fuzzy subsets of $U_1,\ldots,U_n$, respectively, and that we wish to compute the value of Q given that the values of $Q_1,\ldots,Q_n$ are $g_1,\ldots,g_n$.

Let $R(g_1,\ldots,g_n)$ denote the result of the substitution and hence the desired value of Q, and let G denote the cartesian product

$$(4.14)\qquad G = g_1\times\ldots\times g_n .$$

Then, the mapping rule may be expressed compactly as[12]

$$(4.15)\qquad R(g_1,\ldots,g_n) = \tilde{R}\circ G$$

---

[12] This mapping rule may be viewed as an extension to a fuzzy relation of the mapping rule employed in such query languages as SQUARE and SEQUEL [15], [16].

where $\circ$ denotes the composition (see A60) of the (n+1)-ary fuzzy relation $\tilde{R}$ with the n-ary fuzzy relation G.

In more explicit terms, the right-hand member of (4.15) is a fuzzy subset of U which may be computed as follows.

Assume for simplicity that $U_1, \ldots, U_n, V$ are finite sets which may be expressed in the form (+ denotes the union)

(4.16)
$$\begin{aligned} U_1 &= u_1^1 + \ldots + u_{k_1}^1 \\ U_2 &= u_1^2 + \ldots + u_{k_2}^2 \\ &\ldots\ldots\ldots\ldots \\ U_n &= u_1^n + \ldots + u_{k_n}^n \\ V &= v_1 + \ldots + v_k \end{aligned}$$

Now suppose that the g_i are expressed as fuzzy subsets of the U_i by (see A6)

(4.17)
$$\begin{aligned} g_1 &= \gamma_1^1 u_1^1 + \ldots + \gamma_{k_1}^1 u_{k_1}^1 \\ &\ldots\ldots\ldots\ldots \\ g_n &= \gamma_1^n u_1^n + \ldots + \gamma_{k_n}^n u_{k_n}^n \end{aligned}$$

so that

(4.18)
$$G = \sum_I \gamma_{i_1}^1 \wedge \gamma_{i_2}^2 \wedge \ldots \wedge \gamma_{i_n}^n / u_{i_1}^1 u_{i_2}^2 \ldots u_{i_n}^n$$

where I denotes the index sequence $(i_1, \ldots, i_n)$, with $1 \le i_1 \le k_1$, $1 \le i_2 \le k_2, \ldots, 1 \le i_n \le k_n$; $u_{i_1}^1 u_{i_2}^2 \ldots u_{i_n}^n$ is an abbreviation for the n-tuple $(u_{i_1}^1, u_{i_1}^2, \ldots, u_{i_n}^n)$, and $\gamma_{i_1}^1 \wedge \gamma_{i_2}^2 \wedge \ldots \wedge \gamma_{i_n}^n$ is the grade of membership of the n-tuple $u_{i_1}^1 u_{i_2}^2 \ldots u_{i_n}^n$ in the n-ary fuzzy relation G.

By the definition of composition, the composition of $\tilde{R}$ with G may be expressed as the projection on $U_1 \times \ldots \times U_n$ of the intersection of $\tilde{R}$ with the cylindrical extension of G. Thus,

(4.19) $$\tilde{R} \circ G = \mathrm{Proj}_{U_1 \times \dots \times U_n} (\tilde{R} \cap \bar{G})$$

where $\bar{G}$ is given by

(4.20) $$\bar{G} = \sum_{(I,i)} \gamma^1_{i_1} \wedge \gamma^2_{i_2} \wedge \dots \wedge \gamma^n_{i_n} / u^1_{i_1} u^2_{i_2} \dots u^n_{i_n} v_i \,.$$

In this expression, (I,i) denotes the index sequence $(i_1,\dots,i_n,i)$, with $1 \le i \le k$, and $u^1_{i_1} u^2_{i_2} \dots u^n_{i_n} v_i$ is an abbreviation for the $(n+1)$-tuple $(u^1_{i_1}, u^2_{i_2}, \dots, u^n_{i_n}, v_i)$.

Now suppose that the computation of the right-hand member of (4.13) yields $\tilde{R}$ in the form

(4.21) $$\tilde{R} = \sum_{(I,i)} \mu_{(I,i)} / u^1_{i_1} u^2_{i_2} \dots u^n_{i_n} v_i \,.$$

Then, the intersection of $\tilde{R}$ with $\bar{G}$ is given by

(4.22) $$\tilde{R} \cap \bar{G} = \sum_{(I,i)} \gamma^1_{i_1} \wedge \gamma^2_{i_2} \wedge \dots \wedge \gamma^n_{i_n} \wedge \mu_{(I,i)} / u^1_{i_1} \dots u^n_{i_n} v_i$$

and the projection[13] of $\tilde{R} \cap \bar{G}$ on $U_1 \times \dots \times U_n$ -- and hence the composition of $\tilde{R}$ and G -- is expressed by

(4.23) $$\tilde{R} \circ G = \sum_{(I,i)} \gamma^1_{i_1} \wedge \gamma^2_{i_2} \wedge \dots \wedge \gamma^n_{i_n} \wedge \mu_{(I,i)} / v_i$$

where, to recapitulate;

$R(g_1,\dots,g_n) = \tilde{R} \circ G$

= result of substitution of g_i for Q_i, $i = 1,\dots,n$, in R;

$G = g_1 \times \dots \times g_n$;

$\gamma^\lambda_{i_\lambda}$ = grade of membership of u_{i_λ} in g_λ, $\lambda = 1,\dots,n$;

$I \triangleq (i_1,\dots,i_n)$

$(I,i) \triangleq (i_1,\dots,i_n,i)$

---

[13] A convenient way of obtaining the projection is to set $u^1_{i_1} = \dots = u^n_{i_n} = 1$ in the right-hand member of (4.22) and treat the $(n+1)$-tuple $(u^1_{i_1},\dots,u^n_{i_n},v_i)$ as if it were an algebraic product of $u^1_{i_1},\dots,u^n_{i_n},v_i$.

$$\mu_{(i,i)} \triangleq \text{grade of membership of } (u^1_{i_1}, u^2_{i_2}, \ldots, u^n_{i_n}, v_i) \text{ in } \tilde{R}$$

$$\tilde{R} = r_{11} \times \ldots \times r_{1n} \times r_1 + \ldots + r_{m1} \times \ldots \times r_{mn} \times r_m \,.$$

It should be noted that we would obtain the same result by assigning $g_1, \ldots, g_n$ to $Q_1, \ldots, Q_n$ in sequence rather than simultaneously. This is a consequence of the identity

$$(4.24) \qquad \tilde{R} \circ G = (\ldots((\tilde{R} \circ g_1) \circ g_2) \ldots \circ g_n)$$

which in turn follows from the identity

$$(4.25) \qquad \sum_{(I,i)} \gamma^1_{i_1} \wedge \gamma^2_{i_2} \wedge \ldots \wedge \gamma^n_{i_n} \wedge \mu_{(I,i)}/v_i$$

$$= \sum_{(I,i)} [[[\gamma'_{i_1} \wedge \mu_{(I,i)}/u^1_{i_1} u^2_{i_2} \ldots u^n_{i_n} v_i]_{1\, u^1_{i_1}=1} \wedge \gamma^2_{i_2}]_{2\, u^2_{i_2}=1} \wedge \ldots \wedge \gamma^n_{i_n}]_{n\, u^n_{i_n}=1} \,.$$

As a very simple illustration of the mapping operation, assume that

$$n = 2;$$

$$U_1 = U_2 = V = 0 + 0.2 + 0.4 + 0.6 + 0.8 + 1;$$

$\tilde{R}$ is given by

$$(4.26) \quad \tilde{R} = 1/(0,0,0) + 0.8/(0,0,0.2) + 0.7/(0.2,0.2,0) + 0.6/(0.2,0,0) + 0.8/(0.4,0.6,0.4) + 0.8/(0.4,0.2,0) + 0.5/(0.4,0.2,0.4) + 0.6/(0.2,0.6,0.8) + 0.8(0.8,0.8,0.2) + 0.9/(0.8,0.8,1) + 0.8/(0.8,1,0.8) + 0.6/(0.2,0.8,1) + 0.8/(0.6,0.8,1)$$

and

$$(4.27) \qquad g_1 = 0.6/0.4 + 1/0.2$$

$$(4.28) \qquad b_2 = 1/0.6 + 0.8/0.2.$$

Then by (4.18)

$$(4.29) \qquad g = g_1 \times g_2 = 0.6/(0.4,0.6) + 0.6/(0.4,0.2) + 1/(0.2,0.6) + + 0.8/(0.2,0.2)$$

and thus

$$(4.30)\quad R(g_1,g_2) = \tilde{R} \circ g$$
$$= 0.6 \wedge 0.8/0.4 + 0.8 \wedge 0.6/0 + 0.5 \wedge 0.6/0.4 +$$
$$+ 0.6 \wedge 1/0.8 + 0.7 \wedge 0.8/0$$
$$= 0.6/0.4 + 0.7/0 + 0.6/0.8.$$

There are two points related to the computation of $\tilde{R} \circ g$ that are in need of comment. First, if $\tilde{R}$ is sparsely tabulated in the sense that many of the possible n-tuples of values of $Q_1, \ldots, Q_n$ are not in the table, then the interpolation of R by the use of (4.23) may not yield a valid approximation to the answer to Q. And second, the result of substitution of

$$g = r_{i1} \times \ldots \times r_{in}$$

in $\tilde{R}$ would not, in general, be exactly equal to r_i -- as one might expect to be the case. As pointed out in [14], the reason for this phenomenon is the interference between the rows of $\tilde{R}$, which in turn is due to the fact that the fuzzy sets which constitute a column of R are not, in general, disjoint, that is, do not have an empty intersection.

An important assumption that underlies the procedure described in this section is that one has or can obtain a relational tableau which characterizes the dependence of the grade of membership of an object on the linguistic values of its attributes and/or the degree to which it possesses specified properties. The main contribution of the linguistic approach is that it makes it possible to describe this dependence in an approximate manner, using words rather than numbers as values of the relevant variables.

5. Cluster Analysis.

Theory of fuzzy sets was first applied to cluster analysis by E. Ruspini [17]-[19]. More recently, J. Dunn and J. Bezdek have made a number of important contributions to this subject and have described effective algorithms for deriving optimal fuzzy partitions of a given set of sample points [20]-[32].

Viewed within the framework described in Section 2, cluster analysis differs from pattern classification in three essential respects.

First, the point of departure in cluster analysis is not -- as in pattern classification -- an opaque recognition algorithm in U^o which defines a fuzzy subset F of U^o, but a fuzzy similarity relation S^o which is a fuzzy subset of $U^o \times U^o$ and which is characterized by an opaque recognition algorithm R_{op}. Thus, when presented with two objects p and q in U^o, R_{op} yields the degree, $\mu_{S^o}(p,q)$, to which p and q are similar. The function $\mu_{S^o}: U^o \times U^o \rightarrow [0,1]$ is the membership function of the fuzzy relation S^o in U^o.

Second, the problem of cluster analysis includes as a subproblem the following problem in pattern classification.

Let p and q be objects in U^o and let $x \stackrel{\Delta}{=} M(p)$ and $y \stackrel{\Delta}{=} M(q)$ be their correspondents in the space of mathematical objects $U = \{M(p)\}$. The problem is to convert the opaque recognition algorithm R_{op} which acting on p and q yields

(5.1) $$R_{op}(p,q) = \mu_{S^o}(p,q) ,$$

into a transparent recognition algorithm R_{tr} which acting on x and y yields the same result as R_{op}, i.e.,

(5.2) $$R_{tr}(x,y) = R_{op}(p,q)$$
$$= \mu_{S^o}(p,q) .$$

It should be noted that this problem is of the same type as that formulated in Section 2, with the fuzzy subset S^o of $U^o \times U^o$ playing the role of F.

Third, assuming that we have R_{tr} -- which acts on elements of $U \times U$ -- the objective of cluster analysis is to derive from R_{tr} a number, say k, of transparent recognition algorithms $R_{tr_1}, \ldots, R_{tr_k}$ -- acting on elements of U -- such that the fuzzy subsets (fuzzy clusters)

$F_1, \ldots, F_k$ in U defined by $R_{tr_1}, \ldots, R_{tr_k}$, have a property which may be stated as follows.

Fuzzy Affinity Property

Let $x = M(p)$ and $y = M(q)$ be mathematical objects in U corresponding to the objects p and q in U^o. Let $\{F_1, \ldots, F_k\}$ be a collection of well-separated[14] fuzzy subsets of U with membership functions $\mu_1, \ldots, \mu_n$, respectively. Then the F_i are fuzzy clusters induced by S^o if they have the fuzzy affinity property defined below.

(a) Both x and y have high grades of membership in some F_r, $r = 1, \ldots, k \Leftrightarrow (x, y)$ has a high grade of membership in S (the similarity relation induced in U by S^o).

(b) x has a high grade of membership in some F_r, $r = 1, \ldots, k$ and y has a high grade of membership in F_t, $t \neq r \Rightarrow (x, y)$ does not have a high grade of membership in S .

Stated less formally, the fuzzy affinity property implies that (a) if x and y have a high degree of similarity then they have a high grade of membership in some cluster, and vice-versa; and (b) if x and y have high grades of membership in different clusters then they do not have a high degree of similarity. It should be noted that this property of fuzzy clusters is more demanding than that implicit in the conventional definitions in which the degree of similarity of objects which belong to the same cluster is merely required to be greater than the degree of similarity between objects which belong to different clusters. Another point that should be noted is that, if we assumed that the only alternative to the consequent of (b) is "(x,y) has a high grade of membership in S," then (b) would be implied by (a) since the latter consequent would imply that x and y have a high grade of membership in some F_r -- which contradicts the antecedent of (b). Thus, by stating (b) we are tacitly assuming that (x,y) is not restricted to having either "high" or "not high"

[14] By well-separated we mean that if F_r and F_t are distinct fuzzy sets in $\{F_1, \ldots, F_k\}$, then every point of U has a low grade of membership in $F_r \cap F_t$.

grades of membership in S. For example, the grade of membership in S could be "not high and not low."

An important implication of the fuzzy affinity property is the following. Suppose that x and y have high grades of membership in some fuzzy cluster F_r, and that z has a high grade of membership in a different fuzzy cluster, say F_t. Then, by (a) and (b), we have

(5.3) similarity of x and y is high
similarity of y and z is not high
similarity of x and z is not high

which implies that we could not have

(5.4) similarity of x and y is high
similarity of y and z is high
similarity of x and z is not high.

The inconsistency of the assertions in (5.4) is ruled out by the fuzzy transitivity of the similarity relation S which may be stated as[15]

(5.5) similarity of x and z is at least as great as the similarity of x and y or the similarity of y and z.

Thus, if S has the fuzzy transitivity property and the similarities of both x and y and y and z are high, then the similarity of x and z must also be high.

Another point that should be noted is that the fuzzy affinity property does not require that the fuzzy clusters $\{F_1, \ldots, F_k\}$ form a fuzzy partition in the sense of Ruspini. However, the stronger assumption that the F_r form a fuzzy partition makes it possible for Dunn and Bezdek to construct an effective algorithm for deriving from a fuzzy similarity relation a family of fuzzy clusters which form a fuzzy partition.

---

[15] In more precise terms, the transitivity of a fuzzy relation R in U is defined by (see [13])

$$\mu_R(u,v) \geq \vee_w(\mu_R(u,w) \wedge \mu_R(w,v)), \qquad (u,v) \in U \times U$$

where $\mu_R(u,v)$ is the grade of membership of (u,v) in R, and $\vee_w$ is the supremum over $w \in U$.

As described in [26], the Dunn-Bezdek fuzzy ISODATA algorithm may be stated as follows.

Let $\mu_1,\ldots,\mu_k$ denote the membership functions of $F_1,\ldots,F_k$, where the F_i, $i = 1,\ldots,k$, are fuzzy subsets (clusters) of a finite subset, X, of points in U. The fuzzy clusters $F_1,\ldots,F_k$ form a <u>fuzzy k-partition of X</u> if and only if

(5.6) $$\mu_1(x) + \ldots + \mu_k(x) = 1, \qquad x \in X$$

where + denotes the arithmetic sum. The goodness of a fuzzy partition is assumed to be assessed by the criterion functional

(5.7) $$J(\mu) = \min_v \sum_{i=1}^{k} \sum_{x \in X} (\mu_i(x))^2 \|x - v_i\|^2$$

where $\mu \triangleq (\mu_1,\ldots,\mu_k)$, $v = (v_1,\ldots,v_k)$, $v_i \in L$, and $L \triangleq$ vector space with inner product induced norm $\| \|$. Intuitively, the v_i represent the "centers" of $F_1,\ldots,F_k$ and $J(\mu)$ provides a measure of the weighted dispersion of points in X in the relation to the optimal locations of the centers $v_1,\ldots,v_k$.

<u>Step 1:</u> Choose a fuzzy partition $F_1,\ldots,F_k$ characterized by k nonempty membership functions $\mu = (\mu_1,\ldots,\mu_k)$, with $2 \leq k \leq n$.

<u>Step 2:</u> Compute the k weighted means (centers)

(5.8) $$v_i = \frac{\sum_{x \in X} (\mu_i(x))^2 x}{\sum_{x \in X} (\mu_i(x))^2}, \qquad 1 \leq i \leq k$$

where $x \in X \subset L$.

<u>Step 3:</u> Construct a new partition, $\hat{F}_1,\ldots,\hat{F}_k$, characterized by $\hat{\mu} = (\hat{\mu}_1,\ldots,\hat{\mu}_k)$, according to the following rule.

Let $I(x) \triangleq \{1 \leq i \leq k \mid v_i = x\}$. If $I(x)$ is not empty let $\hat{i}$ be the least integer $I(x)$ and put

(5.9) $$\hat{\mu}_i(x) = 1 \quad \text{if} \quad i = \hat{i}$$
$$= 0 \quad \text{if} \quad i \neq \hat{i}$$

for $1 \leq i \leq k$. Otherwise, if $I(x)$ is empty (the usual case), set

(5.10)
$$\hat{\mu}_i(x) = \frac{\dfrac{1}{\|x-v_i\|^2}}{\sum_{j=1}^{k}\left(\dfrac{1}{\|x-v_i\|^2}\right)} .$$

Step 4: Compute some convenient measure, δ, of the defect between μ and $\hat{\mu}$. If $\delta \leq \varepsilon \triangleq$ a specified threshold, then stop. Otherwise go to Step 2.

In a number of papers [20]-[32], Bezdek and Dunn have studied the behavior of this and related algorithms and have established their convergence and other properties. Clearly, the work of Bezdek and Dunn on fuzzy clustering constitutes an important contribution to both the theory of cluster analysis and its practical applications.

Fuzzy Level-Sets

As was pointed out in [13], the conventional hierarchical clustering schemes [33] may be viewed as the resolution of a fuzzy similarity relation into a nested collection of nonfuzzy equivalence relations. To relate this result to the fuzzy affinity property, it is necessary to extend the notion of a level-set as defined in [13] to that of a fuzzy level-set. More specifically, let F be a fuzzy subset of U and let F_α, $0 \leq \alpha \leq 1$, be the α-level subset of U defined by

(5.11)
$$F_\alpha \triangleq \{x \mid \mu_F(x) \geq \alpha\}$$

where μ_F is the membership function of F. We note that F_α -- which is a nonfuzzy set -- may be expressed equivalently as

(5.12)
$$F_\alpha = \mu_F^{-1}([\alpha,1])$$

where μ_F^{-1} is the relation from [0,1] to U which is converse to μ_F, and F_α is the image of the interval $[\alpha,1]$ under this relation -- or, equivalently, multi-valued mapping -- μ_F^{-1}. It is easy to verify that in terms of the membership functions of F_α, F and $[\alpha,1]$, (5.12) translates into

(5.13) $$\mu_{F_\alpha}(x) = \mu_{[\alpha,1]}(\mu_F(x)), \quad x \in U$$

where μ_{F_α} and $\mu_{[\alpha,1]}$ denote the membership (characteristic) functions of the nonfuzzy sets F_α and $[\alpha,1]$, respectively.

Now suppose that α is a fuzzy subset of $[0,1]$ labeled, say, high, with μ_{high} defined by (see A17)

(5.14) $$\mu_{high}(v) = S(v;0.6,0.7,0.8), \qquad 0 \le v \le 1 .$$

When α is a fuzzy subset of $[0,1]$, the fuzzy set $\ge \alpha$ may be expressed as the composition of the nonfuzzy binary relation $\ge$ with the unary fuzzy relation α. Thus, if $\alpha \triangleq$ high, then

(5.15) $$\begin{aligned} \ge \alpha &= \ge \circ\, \alpha \\ &= \ge \circ\, \text{high} \\ &= \text{high} \end{aligned}$$

since the membership function of high is monotone nondecreasing in v. Correspondingly, the expression for the membership function of the fuzzy level set F_{high} becomes (see A73)

(5.16) $$\mu_{F_{high}}(x) = \mu_{high}(\mu_F(x)) .$$

To relate this result to the fuzzy affinity property, we note that if the objects x,y in U have a high degree of similarity, then the ordered pair (x,y) has a high grade of membership in the fuzzy similarity relation S. Thus, by analogy with (5.12), the set of pairs (x,y) in $U \times U$ which have a high grade of membership in S form a fuzzy level-set of S defined by

(5.17) $$S_{high} = \mu_S^{-1}(\mu_{high})$$

or, equivalently,

(5.18) $$\mu_{S_{high}}(x,y) = \mu_{high}(\mu_S(x,y)) .$$

This expression makes it possible to derive in a straight-forward fashion the fuzzy level-set S_{high} from the similarity relation S.

An important property of S_{high} may be stated as the

Proposition. If S is a transitive fuzzy relation, so is S_{high}. The validity of this proposition is readily established by observing that the transitivity of S means that (see (5.5))

(5.19) $$\mu_S(x,y) \geq \vee_z \mu_S(x,z) \wedge \mu_S(z,y), \quad x, y, z \in U .$$

Now, (5.19) implies and is implied by

(5.20) $$\forall z\ (\mu_S(x,y) \geq \mu_S(x,z) \wedge \mu_S(z,y))$$

which in turn implies and is implied by

(5.21) $$\forall z(\mu_S(x,y) \geq \mu_S(x,z) \text{ or } \mu_S(x,y) \geq \mu_S(z,y)).$$

Since μ_{high} is a monotone nondecreasing function, we have

(5.22) $$\mu_S(x,y) \geq \mu_S(x,z) \Rightarrow \mu_{high}(\mu_S(x,y)) \geq \mu_{high}(\mu_S(x,z))$$

and

(5.23) $$\mu_S(x,y) \geq \mu_S(z,y) \Rightarrow \mu_{high}(\mu_S(x,y)) \geq \mu_{high}(\mu_S(z,y))$$

and hence

(5.24) $$\forall z(\mu_{S_{high}}(x,y) \geq \mu_{S_{high}}(x,z) \text{ or } \mu_{S_{high}}(x,y) \geq \mu_{S_{high}}(z,y))$$

which by (5.21) and (5.20) leads to the conclusion that S_{high} is transitive.

Basically, the employment of fuzzy level-sets for purposes of clustering may be viewed as an application of a form of contrast intensification [34] to a fuzzy similarity relation which defines the degrees of similarity of mathematical objects in U . Thus, given a collection of such objects, we can derive S_{high} from S by the use of (5.18) and then apply a Dunn-Bezdek type of fuzzy clustering algorithm to group the given collection of objects into a set of fuzzy clusters $\{F_1, \ldots, F_k\}$.

6. Concluding Remarks.

In the foregoing discussion, we have touched upon only a few of the many basic issues which arise in the application of the theory of fuzzy sets to pattern classification and cluster analysis. Although this is not yet the case at present, it is very likely that in the years ahead it will be widely recognized that most of the problems in pattern classification and cluster analysis are intrinsically fuzzy in nature and that the conceptual framework of the theory of fuzzy sets provides a natural setting both for the formulation of such problems and their solution by fuzzy-algorithmic techniques.

Appendix

Fuzzy Sets -- Notation, Terminology and Basic Properties.

The symbols $U, V, W, \ldots,$ with or without subscripts, are generally used to denote specific universes of discourse, which may be arbitrary collections of objects, concepts or mathematical constructs. For example, U may denote the set of all real numbers; the set of all residents in a city; the set of all sentences in a book; the set of all colors that can be perceived by the human eye, etc.

Conventionally, if A is a fuzzy subset of U whose elements are $u_1, \ldots, u_n$, then A is expressed as

(A1) $$A = \{u_1, \ldots, u_n\} .$$

For our purposes, however, it is more convenient to express A as

(A2) $$A = u_1 + \ldots + u_n$$

or

(A3) $$A = \sum_{i=1}^{n} u_i$$

with the understanding that, for all i, j,

(A4) $$u_i + u_j = u_j + u_i$$

and

(A5) $$u_i + u_i = u_i .$$

As an extension of this notation, a finite fuzzy subset of U is expressed as

(A6) $$F = \mu_1 u_1 + \ldots + \mu_n u_n$$

or, equivalently, as

(A7) $$F = \mu_1/u_1 + \ldots + \mu_n/u_n$$

where the μ_i, $i = 1,\ldots,n$, represent the grades of membership of the u_i in F. Unless stated to the contrary, the μ_i are assumed to lie in the interval $[0,1]$, with 0 and 1 denoting no membership and full membership, respectively.

Consistent with the representation of a finite fuzzy set as a linear form in the u_i, an arbitrary fuzzy subset of U may be expressed in the form of an integral

(A8) $$F = \int_U \mu_F(u)/u$$

in which $\mu_F : U \rightarrow [0,1]$ is the membership or, equivalently, the compatibility function of F; and the integral $\int_U$ denotes the union (defined by (A23)) of fuzzy singletons $\mu_F(u)/u$ over the universe of discourse U.

The points in U at which $\mu_F(u) > 0$ constitute the support of F. The points at which $q_F(u) = 0.5$ are the crossover points of F.

Example A9. Assume

(A10) $$U = a + b + c + d.$$

Then, we may have

(A11) $$A = a + b + d$$

and

(A12) $$F = 0.3a + 0.9b + d$$

as nonfuzzy and fuzzy subsets of U, respectively.

If

(A13) $$U = 0 + 0.1 + 0.2 + \ldots + 1$$

then a fuzzy subset of U would be expressed as, say

(A14) $$F = 0.3/0.5 + 0.6/0.7 + 0.8/0.9 + 1/1.$$

If $U = [0,1]$, then F might be expressed as

$$F = \int_0^1 \frac{1}{1+u^2} / u \tag{A15}$$

which means that F is a fuzzy subset of the unit interval $[0,1]$ whose membership function is defined by

$$\mu_F(u) = \frac{1}{1+u^2} . \tag{A16}$$

In many cases, it is convenient to express the membership function of a fuzzy subset of the real line in terms of a standard function whose parameters may be adjusted to fit a specified membership function in an approximate fashion. Two such functions are defined below.

$$\begin{aligned} S(u;\alpha,\beta,\gamma) &= 0 && \text{for } u \leq \alpha \\ &= 2\left(\frac{u-\gamma}{\gamma-\alpha}\right)^2 && \text{for } \alpha \leq u \leq \beta \\ &= 1 - 2\left(\frac{u-\gamma}{\gamma-\alpha}\right)^2 && \text{for } \beta \leq u \leq \gamma \\ &= 1 && \text{for } u \geq \gamma \end{aligned} \tag{A17}$$

$$\begin{aligned} \pi(u;\beta,\gamma) &= S(u;\gamma-\beta,\gamma-\frac{\beta}{2},\gamma) && \text{for } u \leq \gamma \\ &= 1 - S(u;\gamma,\gamma+\frac{\beta}{2},\gamma+\beta) && \text{for } u \geq \gamma . \end{aligned} \tag{A18}$$

In $S(u;\alpha,\beta,\gamma)$, the parameter β, $\beta = \frac{\alpha+\gamma}{2}$, is the crossover point. In $\pi(u;\beta,\gamma)$, β is the bandwidth, that is the separation between the crossover points of π, while γ is the point at which π is unity.

In some cases, the assumption that μ_F is a mapping from U to $[0,1]$ may be too restrictive, and it may be desirable to allow μ_F to take values in a lattice or, more particularly, in a Boolean algebra. For most purposes, however, it is sufficient to deal with the first two of the following hierarchy of fuzzy sets.

Definition A19. A fuzzy subset, F, of U is of *type* 1 if its membership function, μ_F, is a mapping from U to $[0,1]$; and F is of type n, $n = 2,3,\ldots,$ if μ_F is a mapping from U to the set of fuzzy subsets of type $n-1$. For simplicity, it will always be understood that F is of type 1 if it is not specified to be of a higher type.

Example A20. Suppose that U is the set of all nonnegative integers and F is a fuzzy subset of U labeled small integers. Then F is of type 1 if the grade of membership of a generic element u in F is a number in the interval [0,1], e.g.,

(A21) $$\mu_{\underline{\text{small integers}}}(u) = (1 + (\frac{u}{5})^2)^{-1}, \quad u = 0,1,2,\ldots .$$

On the other hand, F is of type 2 if for each u in U, $\mu_F(u)$ is a fuzzy subset of [0,1] of type 1, e.g., for u = 10,

(A22) $$\mu_{\underline{\text{small integers}}}(10) = \underline{\text{low}}$$

where low is a fuzzy subset of [0,1] whose membership function is defined by, say,

(A23) $$\mu_{\underline{\text{low}}}(v) = 1 - S(v;0,0.25,0.5), \quad v \in [0,1]$$

which implies that

(A24) $$\underline{\text{low}} = \int_0^1 (1 - S(v;0,0.25,0.5))/v .$$

If F is a fuzzy subset of U, then its α-level-set, F_α, is a nonfuzzy subset of U defined by

(A25) $$F_\alpha = \{u \mid \mu_F(u) \geq \alpha\}$$

for $0 < \alpha \leq 1$.

If U is a linear vector space, the F is convex if and only if for all $\lambda \in [0,1]$ and all u_1, u_2 in U,

(A26) $$\mu_F(\lambda u_1 + (1-\mu)u_2) \geq \min(\mu_F(u_1),\mu_F(u_2)) .$$

In terms of the level-sets of F, F is convex if and only if the F_α are convex for all $\alpha \in (0,1]$.[26]

The relation of containment for fuzzy subsets F and G of U is defined by

(A27) $$F \subset G \Leftrightarrow \mu_F(u) \leq \mu_G(u), \quad u \in U .$$

[26] This definition of convexity can readily be extended to fuzzy sets of type 2 by applying the extension principle (see (A70)) to (A26).

Thus, F is a fuzzy subset of G if (A27) holds for all u in U.

Operations on Fuzzy Sets.

If F and G are fuzzy subsets of U, their union, $F \cup G$, intersection, $F \cap G$, bounded-sum, $F \oplus G$, and bounded difference, $F \ominus G$, are fuzzy subsets of U defined by

(A28) $$F \cup G \triangleq \int_U \mu_F(u) \vee \mu_G(u)/u$$

(A29) $$F \cap G \triangleq \int_U \mu_F(u) \wedge \mu_G(u)/u$$

(A30) $$F \oplus G \triangleq \int_U 1 \wedge (\mu_F(u) + \mu_G(u))/u$$

(A31) $$F \ominus G \triangleq \int_U 0 \vee (\mu_F(u) - \mu_G(u))/u$$

where $\vee$ and $\wedge$ denote max and min, respectively. The complement of F is defined by

(A32) $$F' = \int_U (1 - \mu_F(u))/u$$

or, equivalently,

(A33) $$F' = U \ominus F .$$

It can readily be shown that F and G satisfy the identities

(A34) $$(F \cap G)' = F' \cup G'$$

(A35) $$(F \cup G)' = F' \cap G'$$

(A36) $$(F \oplus G)' = F' \ominus G$$

(A37) $$(F \ominus G)' = F' \oplus G$$

and that F satisfies the resolution identity

(A38) $$F = \int_0^1 \alpha F_\alpha$$

where F_α is the α-level-set of F; αF_α is a set whose membership function is $\mu_{\alpha F_\alpha} = \alpha \mu_{F_\alpha}$, and $\int_0^1$ denotes the union of the αF, with $\alpha \in (0,1]$.

Although it is traditional to use the symbol $\cup$ to denote the union of nonfuzzy sets, in the case of fuzzy sets it is advantageous to use the symbol + in place of $\cup$ where no confusion with the arithmetic sum can result. This convention is employed in the following example, which is intended to illustrate (A28), (A29), (A30), (A31) and (A32).

Example A39. For U defined by (A10) and F and G expressed by

(A40) $$F = 0.4a + 0.9b + d$$

(A41) $$G = 0.6a + 0.5b$$

we have

(A42) $$F + G = 0.6a + 0.9b + d$$

(A43) $$F \cap G = 0.4a + 0.5b$$

(A44) $$F \oplus G = a + b + d$$

(A45) $$F \ominus G = 0.4b + d$$

(A46) $$F' = 0.6a + 0.1b + c.$$

The linguistic connectives and (conjuction) and or (disjunction) are identified with $\cap$ and + , respectively. Thus,

(A47) $$F \text{ and } G \triangleq F \cap G$$

and

(A48) $$F \text{ or } G \triangleq F + G.$$

As defined by (A47) and (A48), and and or are implied to be noninteractive in the sense that there is no "trade-off" between their operands. When this is not the case, and and or are denoted by and* and or* respectively, and are defined in a way that reflects the nature of the trade-off. For example, we may have

(A49) $$F \text{ and* } G \triangleq \int_U \mu_F(u)\mu_G(u)/u$$

(A50) $$F \text{ or* } G \triangleq \int_U (\mu_F(u) + \mu_G(u) - \mu_F(u)\mu_G(u))/u$$

whose + denotes the arithmetic sum. In general, the interactive versions of and and or do not possess the simplifying properties of the connectives defined by (A47) and (A48), e.g., associativity, distributivity, etc.

If α is a real number, then F^α is defined by

(A51) $$F^\alpha \stackrel{\Delta}{=} \int_U (\mu_F(n))^\alpha/u \,.$$

For example, for the fuzzy set defined by (A40), we have

(A52) $$F^2 = 0.16a + 0.81b + d$$

and

(A53) $$F^{\frac{1}{2}} = 0.63a + 0.95b + d.$$

These operations may be used to approximate, very roughly, the effect of the linguistic modifiers very and more or less. Thus,

(A54) $$\text{very } F \stackrel{\Delta}{=} F^2$$

and

(A55) $$\text{more or less } F \stackrel{\Delta}{=} F^{\frac{1}{2}} .$$

If $F_1,\ldots,F_n$ are fuzzy subsets of $U_1,\ldots,U_n$, then the cartesian product of $F_1,\ldots,F_n$ is a fuzzy subset of $U_1 \times \ldots \times U_n$ defined by

(A56) $$F_1 \times \ldots \times F_n = \int_{U_1\times\ldots\times U_n} (\mu_{F_1}(u_1)\wedge \ldots \wedge\mu_{F_n}(u_n))/(u_1,\ldots,u_n).$$

As an illustration, for the fuzzy sets defined by (A40) and (A41), we have

(A57) $$\begin{aligned} F \times G &= (0.4a + 0.9b + d) \times (0.6a + 0.5b) \\ &= 0.4/(a,a) + 0.4/(a,b) + 0.6/(b,a) \\ &\quad + 0.5/(b,b) + 0.6/(d,a) + 0.5/(d,b) \end{aligned}$$

which is a fuzzy subset of $(a + b + c + d) \times (a + b + c + d)$.

Fuzzy Relations

An n-ary fuzzy relation R in $U_1 \times \ldots \times U_n$ is a fuzzy subset of $U_1 \times \ldots \times U_n$. The projection of R on $U_{i_1} \times \ldots \times U_{i_k}$, where $(i_1,\ldots,i_k)$ is a subsequence of $(1,\ldots,n)$, is a relation in

$U_{i_1} \times \ldots \times U_{i_k}$ defined by

(A58) $\text{Proj } R \text{ on } U_{i_1} \times \ldots \times U_{i_k} \overset{\Delta}{=}$

$$\int_{U_{i_1} \times \ldots \times U_{i_k}} \vee_{u_{j_1},\ldots,u_{j_\ell}} \mu_R(u_1,\ldots,u_n)/(u_1,\ldots,u_n)$$

where $(j_1,\ldots,j_\ell)$ is the sequence complementary to $(i_1,\ldots,i_k)$ (e.g., if $n = 6$ then $(1,3,6)$ is complementary to $(2,4,5)$), and $\vee_{u_{j_1},\ldots,u_{j_\ell}}$ denotes the supremum over $U_{j_1} \times \ldots \times U_{j_\ell}$.

If R is a fuzzy subset of $U_{i_1},\ldots,U_{i_k}$, then its cylindrical extension in $U_1 \times \ldots \times U_n$ is a fuzzy subset of $U_1 \times \ldots \times U_n$ defined by

(A59) $$\bar{R} = \int_{U_1 \times \ldots \times U_n} \mu_R(u_{i_1},\ldots,u_{i_k})/(u_1,\ldots,u_n)$$

In terms of their cylindrical extensions, the composition of two binary relations R and S (in $U_1 \times U_2$ and $U_2 \times U_3$, respectively) is expressed by

(A60) $$R \circ S = \text{Proj } \bar{R} \cap \bar{S} \text{ on } U_1 \times U_3$$

where $\bar{R}$ and $\bar{S}$ are the cylindrical extensions of R and S in $U_1 \times U_2 \times U_3$. Similarly, if R is a binary relation in $U_1 \times U_2$ and S is a unary relation in U_2, their composition is given by

(A61) $$R \circ S = \text{Proj } R \cap \bar{S} \text{ on } U_1 .$$

Example A62. Let R be defined by the right-hand member of (A57) and

(A63) $$S = 0.4a + b + 0.8d.$$

Then

(A64) $$\text{Proj } R \text{ on } U_1 \; (\overset{\Delta}{=} a + b + c + d) = 0.4a + 0.6b + 0.6d$$

and

(A65) $$R \circ S = 0.4a + 0.5b + 0.5d .$$

Linguistic Variables

Informally, a linguistic variable, χ , is a variable whose values are words or sentences in a natural or artificial language. For example, if age is interpreted as a linguistic variable, then its term-set, $T(\chi)$, that is, the set of linguistic values, might be

(A66) $$T(\text{age}) = \text{young} + \text{old} + \text{very young} + \text{not young} + \text{very old} + \text{very very young} + \text{rather young} + \text{more or less young} + \ldots$$

where each of the terms in T(age) is a label of a fuzzy subset of a universe of discourse, say $U = [0,100]$.

A linguistic variable is associated with two rules: (a) a syntactic rule, which defines the well-formed sentences in $T(\chi)$; and (b) a semantic rule, by which the meaning of the terms in $T(\chi)$ may be determined. If X is a term in $T(\chi)$, then its meaning (in a denotational sense) is a subset of U . A primary term in $T(\chi)$ is a term whose meaning is a primary fuzzy set, that is, a term whose meaning must be defined a priori, and which serves as a basis for the computation of the meaning of the non-primary terms in $T(\chi)$. For example, the primary terms in (A66) are young and old, whose meaning might be defined by their respective compatibility functions μ_{young} and μ_{old}. From these then, the meaning -- or, equivalently, the compatibility functions -- of the non-primary terms in (A66) may be computed by the application of a semantic rule. For example, employing (A54) and (A55) we have

(A67) $$\mu_{\text{very young}} = (\mu_{\text{young}})^2$$

(A68) $$\mu_{\text{more or less old}} = (\mu_{\text{old}})^{\frac{1}{2}}$$

(A69) $$\mu_{\text{not very young}} = 1 - (\mu_{\text{young}})^2 \ .$$

The Extension Principle

Let g be a mapping from U to V . Thus,

(A70) $$v = g(u)$$

where u and v are generic elements of U and V , respectively.

Let F be a fuzzy subset of U expressed as

(A71) $$F = \mu_1 u_1 + \dots + \mu_n u_n$$

or, more generally,

(A72) $$F = \int_U \mu_F(u)/u \ .$$

By the extension principle, the image of F under G is given by

(A73) $$g(F) = \mu_1 g(u_1) + \dots + \mu_n g(u_n)$$

or, more generally,

(A74) $$g(F) = \int_U \mu_F(u)/g(u) \ .$$

Similarly, if g is a mapping from $U \times V$ to W and F and G are fuzzy subsets of U and V , respectively, then

(A75) $$f(F,G) = \int_W (\mu_F(u) \wedge \mu_G(v))/g(u,v).$$

Example A76. Assume that g is the operation of squaring. Then, for the set defined by (A14), we have

(A77) $$g(0.3/0.5 + 0.6/0.7 + 0.8/0.9 + 1/1)$$

$$= 0.3/0.25 + 0.6/0.49 + 0.8/0.81 + 1/1 \ .$$

Similarly, for the binary operation $\vee$ ($\triangleq$ max), we have

(A78) $$(0.9/0.1 + 0.2/0.5 + 1/1) \vee (0.3/0.2 + 0.8/0.6)$$

$$= 0.3/0.2 + 0.2/0.5 + 0.8/1 + 0.8/0.6 + 0.2/0.6.$$

It should be noted that the operation of squaring in (A77) is different from that of (A51) and (A52).

References

1. L. A. Zadeh, Outline of a New Approach to the Analysis of Complex Systems and Decision Processes, IEEE Trans. on Systems, Man and Cybernetics SMC-3 (1973), 28-44.

2. L. A. Zadeh, The Concept of a Linguistic Variable and Its Application to Approximate Reasoning, Information Sciences, Part I, 8 (1975), 199-249; Part II, 8 (1975), 301-357; Part III, 9 (1975), 43-80.

3. L. A. Zadeh, Fuzzy Logic and Approximate Reasoning (In Memory of Grigore Moisil), Synthese 30 (1975), 407-428.

4. R. E. Bellman and L. A. Zadeh, Local and Fuzzy Logics, Memorandum No. ERL-M584, Electronics Research Laboratory, University of Califormia, Berkeley, 1976. (To appear in the Proceedings of the International Symposium on Multi-Valued Logic, University of Indiana, 1975).

5. U. Grenander, Pattern Synthesis: Lectures in Pattern Theory (vol. 1), Springer-Verlag, New York, 1976.

6. K. S. Fu, Syntactic Methods in Pattern Recognition, Academic Press, New York, 1974.

7. E. T. Lee and L. A. Zadeh, Note on Fuzzy Languages, Information Sciences 1 (1969), 421-434.

8. R. E. Bellman, R. Kalaba and L. A. Zadeh, Abstraction and Pattern Classification, J. Math. Anal. and Appl. 13 (1966), 1-7.

9. M. Minsky and S. Papert, Perceptrons, M.I.T. Press, Cambridge, 1969.

10. M. A. Aizerman, E. M. Braverman and L. I. Rozonoer, Method of Potential Functions in the Theory of Learning Machines, Science Press, Moscow, 1974.

11. A. G. Arkad'ev and E. M. Braverman, Computers and Pattern Recognition, Thompson Book Co., Washington, D. C., 1967.

12. V. H. Vapnik and A. Ya. Chervonenkis, Theory of Pattern Recognition, Science Press, Moscow. 1974.

13. L. A. Zadeh, Similarity Relations and Fuzzy Orderings, Information Sciences 3 (1971), 177-200.

14. L. A. Zadeh, Calculus of Fuzzy Restrictions, Proc. U. S.-Japan Seminar on Fuzzy Sets and Their Applications, L. A. Zadeh, K. S. Fu, K. Tanaka, M. Shimura (eds.), Academic Press, New York, 1975, 1-39.

15. R. F. Boyce, D. D. Chamberlin, W. F. King III and M. M. Hammer, Specifying Queries as Relational Expressions: SQUARE, IBM Research Report RJ129, 1973.

16. D. D. Chamberlin and R. F. Boyce, SEQUEL: A Structured English Query Language, Proc. ACM SIGFIDT Workshop on Data Description, Access and Control, 1975. (Also see IBM Research Report RJ1318, 1973.)

17. E. R. Ruspini, A New Approach to Clustering, Information and Control 15 (1969), 22-32.

18. E. R. Ruspini, Numerical Methods for Fuzzy Clustering, Information Sciences 2 (1970), 319-350.

19. E. R. Ruspini, New Experimental Results in Fuzzy Clustering, Information Sciences 6 (1973), 273-284.

20. J. C. Bezdek, Fuzzy Mathematics in Pattern Classification, Ph.D. Dissertation, Center for Applied Mathematics, Cornell University, Ithaca, New York, 1973.

21. J. C. Bezdek, Numerical Taxonomy with Fuzzy Sets, J. Math. Biology 1 (1974), 57-71.

22. J. C. Dunn, Some Recent Investigations of a Fuzzy Partitioning Algorithm and Its Application to Pattern Classification Problems, Center for Applied Mathematics, Cornell University, Ithaca, New York, 1974.

23. J. C. Dunn, A Fuzzy Relative of the ISODATA Process and Its Use in Detecting Compact Well Separated Clusters, J. of Cybernetics 3 (1974), 32-57.

24. J. C. Dunn, Well Separated Clusters and Optimal Fuzzy Partitions, Center for Applied Mathematics, Cornell University, Ithaca, New York, 1974.

25. J. C. Dunn, A Graph-Theoretic Analysis of Pattern Calssification via Tamura's Fuzzy Relation, IEEE Trans. on Systems, Man and Cybernetics SMC-4 (1974), 310-313.

26. J. C. Bezdek and J. C. Dunn, Optimal Fuzzy Partitions: A Heuristic for Estimating the Parameters in a Mixture of Normal Distributions, IEEE Trans. on Computers C-24 (1975), 835-838.

27. J. C. Bezdek, Cluster Validity with Fuzzy Sets, J. of Cybernetics 3 (1974), 58-73.

28. J. C. Bezdek and J. D. Harris, Convex Decompositions of Fuzzy Partitions, Departments of Mathematics, Utah State University, Logan, Utah and Marquette University, Milwaukee, Wisconsin, 1976.

29. J. C. Bezdek, A Physical Interpretation of Fuzzy ISODATA, IEEE Trans. on Systems, Man and Cybernetics SMC-6 (1976), 387-390.

30. J. C. Bezdek, Mathematical Models for Systematics and Taxonomy, Proc. 8th International Conference on Numerical Taxonomy, G. Estabrook (ed.) Freeman Co., San Francisco, 1975, 143-164.

31. J. C. Bezdek, Feature Selection for Binary Data: Medical Diagnosis with Fuzzy Sets, Proc. National Computer Conference, S. Winkler (ed.) AFIPS Press, Montvale, N.J., 1976, 1057-1058.

32. J. C. Bezdek and P. F. Castelaz, Prototype Classification and Feature Selection with Fuzzy Sets, Department of Mathematics Utah State University, Logan, Utah and Department of Electrical Engineering, Marquette University, Mulwaukee, Wisconsin, 1976.

33. S. C. Johnson, Hierarchical Clustering Schemes, Psychometrica 32 (1967), 241-254.

34. L. A. Zadeh, A Fuzzy-Set-Theoretic Interpretation of Linguistic Hedges, J. of Cybernetics 2 (1972), 4-34.

35. R. Sokal and P. Sneath, Principles of Numerical Taxonomy, Freeman Co., San Francisco, 1963.

36. Methodologies of Pattern Recognition, S. Watanabe (ed.), Academic Press, New York, 1969.

37. R. Jardine and R. Sibson, Mathematical Taxonomy, Wiley, New York, 1971.

38. K. Fukunaga, Introduction to Statistical Pattern Recognition, Academic Press, New York, 1972.

39. R. Duda and P. Hart, Pattern Classification and Scene Analysis, Wiley-Interscience, New York, 1973.

40. J. Tou and R. Gonzales, Pattern Recognition Principles, Addison-Wesley, Reading, Mass., 1974.

41. E. A. Patrick, Fundamentals of Pattern Recognition, Prentice-Hall, Englewood Cliffs, N.J., 1972.

42. L. M. Uhr, Pattern Recognition, Learning and Thought, Prentice-Hall, Englewood Cliffs, N.J., 1973.

43. W. Meisel, Computer-Oriented Approaches to Pattern Recognition, Academic Press, New York, 1972.

44. G. Nagy, State of the Art in Pattern Recognition, Proc. IEEE 56 (1969), 836-862.

45. Y. C. Ho and A. K. Agrawala, On Pattern Classification Algorithms: Introduction and Survey, Proc. IEEE 56 (1968), 2101-2114.

46. L. Kanal, Patterns in Pattern Recognition: 1968-1974, IEEE Trans. on Information Theory IT-20 (1974), 697-722.

47. A. Schroeder, Recognition of Components of a Mixture, Ph.D. Dissertation, University of Paris VI, 1974.

48. E. Diday, New Methods and New Concepts in Automatic Classification and Pattern Recognition, Ph. D. Dissertation, University of Paris VI, 1972.

49. V. I. Loginov, Probability Treatment of Zadeh's Membership Functions and Their Use in Pattern Recognition, Engineering Cybernetics (1966), 68-69.

50. C. L. Chang, Fuzzy Sets and Pattern Recognition, Ph. D. Dissertation, Department of Electrical Engineering, University of California, Berkeley, 1967.

51. W. G. Wee, On a Generalization of Adaptive Algorithms and Applications of the Fuzzy Set Concept to Pattern Calssification, Technical Report 67-7, Department of Electrical Engineering, 1967.

52. R. H. Flake and B. L. Turner, Numerical Classification for Taxonomic Problems, J. Theo. Biol. 20 (1968), 260-270.

53. A. N. Borisov and E. A. Kokle, Recognition of Fuzzy Patterns by Feature Analysis, Cybernetics and Diagnostics, No. 4, Riga, U.S.S.R., 1970.

54. I. Gitman and M. D. Levine, An Algorithm for Detecting Unimodal Fuzzy Sets and Its Applications as a Clustering Technique, IEEE Trans. on Computers C-19 (1970), 583-593.

55. S. Otsuki, A Model for Learning and Recognizing Machine, Information Processing 11 (1970), 664-671.

56. S. K. Chang, Automated Interpretation and Editing for Fuzzy Line Drawings, Proc. Spring Joint Conf. (1971), 393-399.

57. S. Tamura, S. Niguchi and K. Tanaka, Pattern Classification Based on Fuzzy Relations, IEEE Trans. on Systems, Man and Cybernetics SMC-1 (1971), 937-944.

58. S. Tamura, Fuzzy Pattern Classification, Proc. Symp. on Fuzziness in Systems and Its Processing (1971), Tokyo.

59. E. T. Lee, Proximity Measures for the Classification of Geometric Figures, J. of Cybernetics 2 (1972), 43-59.

60. M. Shimura, Application of Fuzzy Functions to Pattern Calssification, Trans. IECE 55-d (1972), 218-225.

61. M. Sugeno, Evaluation of Similarity of Patterns by Fuzzy Integrals, Ann. Conf. Records of SICE, Tokyo, 1972.

62. K, Kotoh and K. Hiramatsu, A Representation of Pattern Classes Using the Fuzzy Sets, Trans. IECE 56-d (1973), 275-282.

63. L. E. Larsen, E. Ruspini, J. J. McNew, D.O. Walter and W. R. Adey, A Test of Sleep Staging Systems in the Unrestrained Chimpanzee, Brain Research 40 (1972), 319-343.

64. P. Siy, Fuzzy Logic and Hard-Written Character Recognition, Ph.D. Dissertation, Department of Electrical Engineering, University of Akron, Ohio, 1972.

65. R. K. Ragade, A Multiattribute Perception and Classification of Visual Similarities, Systems Res. and Planning Papers, S-001-73, Bell Northern Research, Canada, 1973.

66. C. C. Negoita, On the Application of the Fuzzy Sets Separation Theorem for Automatic Classification in Information Retrieval Systems, Information Sciences 5 (1973), 279-286.

67. M. Sugeno, Constructing Fuzzy Measure and Grading Similarity of Patterns by Fuzzy Integrals, Trans. SICE 9 (1973), 359-367.

68. Y. Noguchi, Pattern Recognition Systems Based on the Feature Extraction Technique, Report No. 739, Electrotechnical Laboratory, Tokyo, 1973.

69. B. Conche, A Method of Classification Based on the Use of a Fuzzy Automaton, University of Paris - Dauphine, 1973.

70. N. Malvache, Analysis and Identification of Visual Systems in Humans, Ph.D. Dissertation, University of Lille, Lille, 1973.

71. N. Okada and T. Tamachi, Automated Editing of Fuzzy Line Drawings for Picture Description, Trans. IECE 57-a (1974), 216-223.

72. M. Woodbury and J. Clive, Clinical Pure Types vs. A Fuzzy Partition, J. of Cybernetics 4 (1974), 111-120.

73. P. Siy and C. S. Cheu, Fuzzy Logic for Handwritten Numerical Character Recognition, IEEE Trans. on Systems, Man and Cybernetics SMC-4 (1974), 570-575.

74. M. G. Thomason, Finite Fuzzy Automata, Regular Fuzzy Languages, and Pattern Recognition, Department of Electrical Engineering, Duke University, Durham, N.C., 1974.

75. A. Kaufmann, Introduction to the Theory of Fuzzy Subsets (vol. III): Applications to Classification, Pattern Recognition, Automata and Systems, Masson, Paris, 1975.

76. C. V. Negoita and D. A. Ralescu, Theory of Fuzzy Sets and Its Applications, Wiley, New York, 1975.

77. T. Pavlidis, Fuzzy Representations as Means of Overcoming the Over-Commitment of Segmentation, Proc. Conf. On Computer Graphics, Pattern Recognition and Data Structures, Los Angeles, Calif., 1975.

78. A. Rosenfeld, Fuzzy Graphs, Proc. U.S.-Japan Seminar on Fuzzy Sets and Their Applications, L. A. Zadeh, K. S. Fu, K. Tanaka and M. Shimura (eds.), Academic Press, New York, 1975, 77-95.

79. E. T. Lee, Shape-Oriented Chromosome Identification, IEEE Trans. on Systems, Man and Cybernetics SMC-5 (1975) 629-632.

80. R. T. Yeh and S. Y. Band, Fuzzy Relations, Fuzzy Graphs, and Their Applications to Clustering Analysis, Proc. U.S.-Japan Seminar on Fuzzy Sets and Their Applications, L. A. Zadeh, K. S. Fu, K. Tanaka and M. Shimura (eds.), Academic Press, New York, 1975, 125-149.

81. G. F. DePalma and S. S. Yau, Fractionally Fuzzy Grammars with Application to Pattern Recognition, Proc. U.S.-Japan Seminar on Fuzzy Sets and Their Applications, L. A. Zadeh, K. S. Fu, K. Tanaka and M. Shimura (eds.), Academic Press, New York, 1975, 329-351.

82. M. Shimura, An Approach to Pattern Recognition and Associative Memories Using Fuzzy Logic, Proc. U.S.-Japan Seminar on Fuzzy Sets and Their Applications, L. A. Zadeh, K. S. Fu, K. Tanaka and M. Shimura (eds.), Academic Press, New York, 1975, 449-476.

83. R. L. Chang and T. Pavlidis, Fuzzy Decision Trees, Technical Report No. 203, Department of Electrical Engineering and Computer Science, Princeton University, Princeton, 1976.

84. S. Sugeno, Fuzzy Systems and Pattern Recognition, Workshop on Discrete Systems and Fuzzy Reasoning, Queen Mary College, University of London, 1976.

85. H. Bremermann, Pattern Recognition by Deformable Prototypes, in Structural Stability, the Theory of Catastrophes and Applications in the Sciences, Springer Notes in Mathematics 25 (1976), 15-57.

86. L. A. Zadeh, A Fuzzy-Algorithmic Approach to the Definition of Complex or Imprecise Concepts, Int. Jour. Man-Machine Studies 8 (1976), 249-291.

Research Supported by the U. S. Army Research Office Contract DAHCO4-75-G0056.

Computer Science Division,
Department of Electrical Engineering and
Computer Sciences and the
Electronics Research Laboratory,
University of California,
Berkeley, California 94720.

Discrimination, Allocatory and Separatory, Linear Aspects

Seymour Geisser

1. Introduction.

The classification of objects is one of the hoariest and consequently not the least primitive of scientific enterprises - certainly considerably removed from preciser mechanisms directed towards the explanation and prediction of natural and social phenomena. Briefly, it attemps to sort out in some sensible manner objects belonging to two or more labeled classes. When this involves a parsimonious and efficient criterion of choice based on related manifest attributes, we are in the realm of Discrimination.

An early recorded instance appears in the biblical book, Judges, XII, 5-6. A clan of Israelites from Gilead held the fords over the Jordan to prevent the defeated troops of Ephraim, another Israelite tribe, from crossing the river. The Ephraimites sharing race, language, customs and dress were apparently indistinguishable in all respects from the Gileadites. Seizing upon a dialectical variation as an efficient sorting device, the guards made those attempting the ford pronounce the word "Shibboleth". Upon hearing "Sibboleth" they were fairly certain of apprehending an Ephraimite.

It is quite likely that their errors of classification were no greater than many of our current weather "classifiers" aided by a modern computer, who base forecasts of snow on a large number of precisely determined variables. This is of course a situation where the label has in fact not yet occurred but is predictive as opposed to the previous retrodictive case.

Often in the latter case the latent label of a new object can only be ascertained with certainty by prodigeous technical effort which may even involve the destruction or alteration of the object rendering it useless for further inquiry. Other cases may require an inordinate amount of time and patience until the label eventually reveals itself. Hence the utilization of easily assessed related attributes may be of invaluable aid in a study if only for reasons of economics and prudence.

There is also a natural hierarchy in terms of how these problems can be organized. In the least informative situations, the number of classes as well as the labels are unknown, and it is hoped that clues to both these entities will be disclosed by some set of appropriate manifest attributes. Here the basic problem is determining the number of classes and of forming clusters. In more informative cases the number of classes or populations is known or specified. Further knowledge is often also presumed concerning certain aspects of the attribute distributions.

For the sake of clarity we set down the general problem as follows: There are populations (or patterns) π_j, $j = 1, \ldots, r$, with r known or unknown and π_j possibly specified by moments or by a distribution function $F_j(\cdot \mid \theta_j)$, whose form may be known or unknown, and θ_j is the jth set of known or unknown parameters. There may be certain relationships among the π_j, as well as subpopulations π_{ji}. Further there are two sets of observations, the first denoted by X and the second by U, (either set may be empty). Each of the observations belonging to X is such that its population origin or label is known with certitude, but the labels of those belonging to U are not. These may have some prior probabilities attached to them before they are observed and one object of the endeavor is to determine their origin in some optimal manner. Here allocation is the goal. A second goal, which may be primary in certain studies, is basically descriptive (graphical, algebraic or some other qualitative form), and involves initially the disclosure of the manifest differential features of the patterns, populations or potential populations under scrutiny. The purposes of the first are action oriented, predictive or retrodictive while the latter is more in the realm of the speculative in terms of possibly throwing some light on scientific or social issues.

In the first case one attempts to derive some rule which optimally allocates new observations while in the second instance one tends to focus on functions (discriminants) which maximally distinguish or separate the populations. An appropriate allocatory procedure requires prior probabilities of an observation belonging to one or another population or estimates thereof. Often they are not obtainable and one tacitly assumes that these prior probabilities are equal. In many cases this is tantamount to using a separatory function as an allocator and the two original distinct goals tend to fuse or become blurred. Allocatory optimality is basically definable only when stringent assumptions are met while in vague situations a separatory function may sometimes usefully serve as an allocator. Conversely, allocatory notions may also be used to define a separatory function.

Discrimination, in its modern guise, was founded by R. A. Fisher (1936). He derived those linear functions of the class of all linear functions that best separated populations (actually samples) in terms of maximizing a certain distance function depending on only the first two moments.

Since then, linear discriminants have played an important role in the theory. From other points of view it was also found that linear theory was preeminent in one of the most useful of distributions, the multivariate normal, Wald (1944), Welch (1939).

In this paper we shall present not only an exposition of linear discrimination but shall also attempt to give a coherent discussion of its twin goals - allocation and separation.

In the next few sections we review linearity in the multivariate normal case, discuss the extent to which linearity is optimal and indicate the actual use of linear discriminants. This is followed by a section in which the distributional assumptions are dropped and the thrust is on the separation of populations via linear functions. An incidental feature is that some of the basic results are derived algebraically in a manner which differs from customary derivations. The penultimate section indicates some possible applications of sample reuse procedures to linear discriminants.

2. Multivariate Normal Case -- Allocation and Reduction.

Suppose thare are p-dimensional multivariate populations $\pi_1, \ldots, \pi_r$ with vector means $\mu_1, \ldots, \mu_r$ and common positive definite covariance matrix Σ. One is interested in allocating a new p-dimensional observation u to one of these various populations in some optimal fashion. Assuming u has prior probability q_i of belonging to π_i, $\sum_{i=1}^{r} q_i = 1$, then the optimal method for multivariate normal populations with regard to total posterior probability of correct classification (PCC), c.f. Anderson (1958) is to allocate u to that π_i for which

$$w_i(p) = \log q_i - \tfrac{1}{2} D_1^2(p), \quad i = 1, \ldots, r \tag{2.1}$$

is a maximum where

$$D_i^2(p) = (u - \mu_i)' \Sigma^{-1} (u - \mu_i), \tag{2.2}$$

the Mahalanobis distance. This is the solution which allocates u to that π_i which has maximum posterior probability since $w_i(p)$ is easily shown to be a monotone function of $\Pr[\pi_i | u]$, the posterior probability that u is from π_i.

It is sometimes of interest to determine whether we can transform linearly the set of p variables into $k \leq p$ variables and preserve the allocation in k dimensions. Let $y = Cu$, $\eta_i = C\mu_i$, $\Omega = C\Sigma C'$, for C, a $k \times p$ matrix of rank $k \leq p$, and

$$D_i^2(k) = (y - \eta_i)' \Omega^{-1} (y - \eta_i) = (u - \mu_i)' C' (C\Sigma C')^{-1} C (u - \mu_i), \tag{2.3}$$

the corresponding distance in k dimensions.

Assume $\beta = \sum_{i=1}^{r} (\mu_i - \bar{\mu})(\mu_i - \bar{\mu})'$ where $\bar{\mu} = r^{-1}\sum_{i=1}^{r} \mu_i$ and β is of rank $r - v \leq p$, noting that when $\mu_1, \ldots, \mu_r$ are linearly independent $v = 1$. Since β is a p.s.d. matrix there exists a Λ such that, $\beta = \Lambda\Lambda'$, where Λ is $p \times r - v$. If we let $C = P'\Lambda'\Sigma^{-1}$ where $k = r - v$ and P, the $r\text{-}v \times r - v$ orthogonal matrix such that $P'\Lambda'\Sigma^{-1}\Lambda P$ is $\text{diag}(\delta_1, \ldots, \delta_{r-v})$ where δ_j are the non-zero roots of $\Sigma^{-1}\beta$ in descending order, then

$$(2.4) \qquad D_i^2(k) = (u - \mu_i)'\Sigma^{-1}\Lambda[\Lambda'\Sigma^{-1}\Lambda]^{-1}\Lambda'\Sigma^{-1}(u - \mu_i) .$$

Further by adding and subtracting $\bar{\mu}$ in $u - \mu_i$ and noting that $\mu_i - \bar{\mu}$ is in the vector space generated by $\Lambda\Lambda'$ it is easily shown that for all i

$$(2.5) \quad w_i(p) - w_i(k) = D_i^2(p) - D_i^2(k) = (u - \bar{\mu})'[\Sigma^{-1} - \Sigma^{-1}\Lambda[\Lambda'\Sigma^{-1}\Lambda]^{-1}\Lambda'\Sigma^{-1}](u - \bar{\mu})$$

and hence is independent of i. Therefore allocation of y by means of the maximum $w_i(k)$ is equivalent to the original allocation of u, thus verifying that $C = P'\Lambda'\Sigma^{-1}$ is a solution that preserves the original allocation. The new set of coordinates y are referred to as the complete set of linear multiple discriminants and they contain all of the discriminatory power of the original set of coordinates. The set y is an orthogonal set and forms a basis for all other solutions $y^* = Ry$ where R is any real nonsingular $k \times k$ matrix. On the other hand if $k < r - v$, the allocation by y, the transform of u, will not be the same as the allocation by u for all u, as can easily be verified.

The total probability that u will be correctly allocated by the procedure is

$$(2.6) \qquad q(p) = \sum_{i=1}^{r} q_i \Pr[u \in R_i | \pi_i]$$

where R_i is the region given by those u satisfying $\underset{j}{\text{Max}}\, w_j(p) = w_i(p)$, and is a maximum with respect to all possible procedures. If $y = P'\Lambda'\Sigma^{-1} u$ then it is clear that

$$(2.7) \qquad q(p) = q(r\text{-}v) = \sum_{i=1}^{r} q_i \Pr[y \in R_i^* \mid \pi_i]$$

where R_i^* is given by those y satisfying $\underset{j}{\text{Max}}\ w_j(r-v) = w_i(r-v)$. To simplify matters let $q_i = r^{-1}$ for all i so that

$$q(p) = q(r-v) = r^{-1}\sum_{i=1}^{r} \Pr[y \in R_i^* \mid \pi_i] \tag{2.8}$$

then R_i and R_i^* are given by u and y which minimize $D_i^2(p)$ and $D_i^2(r-v)$ respectively. When $k < r - v$ and we use the procedure, i.e., minimizing $D_i^2(k)$ of (2.3) where $y = Cu$, then $q(k) < q(r-v)$ by continuity arguments. On the other hand it might be conjectured that the best one can do with respect to maximizing $q(k)$ is to let $C = P'_{(k)}\Lambda'\Sigma^{-1}$ where $P_{(k)} = (P_1, \ldots, P_k)$ is the matrix of the first k columns of P, i.e., P_i is the invariant vector associated with the ith largest root of $\Lambda'\Sigma^{-1}\Lambda$, or equivalently ΛP_i is the invariant vector associated with the identical root of $\beta\Sigma^{-1}$. This conjecture is in general false whenever $r - v \geq 2$ if we wish to maximize $q(k)$, as almost any precisely calculated example will show. But from another point of view, i.e., optimizing on separatory criteria which we shall discuss in Section 5, it can be best.

A further note of caution should be introduced to the effect that the PCC is only of value in assessing the discriminatory power of the manifest variables at hand prior to the observation of u. Once a set of such variables is determined and a particular u observed, the only relevant factor is the posterior probability, when calculable, that u belongs to one or another of $\pi_1, \ldots, \pi_r$,

$$\Pr[\pi_j \mid u] = q_j f_j(u) \Big/ \sum_{i=1}^{r} q_i f_i(u) \tag{2.9}$$

where $f_j(\cdot)$ represents in general the probability function associated with π_j.

We now present a numerical demonstration of the falsity of the conjecture. Suppose we ask for a single linear combination that will maximize the PCC assuming the r populations all have equal prior probability r^{-1}, blurring the distinction between allocation and separation. Then $c'u$, under π_j is univariate normal with mean $c'\mu_j$ and variance

$c'\Sigma c$. Then $z = c'u/\sqrt{c'\Sigma c}$ is, under π_j, $N(\eta_j, 1)$ where $\eta_j = c'\mu_j/\sqrt{c'\Sigma c}$. Hence we can calculate the maximal probability of correct classification for any c such that

$$\text{PCC} = 2r^{-1} \sum_{i=1}^{r-1} \Phi\left(\frac{\eta_{(i)} - \eta_{(i+1)}}{2}\right) + (2 - r)r^{-1} \tag{2.10}$$

where Φ is the distribution function of a standardized normal variate and $\eta_{(i)}$ are the ordered values of η_i such that $\eta_{(1)} \geq \eta_{(2)} \geq \cdots \geq \eta_{(r)}$. Maximization of the PCC with respect to c is troublesome, but it can be shown that the c that maximizes PCC is not necessarily the vector associated with the largest root of $\beta\Sigma^{-1}$ as one might initially suspect. Such a suspicion of course would arise from the fact that this vector does maximize the variation amongst the η_i. While this variation is contributory, the PCC is also quite sensitive to the spacing amongst the $\eta_{(i)}$.

The following example demonstrates these facts: Let u be a 3×1 vector with means under π_1, π_2, and π_3, respectively

$$\mu_1' = (1, 0, 0), \quad \mu_2' = (0, 1, -1), \quad \mu_3' = (-1, -1, 1) \quad \text{and} \quad \Sigma = I .$$

Then $\beta\Sigma^{-1} = \beta$ and

$$\beta = \sum_{j=1}^{3} (\mu_j - \bar{\mu})(\mu_j - \bar{\mu})' = \begin{pmatrix} 2 & 1 & -1 \\ 1 & 2 & -2 \\ -1 & -2 & 2 \end{pmatrix}$$

with characteristic roots $3 + \sqrt{3}$, 0, $3 - \sqrt{3}$. The normed vector associated with the largest root is $c_1' = (6 - 2\sqrt{3})^{-\frac{1}{2}}(\sqrt{3} - 1, 1, -1)$. Using $c_1'u$ we find that $(\eta_{(1)}, \eta_{(2)}, \eta_{(3)}) = (6 - 2\sqrt{3})^{-\frac{1}{2}}(2, \sqrt{3} - 1, -1 - \sqrt{3})$ and compute the PCC to be .67757. On the other hand the simple normed vector $c_0' = (0, -\frac{1}{\sqrt{2}}, \frac{1}{\sqrt{2}})$ yields $(\eta_{(1)}, \eta_{(2)}, \eta_{(3)}) = (\sqrt{2}, 0, -\sqrt{2})$, equally spaced, and results in a PCC of .68033, which is just a trifle larger than that attained by the vector associated with the maximum root of β. In actual fact the normed vector $c^* = (.173, .697, -.697)$ leads to $(\eta_{(1)}, \eta_{(2)}, \eta_{(3)}) = (1.394, .173, -1.567)$ and yields a PCC of .69139 which is the maximum attainable here for a single linear combination.

It is well known that the dispersion $\sum_{i=1}^{r}(\eta_i - \bar{\eta})^2$ attains its maximum, $3 + \sqrt{3} = 4.732$ in this case, when the vector associated with the largest root is utilized, while the same measure of dispersion for the vector c_0 is 4 - considerably less, and for the vector c^* we obtain 4.42.

Another way of viewing this problem is to realize that we are basically maximizing two quite different functions of the ordered values of $\eta_j = c'\mu_j/\sqrt{c'\Sigma c}$, $j = 1,\ldots,r$ with respect to the arbitrary vector c . One function is given by (2.10) while the other is

$$\sum_{j=1}^{r}(\eta_{(j)} - \bar{\eta})^2 . \tag{2.11}$$

That the characteristic vector associated with the largest root maximizes (2.11) results from the fact that (2.11) is invariant with regard to the ordering of the η_j so that from the definition of η_j we obtain

$$\sum_{j=1}^{r}(\eta_{(j)} - \bar{\eta})^2 = \sum_{j=1}^{r}(\eta_j - \bar{\eta})^2 = \frac{c'\beta c}{c'\Sigma c} . \tag{2.12}$$

As is well known, the quantity on the right of (2.12) is maximized when c is set equal to the characteristic vector associated with the largest root of $\beta\Sigma^{-1}$. Hence there is really no reason to expect the same solution for both cases.

The general problem of maximizing PCC with arbitrary q_i and k does not seem susceptible to explicit solution nor even be numerically tractable except under rather special circumstances. Some results and problems in this regard are discussed in a paper by Guseman, Peters and Walker (1975). There are also some interesting mathematical questions that require investigation in terms of the amount of loss incurred in PCC when one uses the vectors associated with the invariant roots. It is anticipated that this loss decreases as k increases and for any given k depends heavily on the closeness of the invariant roots.

We note that when $r = 2$, the optimal allocatory procedure yields the single linear discriminant

$$(2.13) \qquad U^* = (u - \tfrac{1}{2}(\mu_1 + \mu_2))' \Sigma^{-1}(\mu_1 - \mu_2) + \log \frac{q_1}{q_2}$$

such that $U^* > 0$ assigns u to π_1 and $U^* \leq 0$ assigns u to π_2. Insertion of the usual estimates for μ_1, μ_2 and Σ when they are unknown and estimable from data yields the plug-in rule

$$V^* = (u - \tfrac{1}{2}(\hat{\mu}_1 + \hat{\mu}_2)' \hat{\Sigma}^{-1}(\hat{\mu}_1 - \hat{\mu}_2) + \log \frac{q_1}{q_2}$$

with $V^* > 0$ assigning u to π_1 and to π_2 otherwise.

It will be shown, however, that from certain points of view, even in this most structured of cases, linear theory, strictly speaking, may be inappropriate though approximately correct and certainly convenient.

3. The Limits of Linear Theory - Allocatory Aspects.

For the remainder of our discussion we shall restrict ourselves to the two population cases: π_1 and π_2 with density function $f(\cdot\,|\,\theta_1)$ and $f(\cdot\,|\,\theta_2)$ respectively. Optimal allocation for a new observation u with regard to the PCC involves assigning

$$(3.1) \qquad \begin{cases} u \text{ to } \pi_1 & \text{if } \rho(\theta,u) = \dfrac{q_1 f(u\,|\,\theta_1)}{q_2 f(u\,|\,\theta_2)} > 1 \\ u \text{ to } \pi_2 & \text{otherwise} \end{cases}$$

where $\theta = \theta_1 \cup \theta_2$ is the entire set of distinct parameters of the problem. Equivalently any monotonic increasing function of ρ, say $h(\rho)$ for every fixed u will also do, so that any $h(\rho)$ may be denoted as an allocatory population discriminant. "Linear" theory is then surely optimal whenever there exists an $h(\rho)$ which is linear in u, although there are other cases as well, The multivariate normal distribution with equal covariance matrices is an example of the logistic class which always yields a linear population discriminant because of the form of

$$(3.2) \qquad \rho(\theta,u) = e^{\alpha_0 + \alpha' u}$$

where $\alpha' = (\alpha_1,\ldots,\alpha_p)$ and consequently $\log \rho(\theta,u)$ is linear in u.

J. A. Anderson [1973] points out that multivariate independent dichotomous variables as well as several other interesting cases also belong to the logistic class. In fact this type of linearity remains valid for a special case of the general exponential family where θ_i is the set of parameters (β_i, τ) and

$$f(u \mid \theta_i) = g(\beta_i, \tau) h(u, \tau) e^{\beta_i' u} \tag{3.3}$$

where β_i is a p-dimensional vector and τ is a set of extraneous parameters. But there are also other possibilities for linearity, e.g., two multivariate "student" distributions that differ in their location but have the same covariance matrices and equal prior probabilities Geisser (1966]. Here the rule (3.1) is equivalent to a rule linear in u derived from a positive root of $\rho(\theta, u)$. The rules conform exactly even though the positive root of $\rho(\theta, u)$ is nonlinear. For a slightly wider class of which the above is a special case see Enis and Geisser [1974]. Exact linear theory is then only strictly appropriate for restricted sets of distributional assumptions though somewhat wider than the logistic family. However, it is generally hoped that it will give reasonably robust, if less than optimal, solutions to many other cases. There are situations, however, where it certainly should not be applied, e.g., where two normal populations have the same mean but differ in their covariance matrices. Here linear discriminants will be quite inappropriate. This model reflects to a degree the situation arising in discriminating between fraternal and nonfraternal twins, see e.g., Richter and Geisser [1960], Okamoto [1961], Geisser and Desu [1968], Desu and Geisser [1973], Geisser [1973a].

However, except for special situations as just described, it is usually assumed or piously hoped that linearity will be at least a not unreasonable first approximation. By this is implied that the rule (3.1) can be replaced by a rule linear in u without great loss. For a contrary view in taxonomy see Reyment [1973]. In the classical frequential paradigm often an estimate of α is plugged into (3.2) while α_0 is resolved into its constituent sum $\log q_1 q_2^{-1} + \alpha_0^*$ with an estimate for α_0^* plugged in and $\log q_1 q_2^{-1}$ assumed to be particular value, often 0 for convenience resulting in a discriminatory blur. Sometimes $q_1 q_2^{-1}$ is derived from a model, Geisser [1973a], or estimated from previous data or from the data at hand, when the situation permits Geisser [1964]. Now when α_0 and α are known any $h(\rho)$, of course, will do as well. However, depending on which $h(\rho)$ and what is used for its estimation when the parameter values are only estimable from data, the sample discriminant or rule for allocation will in general vary. One way around this is to use maximum likelihood or any other estimator which will preserve the invariance of the rule. For a discussion of some of these and related points see Geisser [1969, 1970] and Desu and Geisser [1973]. To do otherwise requires that the statistician decide on whether the rule is paramount or the estimation of a particular discriminatory function $h(\rho)$ is crucial. Of course for large samples the discrepancy may be quite neglible.

However, as was noted, the logistic model itself encompasses a variety of possible distributional assumptions. While presumably robust for its class when its parameters are estimated it is not expected to yield as efficient a procedure when compared to one that is based on the true member of the class. For a logistic and normal comparison see Efron [1975].

Another classical approach, Wald [1944], Anderson [1958, 141-2], is via the testing of hypotheses. Here one computes the likelihood ratio test of the hypothesis that the new observation belongs to either of the two populations under scrutiny. More specifically if X_i is the set of observations known to be from π_i, then

$$\lambda = \frac{\underset{\theta}{\text{Max}}\ f(X_1|\theta_1)\, f(X_2|\theta_2)\, f(u|\theta_1)}{\underset{\theta}{\text{Max}}\ f(X_1|\theta_1)\, f(X_2|\theta_2)\, f(u|\theta_2)} \tag{3.4}$$

and u is assigned to π_1 if $\lambda > \dfrac{q_2}{q_1}$, π_2 otherwise. Hence $q_1 q_2^{-1}\lambda$ may be termed the likelihood ratio allocatory discriminant.

For the multivariate normal case with equal covariance matrices,

$$\lambda = \left[\frac{1 + N_2\nu^{-1}(N_2+1)^{-1}(u-\bar{x}_2)'S^{-1}(u-\bar{x}_2)}{1 + N_1\nu^{-1}(N_1+1)^{-1}(u-\bar{x}_1)'S^{-1}(u-\bar{x}_1)}\right]^{(\nu+3)/2} \tag{3.5}$$

where $\bar{x}_i$ is the sample mean of N_i independent observations represented by X_i and known to have originated from π_i and S is the usual unbiased estimate of Σ with $\nu = N_1 + N_2 - 2$ degrees of freedom, $\nu > p$.

It is interesting to note that it is no longer necessarily possible to recover a linear discriminant from this procedure except under the rather restrictive assumption that $q_1^{2/(\nu+3)}N_2(N_2+1)^{-1} = q_2^{2/(\nu+3)}N_1(N_1+1)^{-1}$. Of course satisfaction is guaranteed if both $q_1 = q_2$ and $N_1 = N_2$. Although this may be disconcerting, it is not surprising as the thrust here is essentially on a rule (or test) rather than on the estimation of a true underlying linear population discriminant. Although the likelihood ratio discriminant for this paradigm is equivalent to a rule based on a quadratic discriminant it approaches linearity for large N_1 and N_2 so that for large enough samples there will be virtually little difference between it and the "usual" plug-in estimate (rule)

$$V^* = [u - \tfrac{1}{2}(\bar{x}_1 + \bar{x}_2)]'\, S^{-1}(\bar{x}_1 - \bar{x}_2) + \log \frac{q_1}{q_2} \tag{3.6}$$

for the true population discriminant (rule)

$$U^* = [u - \tfrac{1}{2}(\mu_1 + \mu_2)]\, \Sigma^{-1}(\mu_1 - \mu_2) + \log \frac{q_1}{q_2}\ . \tag{3.7}$$

The rule indicated by (3.5) was shown to be an admissible Bayes rule by Kiefer and Schwartz [1965] and also Das Gupta (1965] for this allocation problem. However, the proper prior distribution which is utilized to prove admissibility is one that most Bayesians would consider grossly deficient in that it depends on the sum of the sample sizes and only assigns non-zero density to functions of Σ, μ_1, μ_2 in a space restricted to p-dimensions whereas the set (Σ, μ_1, μ_2) ostensibly contains $(1/2)p(p+5)$ parameters. This does not say too much for the Bayes admissible character of the rule. Whether a proper prior can be obtained which does not have these drawbacks is an open question but it is not likely.

Another Bayesian derivation given by Geisser [1964] uses the simple improper prior density

$$g(\Sigma^{-1}, \mu_1, \mu_2) \propto |\Sigma|^{\frac{p+1}{2}}, \tag{3.8}$$

and also results in a quadratic rule in general. Hence V^* is not recoverable for arbitrary values of N_1, N_2, p, q_1 and q_2, Geisser [1966], but it is recoverable except for an additive constant depending on a particular relationship existing among these values, Enis and Geisser [1974]. It is only fully recoverable for the special case $N_1 = N_2$ and $q_1 = q_2$. Hence on a strictly allocatory basis the linear discriminant V has not been found to be admissible.

A semi-Bayesian justification, Geisser [1967], based on the aforementioned improper prior, focuses on the Bayesian estimation of U^*, rather than on allocation. This approach yields for the posterior expectation of U^*

$$E(U^*|u) = V^* + \frac{p}{2}(N_2^{-1} - N_1^{-1}) \tag{3.9}$$

which all but recovers the linear rule (5.6) and completely so whenever $N_1 = N_2$. Elaborations of the use of this method are presented by Enis and Geisser [1970]. Another Bayesian approach Enis and Geisser [1974], which stresses linearity also yields results close to the rule V^*. Here one determines that linear function which maximizes the PCC with respect to the predictive distribution of the observation to be classified. Here an allocatory notion is utilized as the separatory criterion with discriminants restricted to a linear class. This attempts to "optimally" compromise the allocatory needs with the desirability of linearity.

In both normal and non-normal applications, V^* is often utilized although it is not clear whether this emanates from the fact that the normal population discriminant U^* is linear and both allocatory and separatory and V^* is a good estimate of U^*, or that V^* for $q_1 = q_2$ can be derived as the "best" separatory linear discriminant in a distribution-free setting utilizing the sample, Fisher [1936]. Basically it appears that for many less sophisticated users of the technique it is both the simplicity of linearity combined with the authority of Fisher that is compelling. At any rate, there seems to be a bias in applications (as well as theory) for focusing on linear discriminants rather than a quest for overall optimal allocation irrespective of the goal. One has only to peruse the discriminatory literature to observe that almost all applications are linear and much theory devoted to the "improvement" of linear estimates of linear discriminants. We also note that even for the particular normal distribution setup discussed here there has as yet not been any completely frequentist rule that guarantees optimal allocation when the parameters are unknown nor a Bayesian rule which yields V^* for all values of q_1, q_2, N_1 and N_2. On the other hand, when allocation is actually not the goal, lineaity may be inherently more useful (certainly descriptively) because of its simplicity in discussing certain issues, and in the normal case both frequentist and Bayesian estimation procedures will yield linear sample discriminants.

4. Using Linear Discriminants- Normal Case.

As in the previous section let X_i, $i = 1,2$ represent a set of N_i observations known to be from π_i, a $N(\mu_i, \Sigma)$ population. The object is to optimally allocate a new observation u which has prior probability q_i of being from π_i. We then assign a prior density $g(\mu_1, \mu_2, \Sigma)$ to the unknown set of parameters. Hence

$$R = \frac{\Pr[\pi_1 | u]}{\Pr[\pi_2 | u]} = \frac{q_1 f(u | X, \pi_1]}{q_2 f(u | X, \pi_2]} \tag{4.1}$$

where $X = (X_1, X_2)$ and

$$f(u | X\ \pi_i) = \int f(u | \mu_i, \Sigma)\, p\,(\mu_1, \mu_2, \Sigma | X) d\mu_1 d\mu_2 d\,\Sigma\ , \tag{4.2}$$

the predictive density of a future observation where

$$p(\mu_1, \mu_2, \Sigma | X) \propto L(\mu_1, \mu_2, \Sigma | X)\ g(\mu_1, \mu_2, \Sigma)\ . \tag{4.3}$$

This then provides the solution for the allocation of the next observation and can be used on all further observations. This latter use is frequently not optimal as the predictive distribution very often indicates the dependence of the new observations, even though they are independent when conditioned on the set of parameters, and here it would be utilized as if they were independent (for the optimal solution see Geisser (1966)). At any rate the solution is optimal for the next observation u. Only modest loss is to be expected by using this marginal allocation scheme rather than the joint or sequential optimal scheme. However one is in quandary as how to calculate a joint prior distribution for μ_1, μ_2 and Σ that realistically reflects prior knowledge one may have about them. One way out of this dilemma is to use the improper prior $g(\mu_1, \mu_2, \Sigma^{-1}) \alpha |\Sigma|^{\frac{p+1}{2}}$ which tends to minimize the effect of the prior distribution. The results for this case were given by Geisser (1964) and yields for the posterior probability ratio

$$R = \tag{4.4}$$

$$\frac{q_1}{q_2}\left(\frac{N_1}{N_2}\right)^{\frac{p}{2}}\left(\frac{N_1+1}{N_2+1}\right)^{\frac{N_1+N_2-1-p}{2}}\left[\frac{(N_2+1)(N_1+N_2-2)+N_2(u-\bar{x}_2)'S^{-1}(u-\bar{x}_2)}{(N_1+1)(N_1+N_2-2)+N_1(u-\bar{x}_1)'S^{-1}(u-\bar{x}_1)}\right]^{\frac{N_1+N_2-1}{2}}$$

so that when $R > 1$ assign u to π_1 and to π_2 otherwise. This rule is in general quadratic and is linear only for very special cases among which is $N_1 = N_2$ and $q_1 = q_2$, but tends to linearity as the sample sizes increase.

All evidence to date indicates that this procedure is superior for allocation to the plug-in rule V^* of (3.6). In this regard admissibility was previously discussed. From the point of view of density estimation the predictive density, as generally suggested in Geisser (1971), is shown by Aitchison (1975) to be a better estimate of the true density in this case than what results from plugging in the maximum likelihood estimates into the known normal density (which is basically the rule V^*) by a "frequentist" goodness of fit criterion based on the Kullback-Leibler (1951) directed measure of divergence.

On the other hand the use of V (V^* with $q_1 = q_2$) as a separatory function can be made compelling or approximately so even when based on probabilistic criterion of the kind discussed in (2.10), Enis and Geisser (1974). Neither in its form nor its interpretation, is (4.4) very appealing for separatory purposes, while using V as a separatory function seems to be very attractive for many applications.

If one then were satisfied with V as a separatory function and decided to use V^* as well in the allocatory mode as in most applications, what can we say about its properties, i.e., how good an allocator is it. Before answering this let us examine the allocatory prowess of U^* when the parameters are known--the best possible situation. Then, letting θ stand for μ_1, μ_2 and Σ,

(4.5) $$PCC = \gamma(\theta) = q_1\gamma_1(\theta) + q_2\gamma(\theta)$$

where

(4.6) $$\begin{cases} \gamma_1(\theta) = \Pr[U^* > 0 | \pi_1, \theta] \\ \gamma_2(\theta) = \Pr[U^* < 0 | \pi_2, \theta] \end{cases},$$

$\gamma_i(\theta)$ being the probability of U^* correctly classifying an observation emanating from π_i. It is easily shown, Geisser (1967), that

(4.7) $$\begin{cases} \gamma_1(\theta) = 1 - \Phi(\tau_1) \\ \gamma_2(\theta) = \Phi(\tau_2) \end{cases}$$

where

(4.8) $$\begin{cases} \tau_1 = (\log q_2 q_1^{-1} - \tfrac{1}{2}\alpha)/\alpha^{\frac{1}{2}}, \quad \tau_2 = (\log q_2 q_1^{-1} + \tfrac{1}{2}\alpha)/\alpha^{\frac{1}{2}} \\ \alpha = (\mu_1 - \mu_2)' \Sigma^{-1} (\mu_1 - \mu_2) . \end{cases}$$

Hence a "plug-in" estimate of the best one can do is $\gamma(\hat{\theta}) = q_1(1 - \Phi(\hat{\tau}_1)) + q_2\Phi(\hat{\tau}_2)$ and $\hat{\tau}_1$ and $\hat{\tau}_2$ are estimated by employing $Q = (\bar{x}_1 - \bar{x}_2)'S^{-1}(\bar{x}_1 - \bar{x}_2)$ as an estimate for α where μ_1, μ_2 and Σ are unknown. For a Bayesian estimate of $\gamma(\theta)$ which employs $E_\theta \gamma(\theta) = \bar{\gamma}$, see Geisser (1967, 1970). It must be noted that this is an estimate of the best that can be done in terms of $\gamma(\theta)$ and not an estimate of what may be achieved with a given V^* when the parameters are unknown. When one actually uses V^* then we have the conditional or Actual PCC(APCC)

$$\text{APCC} = \delta(\hat{\theta},\theta) = q_1\delta_1(\hat{\theta},\theta) + q_2\delta_2(\hat{\theta},\theta) \tag{4.9}$$

where

$$\begin{cases} \delta_1(\hat{\theta},\theta) = \Pr(V^* > 0 \mid \pi_1, \hat{\theta},\theta) = 1 - \Phi(\eta_1) \\ \delta_2(\hat{\theta},\theta) = \Pr[V^* < 0 \mid \pi_2, \hat{\theta},\theta) = \Phi(\eta_2) \end{cases} \tag{4.10}$$

where

$$\eta_i = \{(\tfrac{1}{2}(\bar{x}_1+\bar{x}_2)-\mu_i)'S^{-1}(\bar{x}_1-\bar{x}_2) + \log q_2 q_1^{-1}\}/[(\bar{x}_1-\bar{x}_2)'S^{-1}\Sigma S^{-1}(\bar{x}_1-\bar{x}_2)]^{\frac{1}{2}} \tag{4.11}$$

$i = 1, 2$. A naive estimate $\hat{\delta}(\hat{\theta},\theta) = \delta(\hat{\theta},\hat{\theta})$ turns out to be $\gamma(\hat{\theta})$, so that the estimate for APCC is the same as for the PCC which is most unsatisfactory and has led to much effort by frequentists in attempting to correct that estimate of APCC or its peculiar companion E[APCC] = $E_{\hat{\theta}}[\delta(\hat{\theta},\theta)]$, Hills (1966). In fact for a long time the various possible probabilities of correct classification were confused and the subject in somewhat of a chaotic state until Hills (1966) presented a careful analysis of the various frequentist allocation error rates. For some further remarks on this propagation of error see Geisser (1969, 1970). From a Bayesian point of view the problem as such completely disappears by using as estimators for $\delta_i(\hat{\theta},\theta)$ its posterior expectation $\bar{\delta}_i = E_\theta(\delta_i(\hat{\theta},\theta))$ which yields Geisser (1967).

$$\begin{cases} \bar{\delta}_1 = \Pr\left[t_{N_1+N_2-1-p} > \dfrac{(\log q_2 q_1^{-1} - \frac{1}{2}Q)(N_1+N_2-1-p)^{\frac{1}{2}} N_1^{\frac{1}{2}}}{[(N_1+N_2-2)(N_1+1)Q]^{\frac{1}{2}}} \right] \\ \bar{\delta}_2 = \Pr\left[t_{N_1+N_2-1-p} < \dfrac{(\log q_2 q_1^{-1} + \frac{1}{2}Q)(N_1+N_2-1-p)^{\frac{1}{2}} N_2^{\frac{1}{2}}}{[(N_1+N_2-2)(N_2+1)Q]^{\frac{1}{2}}} \right] \end{cases} \tag{4.12}$$

where $t_{N_1 + N_2 - 1 - p}$ is the student "t" random variable with $N_1 + N_2 - 1 - p$ degrees of freedom. Clearly then $\bar{\delta} < \bar{\gamma}$ since $\delta(\theta) < \gamma(\theta)$, as required. Actually the Bayesian estimate of $\gamma(\theta)$ is rather difficult to compute explicitly but $\bar{\gamma} \doteq \gamma(\hat{\theta})$, for a better approximation see Geisser (1970). Note also that $\bar{\delta} < \gamma(\hat{\theta})$. At any rate a clear interpretation emerges -- $\bar{\gamma}$ or $\gamma(\hat{\theta})$ is an estimate of what potentially could be achieved with sample sizes very much larger than those in hand whilst $\bar{\delta}$ is what is actually achievable with the data in hand. In other words if say $\bar{\gamma}$ is large enough, then the discriminatory variables are satisfactory. However the user may not be satisfied with an appreciably lower value of $\bar{\delta}$. But then the remedy is clear, one needs larger sample sizes until $\bar{\delta}$ is close enough to $\bar{\gamma}$ to be satisfactory. If $\bar{\gamma}$ is not large enough to suit the purposes of the allocation then one must find other discriminatory variables.

While $\bar{\delta} = q_1\bar{\delta}_1 + q_2\bar{\delta}_2$ is an estimate of the APCC, it turns out that $\bar{\delta}$ is exactly the predictive probability of correct classification (PPCC) using V^*, though not optimal unless $q_1 = q_2$ and $N_1 = N_2$. Even when these conditions do not hold it should not be too far from optimal as it approaches optimality for large N_1 and N_2. Further $\bar{\delta}$ is also a useful guide in determining which variables may be omitted in measuring future observations. For example it can happen for economic or other reasons that only a subset of r of the p original variables can be utilized for allocating future observations. Then one could compute $\bar{\delta}$ weighted by an appropriate cost or utility factor for each subset of r out of the p variables in order to make an optimal determination.

At any rate this approach yields sensible answers when one uses the usual linear discriminant for allocation. Slight improvements can be made by some adjustment of V within the Bayesian framework as noted by Geisser (1967) and Enis and Geisser (1974), but it's not likely the effect will be significant. Extension to $r > 2$ populations throughout or q_i unknown presents no intrinsic difficulty, see Geisser (1964, 1967).

5. Maximizing Measures of Spread for Linear Discriminatory Forms.

When there are no appropriate distributional assumptions one can proceed by both choosing a class $\mathcal{S}(u)$ of discriminatory functions, linear, quadratic, etc. and defining either a distance between any two populations, Fisher [1936], or a more general measure of spread amongst all of the populations. A minimal set of "best" discriminants then presumably would be selected from all of those solutions that maximize the spread with respect to the parameters of the discriminatory functions given the constraints under consideration. These discriminants then can be used to completely characterize the differential aspects of the populations with respect to the manifest variables.

Let us further assume all of the distributions of the r populations are roughly the same in that they enjoy approximately the type of clustering and symmetry about their mean vectors exhibited by a set of multivariate normal densities with equal covariance matrices. Basically then the important differences are in the location of these central vectors. Fisher (1936) then found it sensible for r populations, to find the set of linear combinations $c'u$ which maximized pairwise the distance functions $\{c'(\mu_i - \mu_j)\}^2/c'\Sigma c$, $i \neq j = 1,\ldots,r$ where Σ was assumed to be the common covariance matrix. This generates the optimal reduced set of linear discriminants previously obtained where multivariate normal theory was assumed. The technique used by Fisher was essentially differentation with Lagrange multipliers. An alternate geometric derivation is given by Dempster [1969].

There are other methods of obtaining these linear discriminants, which involve maximizing some measure of spread, Wilks (1962). The technique used for the maximization by Wilks also involved Lagrange multipliers. We now present an alternate derivation, Geisser (1973b) which is completely algebraic and somewhat more general. Again, suppose there are r p-dimensional multivariate populations with mean $\mu_1,\ldots,\mu_r$ and common positive definite covariance matrix Σ. Further let β, of rank $r - v < p$, be defined as previously in Section 2.

Assume that $g(\beta\Sigma^{-1}) = g(\delta_1,\ldots,\delta_{r-v})$ is any scalar measure of the spread of these r populations that is increasing in the non-zero roots of $\beta\Sigma^{-1}$, $\delta_i \geq \ldots \geq \delta_{r-v} > 0$. Suppose further we transform these r p-dimensional populations into a $k \leq p$ space by a real transformation matrix $C_{k\times p}$ which is of rank k. Hence $\eta = C\mu_i$, $i = 1,\ldots,r$, $\Omega = C\Sigma C'$, $\Gamma = C\beta C'$ and the measure of spread in k dimensions is $g_k(\Gamma\Omega^{-1})$, i.e., the same scalar function of the non-zero roots $d_1 \geq \ldots \geq d_t > 0$ of $\Gamma\Omega^{-1}$ where $t = \min(k, r-v)$. Then we shall show that

$$\text{(5.1)} \qquad \underset{C}{\text{Max}}\ g_k(\Omega^{-1}\Gamma) = g(\delta_1,\ldots,\delta_t).$$

As the maximum spread is attained for $k = r-v$, there is no interest in the discriminatory situation in considering $k > r-v$. An orthogonal basis solution for C, when $k \leq r-v$, would then be

$$\text{(5.2)} \qquad C = P'_{(k)}\Lambda'\Sigma^{-1}$$

where $P_{(k)}$ is as previously defined. Consequently ΛP_j is the characteristic vector associated with the jth largest root of $\beta\Sigma^{-1}$. Hence the conjecture made previously in Section 2 has a basis in fact if optimization depends on maximizing every scalar measure of spread which is an increasing function of the non-zero roots of $\beta\Sigma^{-1}$. One can also define the fraction of total loss sustained in the measure of spread when $k \leq r - v$ as

$$\text{(5.3)} \qquad L = \frac{g(\delta_1,\ldots,\delta_{r-v}) - g(\delta_1,\ldots,\delta_k)}{g(\delta_1,\ldots,\delta_{r-v})}.$$

For example, if we are using either the "Hotelling" or "Wilks" measure of spread:

$$\text{(5.4)} \qquad g_H = \text{Tr}\Sigma^{-1}\beta = \sum_{i=1}^{r-v}\delta_i, \quad g_W = |I + \beta\Sigma^{-1}| = \prod_{i=1}^{r-v}(1+\delta_i)$$

then

$$\text{(5.5)} \qquad L_H = \sum_{i=k+1}^{r-v}\delta_i \Big/ \sum_{i=1}^{r-v}\delta_i, \quad L_W = 1 - \prod_{i=k+1}^{r-v}(1+\delta_i)^{-1}.$$

The algebraic derivation of the aforementioned results is basically an application of the following matrix theorem.

Theorem: Let Z be a real $p \times m$ matrix of rank $s = \min(p,m)$ and E_k be the class of $p \times p$ real symmetric idempotent matrices of rank k. Then for all $F \in E_k$ the maximum attainable values of the first t ordered roots a_i of $Z'FZ$ are α_i, $i = 1,\ldots,t$, $t = \min(k,m)$, where the α_i's are the non-zero ordered roots of $Z'Z$. Further, the totality of solutions for F, where the maximum values of the roots are attained is given by

$$F_0 = \begin{cases} Y_{(k)} D_k^{-1} Y'_{(k)} & \text{for } k \le m \\ Z(Z'Z)^{-1} Z' + G_{k-m} & \text{for } k \ge m \end{cases} \tag{5.6}$$

where $Y_{(k)} = (Y_1,\ldots,Y_k)$, represents the first k columns of $Y = ZP$, and P is the orthogonal matrix such that $P'Z'ZP = D_m$ where

$$D_j = \begin{pmatrix} \alpha_1 & 0 & \cdots & 0 \\ 0 & \ddots & & \vdots \\ \vdots & & \ddots & \\ 0 & \cdots & 0 & \alpha_j \end{pmatrix}$$

and G_{k-m} is any idempotent matrix of rank $k-m$ orthogonal to Z.

Proof:

Given the definitions of P and Y above then

$$\begin{aligned} D_m &= P'Z'ZP = P'Z'FZP + P'Z'(I-F)ZP \\ D_m &= Y'Y = Y'FY + Y'(I-F)Y . \end{aligned} \tag{5.7}$$

Hence the roots of $Z'FZ$ are the roots of $Y'FY$ and by virtue of (5.7) the ordered roots of $Y'FY$, $a_1 \ge \cdots \ge a_m \ge 0$, are not greater than the ordered roots of $Y'Y$, i.e., $\alpha_i \ge a_i$, $i = 1,\ldots,m$, Bellman [1960, p. 113].

Now let Q be a $p \times p$ orthogonal matrix such that $Q = (Y_{(s)} D_s^{-\frac{1}{2}}, Q_2)$ where $Y_{(s)}$ consists of the first s columns of Y. Then

$$Y'FY = Y'QQ'FQQ'Y = Y'QF^*Q'Y$$

where F^* is obviously idempotent since F is assumed to be. Further

(5.8)
$$Y'Q = \begin{array}{c} \\ s \\ m-s \end{array}\!\!\begin{array}{c} \begin{array}{cc} s & p-s \end{array} \\ \left(\begin{array}{c|c} D_s^{\frac{1}{2}} & 0 \\ \hline 0 & 0 \end{array}\right) \end{array}$$

so that

(5.9)
$$Y'FY = Y'QF^*Q'Y = \begin{array}{c} \\ s \\ m-s \end{array}\!\!\begin{array}{c} \begin{array}{cc} s & m-s \end{array} \\ \left(\begin{array}{c|c} D_s^{\frac{1}{2}}F_{11}^*D_s^{\frac{1}{2}} & 0 \\ \hline 0 & 0 \end{array}\right) \end{array}$$

And F_{11}^* is the $s \times s$ matrix in the upper left hand corner of F^*. Now the maximum rank of $Y'FY$ is $t = \min(k,n)$ or $t = \min(k,s)$. Hence all solutions, for which $a_i = \alpha_i$, $i = 1,\ldots,t$ are such that the rank of $Y'FY$ must be t. Further the first t diagonal elements of F_{11}^* are then 1 since $\sum_{i=1}^{t} \alpha_i = \sum_{i=1}^{s} \alpha_i f_{ii}^*$, $0 \leq f_{ii}^* \leq 1$ and $\sum_{i=1}^{p} f_{ii}^* = k$ must be satisfied. This implies that the off diagonal elements in those rows and columns are zero since we are dealing with idempotent matrices. Therefore, all solutions for F^* are

$$F_0^* = \begin{array}{c} \\ k \\ p-k \end{array}\!\!\begin{array}{c} \begin{array}{cc} k & p-k \end{array} \\ \left(\begin{array}{c|c} I_k & 0 \\ \hline 0 & 0 \end{array}\right) \end{array} \quad \text{for } t = k, \text{ i.e., } k \leq m$$

(5.10) or

$$F_0^* = \begin{array}{c} \\ m \\ p-m \end{array}\!\!\begin{array}{c} \begin{array}{cc} m & p-m \end{array} \\ \left(\begin{array}{c|c} I_m & 0 \\ \hline 0 & G \end{array}\right) \end{array} \quad \text{for } t = m, \text{ i.e., } m \leq k,$$

where G is any idempotent $p-m \times p-m$ matrix of rank $k-m$. Hence the totality of solutions for F are $F_0 = QF_0^*Q'$, so that

(5.11) $$F_0 = Y_{(k)}D_k^{-1}Y'_{(k)} \qquad \text{for } k \leq m$$

which is unique if $\alpha_1, \ldots, \alpha_k$ are distinct and

(5.12) $$F_0 = YD_m^{-1}Y' + Q_2GQ_2' \qquad \text{for } m \leq k .$$

Note that from (5.12) and $ZP = Y$ that

(5.13) $$F_0 = ZPD_m^{-1}P'Z' + Q_2GQ_2' = Z(Z'Z)^{-1}Z' + Q_2GQ_2 \quad \text{for } m \leq k .$$

Hence set $Q_2GQ_2' = G_{k-m}$. Since G is an arbitrary idempotent matrix of rank $k-m$ and Q_2 is orthogonal to Z being it is orthogonal to Y, then G_{k-m} is an arbitrary idempotent matrix orthogonal to Z and the theorem is established.

As an immediate consequence of the theorem and the fact that $a_i \leq \alpha_i$ we have the following:

Corollary

If $g(Z'FZ) = g(a_1, \ldots, a_t)$ is a scalar non-decreasing function in the roots a_i, then

(5.14) $$\max_{F \in E_k} g(a_1, \ldots, a_t) = g(\alpha_1, \ldots, \alpha_t) .$$

In order to apply the theorem and corollary we first note that the non-zero roots of $\Gamma\Omega^{-1} = C\beta C'(C \Sigma C')^{-1}$ are the same as the non-zero roots of $\Lambda' C'(C \Sigma C')^{-1} C\Lambda$ where $\beta = \Lambda\Lambda'$ and Λ is $p \times r-v$. Set $C \Sigma^{\frac{1}{2}} = H$ where $\Sigma^{\frac{1}{2}}$ is the positive definite symmetric square root of Σ so that the non-zero roots of $\Gamma\Omega^{-1}$ are the same as the non-zero roots of $\Lambda' \Sigma^{-\frac{1}{2}} H'(HH')^{-1} H \Sigma^{\frac{1}{2}} \Lambda$. Set $r-v = m$, $Z = \Sigma^{-\frac{1}{2}}$ and the idempotent matrix $H'(HH')^{-1}H = F$. Hence as by our previous corollary

$$\operatorname*{Max}_{C} g_k(\Gamma\Omega^{-1}) = \operatorname*{Max}_{F} g_k(Z'FZ) = g(\delta_1, \ldots, \delta_t).$$

To find solutions for C we note that there is an orthogonal matrix P such that

$$P'\Lambda'\Sigma^{-1}\Lambda P = \Delta_m = \begin{pmatrix} \delta_1 & & 0 \\ & \ddots & \\ 0 & & \delta_m \end{pmatrix}$$

and in the theorem set $Y = \Sigma^{-\frac{1}{2}}\Lambda P$ and $Y_{(j)} = \Sigma^{-\frac{1}{2}}\Lambda P_{(j)}$ where $P_{(j)} = (P_1, \ldots, P_j)$ is the matrix consisting of the first j columns of P. Note also that ΛP_i is the invariant vector of $\beta\Sigma^{-1}$ corresponding to the root δ_i, $i = 1, \ldots, m$ and the calculation of $Y_{(j)}$ does not depend on Λ. Hence from the theorem

$$F_0 = Y_{(k)} \Delta_k^{-1} Y'_{(k)} \qquad \text{for } k \le r - v . \tag{5.15}$$

From $H'(HH')^{-1}H = F$ we obtain $H = HF$ and noting from (5.15) that $Y'_{(k)}F_0 = Y'_{(k)}$ then $H_0 = Y'_{(k)}$ and $C_0 = Y'_{(k)}\Sigma^{-\frac{1}{2}} = P'_{(k)}\Lambda'\Sigma^{-1}$ for $k \le r-v$, as required

The derivations in this section and in Section 2 have been presented in terms of population parameters. But obviously if sample estimates based on data at hand, $\hat{\beta}$ and $\hat{\Sigma}$ are utilized there need not be, from one point of view, any essential change other than "optimization" now takes place with regard to sample estimates. The problem then is to decide on the "plug-in" estimators. The substitution of unbiased sample moments for the population moments results in the same set of sample discriminants as is used in the normal case where maximum likelihood estimates corrected for bias are utilized. Of course the simplifying assumption that the multivariate populations differ mainly in their locations, and relative to this variations in the dispersion matrices were unimportant as exemplified explicitly in the previously discussed multivariate normal model, was our guide. This latter model gave rise to linear theory in terms of population discriminants and consequently, it is no surprise that focusing on linear theory can yield the same results.

6. Sample Reuse Techniques - Allocatory and Separatory.

When precise distributional assumptions are untenable or abandoned entirely so that theoretical calculations are precluded, one is then obliged to provide some other means for rendering discriminants and assessing their quality. In this section we shall discuss how sample reuse notions can be directed towards an evaluation of discriminants and their further refinement. Again for simplicity let us focus on the two population case. Let the usual single linear discriminant be $V = V(u)$. By substituting each of the p-dimensional observations x_{ij} for $j = 1, \ldots, N_i$ and $i = 1,2$, in V we obtain $V(x_{ij}) = v_{ij}$ or two sets of univariate observations. These may now be plotted on a single axis distinguishing them only by their population origin. If there are a great many of them a histogram for each set is visually informative in indicating the quality the linear separation induced. If V^* is to be used for allocatory purposes then some assessment of the APCC is in order. It has long been clear that the naive assessment of the APCC, by merely calculating $q_1 n_1 N_1^{-1} + q_2 n_2 N_2^{-1}$ where n_i represents the number of x_{ij}'s that are correctly classified by V^*, will be too large - just as in the normal case $\gamma(\hat{\theta})$ was generally too large as discussed in Section 4. A sample reuse technique for correcting this flaw was proposed by Lachenbruch (1965). He proposed calculating $V_{ij}^*(u)$ the linear discriminant with x_{ij} omitted and then computing $V_{ij}^*(x_{ij}) = u_{ij}$ and classifying x_{ij} on the basis of whether u_{ij} exceeded or fell short of 0. An adjusted estimate of APCC, $q_1 n_1' N_1^{-1} + q_2 n_2' N_2^{-1}$, is obtained where n_i' represents the number of x_{ij}'s, $i = 1,2$, correctly classified. Note that if $N_i/(N_1 + N_2)$ is appropriate as an estimator of q_i, when it is unknown, that the estimator for the APCC becomes $(n_1' + n_2')/(N_1 + N_2)$. This is reasonable in situations where the initial sample of size $N = N_1 + N_2$ is drawn at random from $\pi = \pi_1 \cup \pi_2$, so that the random frequency N_i provides information on q_i.

This method may also be of value in the determination of which variables could be eliminated or which subset of r out of the p would be optimal in future measurements as discussed in Section 4.

One could attempt a finer tuning, as it were, by applying the predictive sample reuse (PSR) method, Geisser (1975). A criterion applicable here would be to maximize

(6.1) $$P(u_0) = q_1 n_1(u_0)N_1^{-1} + q_2 n_2(u_0)N_2^{-1}$$

with respect to u_0, where $n_i(u_0)$ represents the number of x_{ii}'s correctly allocated by V_{ij}^* such that x_{ij} is allocated to π_1 if $V_{ij}^*(x_{ij}) = u_{ij} > u_0$ and to π_2 otherwise. One would order the scalar values u_{ij} and find that cutoff point $\hat{u}_0$ which maximizes $P(u_0)$. This can easily be done numerically as it is essentially a counting procedure. Convenient algorithms can be found to shorten the process. While it is also clear that $\hat{u}_0$ need not be unique, it can be made so by arbitrarily selecting a particular one of them, e.g. the maximizer $\hat{u}_0$ closest to zero. Then for future allocation one uses $V^*(u) \gtrless \hat{u}_0$ as the allocatory discriminant. One could also alter the criterion when q_i is unknown and maximize $\hat{P}(u_0) = \hat{q}_1 n_1(u_0)N_1^{-1} + \hat{q}_2 n_2(u_0)N_2^{-1}$. If $N_i(N_1 + N_2)^{-1}$ can be used as an estimator for q_i then the new criterion effectively maximizes the total number of x_{ij}'s correctly classified by $\hat{V}_{ij}^* \gtrless u_0$, and again one would use $\hat{V}^*(u) \gtrless \hat{u}_0$ as the allocator where $\hat{V}_{ij}^*$ and $\hat{V}^*(u)$ are merely V_{ij}^* and $V^*(u)$ respectively with $q_1 q_2^{-1}$ replaced by $N_1 N_2^{-1}$.

An appropriate assessment of the new discriminant would require a two-deep cross-validatory assessment i.e. of the Lachenbruch (1965) type. There are obviously ways of applying the PSR approach to linear discriminants other than the cut-off point allocatory approach illustrated here, e.g. estimating by PSR the linear regression coefficients.

If one is concerned mainly with the reliability of a discriminant in its separatory role, then Mosteller and Tukey (1968) and Lachenbruch and Mickey (1968) suggest jackknifing V. First one calculates the set of pseudo-discriminant functions $V'_{ij} = (N_1+N_2)V-(N_1+N_2-1)V_{ij}$ $j=1,\ldots,N_i$, $i = 1,2$, and then $V' = (N_1 + N_2)^{-1} \sum_{i,j} V'_{ij}$, which is termed the jackknifed discriminant. One can compute the reliability of V' in terms of the variation of the regression coefficients of the V'_{ij}, the individual values averaged to compute V'. Examining the ratio of a regression coefficient in V' to its sample standard error permits a judgement on the significance of its deviation from zero. The main point of this exercise is to assess to some degree the reliability of the jackknifed discriminant function in its separatory role. Again if one decides to use V' (or $V^{*\prime}$) as an allocator one can assess it by using a two-deep cross-validatory approach.

7. Remarks.

I have attempted to carefully delineate two distinct purposes of discriminatory analyses and to examine the linear aspects involved. However linearity has been surveyed, as it were, from a personal (not to be confused with personalistic) point of view, in that much work on the linear aspects of normal parametric discrimination has not been mentioned chiefly because it involves the generation of modest improvements with regard to certain frequency properties by some slight alteration of the linear discriminants. It is my contention that here the Bayesian type of adjustment will yield better results than frequential tinkering. When parametric assumptions are fuzzy or nonexistent, sample reuse methods, which are frequency oriented predictive simulation techniques, should serve.

Finally it must be borne in mind that Discrimination is a technique which is often most useful in the early history or soft stage of a discipline when notions are fuzzy, measurements crude or indirect and relationships vaguely understood at best. Hence it is generally an appreciable improvement of whatever has gone before--theoretical niceties notwithstanding. No doubt Linear Discrimination fulfills the role played by Barnard's (1972) "midwife" in fostering the parturition of pertinent distinctions, probabilistic or classificatory, during the birthpangs of a scientific discipline--but soon abandoned or its focus shifted as the discipline hardens.

References

Aitchison, J. (1975). Goodness of prediction fit. Biometrika, 62, 3, pp. 547-554.

Anderson, J. A. (1973). Logistic discrimination with medical applications, Discriminant Analysis and Applications, edited by T. Cacoullos, Academic Press, New York, pp. 1-16.

Anderson, T. W. (1958). An Introduction to Multivariate Statistical Analysis, John Wiley and Sons, New York.

Barnard, G. A. (1972). The unity of statistics, J. R. Statist. Soc., A., 135, pp. 1-14.

Bellman, R. (1960), Introduction to Matrix Analysis, McGraw-Hill, New York.

Das Gupta, S. (1965), Optimum classification rules for classification into two multivariate normal populations. Ann. Math. Statist., 36, pp. 1174-1184.

Dempster, A. P. (1969), Elements of Continuous Multivariate Analysis, Addison-Wesley, Reading, Massachusetts.

Desu, M. M. and Geisser, S. (1973), Methods and applications of equal-mean discrimination, Discriminant Analysis and Applications, edited by T. Cacoullos, Academic Press, New York, pp. 139-161.

Efron, B. (1975), The efficiency of logistic regression compared to normal discriminant analysis. J. Am. Statist. Assoc. 70, 352, pp. 892-898.

Enis, P. and Geisser, S. (1970), Sample discriminants which minimize posterior squared error loss, South African Statist. J., 4, pp. 85-93.

Enis, P. and Geisser, S. (1974), Optimal predictive linear discriminants, Ann. Statist., 2, 2, pp. 403-410.

Fisher, R. A. (1936), The use of multiple measurements in taxonomic problems, Annals of Eugenics, 7, pp. 179-188.

Geisser, S. (1964), Posterior odds for multivariate normal classification, J. R. Statist. Soc. B, 1, pp. 69-76.

Geisser, S. (1966), Predictive discrimination, Multivariate Analysis, edited by P. Krishnaiah, Academic Press, New York, pp. 149-163.

Geisser, S. (1967), Estimation associated with linear discriminants, Ann. Math. Statist. 38, pp. 807-817.

Geisser, S. (1969), Alternative views of discrimination, Bull. Int. Inst. Statistl, 2, pp. 132-134.

Geisser, S. (1970), Discriminatory practices, Bayesian Statistics, edited by D. Meyer and R. C. Collier, Peacock, Illinois, pp. 57-70.

Geisser, S. (1973a), Multiple birth discrimination, Multivariate Statistical Inference, edited by D. G. Kabe and R. P. Gupta, North-Holland, Amsterdam, pp. 49-55.

Geisser, S (1973b), A note on linear discriminants, Bull. Int. Statist. Inst. 1, pp. 442-448.

Geisser, S. (1975), The predictive sample reuse method with applications, J. Amer. Statist. Assoc., 70, 350, pp. 320-328.

Geisser, S. and Desu, M. M. (1968), Predictive zero-mean uniform discrimination, Biometrika , 55, 3, pp. 519-524.

Guseman, L. F., Peters, B. C., and Walker, H. F. (1975), On minimizing the probability of misclassification for linear feature selection, Annals of Statistics, 3, 3, pp. 661-668.

Hills, M. (1966), Allocation rules and their error rates, J. R. Statist. Soc. B, 28, pp. 1-31.

Kiefer, J. and Schwartz, R. (1965), Admissible Bayes character of T^2, R^2, and other fully invariant tests for classical multivariate normal problems, Ann. Math. Statist., 36, pp. 747-770.

Kullback, S., and Liebler, R. A. (1951), On information and sufficiency, Ann. Math. Statist. 22, pp. 525-540.

Lachenbruch, P. A. (1965), Estimation of error rates in discriminant analysis, unpublished dissertation, University of California at Los Angeles.

Lachenbruch, P. A. and Mickey, M. R. (1968), Estimation of error rates in discriminant analysis. Technometrics, 10, pp. 1-11.

Mostetler, F. and Tukey, J. W. (1968), Data analysis, including statistics, Handbook of Social Psychology, edited by G. Lindzey and E. Aronson, Addison-Wesley, Reading, Massachusetts.

Okamoto, M. (1961), Discrimination for variance matrices, Osaka Math. J., 13, pp. 1-39.

Reyment, R. A. (1973), The discriminant function in systematic biology, Discriminant Analysis and Applications, edited by T. Cacoullos, Academic Press, New York, pp. 311-338.

Richter, D. L. and Geisser, S. (1960), A statistical model for testing treatment effects in the presence of learning, Biometrics, 16, 1, pp. 110-114.

Wald, A. (1944), On a statistical problem arising in the classification of an individual into one of two groups, Ann. Math. Statist. 15, 2, pp. 145-162.

Welch, B. L. (1939), Note on discriminant functions, Biometrika, 31, pp. 218-220.

Wilks, S. S. (1962), Mathematical Statistics, John Wiley and Sons, New York.

This work was supported in part by the U. S. Army Research Office and the Office of Naval Research.

School of Statistics
University of Minnesota
Minneapolis, Minnesota 55455

Discriminant Analysis When Scale Contamination Is Present in the Initial Sample

Susan W. Ahmed
Peter A. Lachenbruch

1. Introduction

The linear discriminant function is used to assign observations to one of two populations. If the underlying distributions of the observations are multivariate normal with the same covariance matrices, the sample discriminant has certain desireable asymptotic properties. If the underlying populations are not normal, the discriminant still has some desireable properties, but is not, in general, the optimal rule in any sense. The amount of loss of discriminating power is a function of how non-normal the distributions are. It is known, however, that even slight non-normality can hurt location estimators considerably. Indeed, Andrews et. al (3) have shown that the sample mean is rarely the location estimator of choice. A variety of non-normal situations can be envisioned: the variables may be Bernouilli, Poisson, continuous non-normal such as might be generated by transforming normal variates by the Johnson system of transformations, or the sampled populations might be contaminated distributions.

This study will investigate one particular aspect of the robustness of the linear discriminant function: robustness against contamination of the initial samples. Very little research has been done on the robustness of the linear discriminant function. Revo (9) and Gilbert (5) have examined the use of the LDF with qualitative variables; Lachenbruch has investigated the effects of misclassification in the initial samples (7); and Lachenbruch, Sneeringer and Revo have studied the robustness of the LDF to certain continuous non-normal distributions (8).

The purpose of this study is to investigate the effect of contamination of one or both of the initial samples. Assume that we wish to classify individuals into one of two groups:

$$\pi_1 : f_1(\underset{\sim}{x}) = N_m(\underset{\sim}{\mu}_1, \underset{\sim}{\Sigma})$$

$$\pi_2 : f_2(\underset{\sim}{x}) = N_m(\underset{\sim}{\mu}_2, \underset{\sim}{\Sigma}) .$$

We know that Fisher's LDF is the optimal solution to this problem. If the parameters $\underset{\sim}{\mu}_1$, $\underset{\sim}{\mu}_2$ and $\underset{\sim}{\Sigma}$ are unknown, we take samples from π_1 and π_2 and use the sample discriminant function.

We say that sample 1 is scale contaminated if it is really a sample from $\pi_1^C : (1-\alpha)\, N_m(\underset{\sim}{\mu}_1, \underset{\sim}{\Sigma}) + \alpha G_m(\underset{\sim}{\mu}_1, \underset{\sim}{\theta})$ rather than π_1, where α is the proportion of contamination $(0 \leq \alpha \leq 1)$, $\underset{\sim}{\theta}$ stands for the parameters of G other than $\underset{\sim}{\mu}_1$, and the contaminating distribution G is assumed to be symmetric, unimodal, centered at $\underset{\sim}{\mu}_1$, and have greater variability and heavier tails than π_1. Then π_1^C is a symmetric long-tailed distribution.

Sample 1 is said to be location contaminated if it is really a sample from $\pi_1^C : (1-\alpha)\, N_m(\underset{\sim}{\mu}_1, \underset{\sim}{\Sigma}) + \alpha G_m(\underset{\sim}{\nu}_1, \underset{\sim}{\Sigma})$ rather than π_1. In this study, we shall consider the particular location contamination model where the contaminating distribution is normal. The density of π_1 is either an asymmetric unimodal distribution or a distribution with one major mode and one minor mode.

A scale contaminated sample can arise as an accidental mixture. Suppose that we wish to discriminate between two groups: π_1 is the population of individuals having disease D and π_2 is the population of individuals free of disease D. The classification into π_1 or π_2 will be made on the basis of m measurements $\underset{\sim}{x}$ thought to be important in determining whether or not an individual has the disease. It is assumed that $\underset{\sim}{x}$ is normally distributed in both populations. Samples will be taken from the diseased and well populations. The sample from π_1 will be drawn from hospital patients and the m measurements on each of the sample individuals will be made in the hospital with hospital instruments. The sample from π_2 will consist of individuals contacted in a home survey. Some of these measurements will be made in the hospital, while others will be made by local nurses or doctors. Suppose that the instruments used by the local nurses and doctors are not as precise as the hospital instruments. Then we have some contamination in the π_2 sample due to the varying precision of the measuring instruments. In the future, when we wish to classify an individual as diseased (π_1) or free of disease (π_2) we will bring him to the hospital and make all measurements with hospital instruments (i.e., we wish to classify him into one of π_1 or π_2, not π_1 or π_2^c). Note that only the sample is contaminated, not the underlying population.

Rather than looking at the way π_2^c is constructed, we can also look at the shape of its density function. It is a long or heavy-tailed distribution. Thus, another example of a scale-contaminated sample is a sample which includes some "gross errors" or "outliers" or in Tukey's words "dirt in the tails". Hampel suggests that the proportion of "gross errors" or "outliers" in data is normally between 0.1% and 10% depending on circumstances, with several percent being the rule rather than the exception.

As an example of location contamination, suppose that we wish to discriminate between individuals with a specified psychiatric disorder (π_1) and normal individuals (π_2). The classification is to be made on the basis of a battery of psychiatric tests. The results of the tests for an individual are represented by the vector $\underset{\sim}{x} = (x_1, \ldots, x_m)$. The vector of test results $\underset{\sim}{x}$ is distributed as $N(\underset{\sim}{\mu}_1, \underset{\sim}{\Sigma})$ in π_1; $\underset{\sim}{x}$ is distributed as $N(\underset{\sim}{\mu}_2, \underset{\sim}{\Sigma})$ in π_2. In order to form the linear discriminant function, samples are drawn from π_1 and π_2. The sample from π_1 is drawn from a group of individuals who have recently been diagnosed by a psychiatrist as having a disorder. The sample from π_2 is drawn from a supposedly normal population. Suppose that among the "normal" sample, there are some undiagnosed cases of the disorder. Then the π_2 sample is location contaminated.

Suppose, then, that although both of the underlying populations are normal, when initial samples are taken, one or both of them are contaminated. What effect does this have on the use of the linear discriminant function? Does the LDF still do a good job of classification? How does the performance of the LDF change as α increases and as G changes? Can the LDF be modified to improve the results with contaminated samples? Suppose we guard against the possibility of contamination of the initial samples by using some "robust" estimation procedures. How do these robust procedures perform when there is no contamination, i.e., when $\alpha = 0$?

Throughout this study, our criterion for evaluating the LDF or any other procedure will be the probability of misclassification, the probability of incorrectly assigning an individual from π_1 to π_2 or assigning an individual from π_2 to π_1. This appears to be a reasonable criterion since our major concern in the classification problem is the correct assignment of an individual to the group of which he is a member.

2. The Asymptotic Probability of Misclassification when the Initial Samples are Contaminated.

The following theorem was proved by Ahmed (1) and is also reported in Ahmed and Lachenbruch (2). Consider the following populations

$$\pi_1 : N_m(\underset{\sim}{0},\underset{\sim}{I})$$

$$\pi_2 : N_m((\delta,0,\ldots,0)',\underset{\sim}{I})$$

$$\pi_1^c : (1-\alpha)N_m(\underset{\sim}{0},\underset{\sim}{I}) + \alpha N_m((\gamma_{11},\gamma_{12},0,\ldots,0)',k_1\underset{\sim}{I})$$

$$\pi_2^c : (1-\beta)N_m((\delta,0,\ldots,0)',\underset{\sim}{I}) + \beta N_m((\gamma_{21},\gamma_{22},\gamma_{23},0,\ldots,0)',k_2\underset{\sim}{I}).$$

This describes the contamination model. We are forced by nature to sample from π_1^c and π_2^c, but wish to assign unknown observations to π_1 and π_2.

Theorem: The asymptotic probabilities of misclassification using the linear discriminant function to discriminate between π_1 and π_2 given contaminated initial samples of size n_1 and n_2 from π_1^c and π_2^c are

$$P_1^C = \Phi\left[\frac{\ln((1-p)/p)}{g_1\delta} - g_2\delta/2\right]$$

$$P_2^C = \Phi\left[\frac{-\ln((1-p)/p)}{g_1\delta} - g_3\delta/2\right]$$

$$P^C = pP_1^C + (1-p)P_2^C$$

where Φ is the cumulative normal distribution, and g_1, g_2 and g_3 are functions of α, β, δ, γ_{ij}, n_1, and n_2. The exact definitions of g_1, g_2, g_3 and a proof of the theorem is given in Ahmed (1).

Some consequences are given here. If $\gamma_{11} = \gamma_{12} = 0$, $\gamma_{21} = \delta$, $\gamma_{22} = \gamma_{23} = 0$, we have pure scale contamination and

$$P_1^C = \Phi\left[\ln(\tfrac{1-p}{p})/g\delta - \delta/2\right]$$

$$P_2^C = \Phi\left[-\ln(\tfrac{1-p}{p})/g\delta - \delta/2\right]$$

where

$$g = \frac{n_1 + n_2 - 2}{n_1 + n_2 - 2 + (k_1-1)(n_1-1) + (k_2-1)(n_2-1)} .$$

Note in particular that if $p = \frac{1}{2}$ there is no asymptotic effect of contamination on the linear discriminant function. This is important for simulation studies as it suggests that the equal a priori probability case is the most favorable one for the linear discriminant function. A second special case arises if $\gamma_{11} = \delta$, $\gamma_{12} = \gamma_{21} = \gamma_{22} = \gamma_{23} = 0$, $k_1 = k_2 = 1$. This is the random initial misclassification model described by Lachenbruch (7).

The asymptotic effects of scale contamination are illustrated in Figures 1-8. Figure 1 illustrates the effect of scale contaminated initial samples on P_1, the probability of misclassification in population 1. The probability of misclassification is plotted as a function of p, the prior probability of population 1. The three graphs in the figure (from top to bottom at $p < .50$) correspond to $\alpha = 0.25$, 0.10, 0.0. Since $P_2^C(p) = P_1^C(1-p)$, i.e., the probability of misclassification in population 2 when the prior probability of population 1 is p is equal to the probability of misclassification in population 1 when the prior probability of population 1 is $1-p$, the corresponding plot of P_2^C is a reflection about $p = .5$ of the plot of P_1^C. We have $P_1^C > P_1$ for $p < \frac{1}{2}$; $P_1^C < P_1$ for $p > \frac{1}{2}$; $P_1^C = P_1$ for $p = \frac{1}{2}$. The relationship is reversed for P_2: $P_2^C < P_2$ for $p < \frac{1}{2}$; $P_2^C > P_2$ for $p > \frac{1}{2}$; $P_2^C = P_2$ for $p = \frac{1}{2}$. Thus, contamination hurts us in classifying individuals who belong to the population having the smaller prior probability, while it improves our ability to classify individuals from the population with larger prior probability. Note, however, that the detrimental effect in the less likely population is greater than the benefical effect in the population with larger prior probability. The above relationship holds, no matter which of the two samples is contaminated, i.e., $(\alpha,\beta) = (0.10, 0)$ gives the same probabilities of misclassification as $(\alpha,\beta) = (0, 0.10)$. In general, scale contamination has a significant effect on the individual probabilities of misclassification, even when the contamination is mild.

The asymptotic effects of scale contamination on P^C are illustrated in Figures 2-5. Since the graphs are symmetric about $p = .5$, only the values of $p < .5$ are given. The overall probability of misclassification is less affected by scale contamination than the individual probabilities. For mild contamination $(k = 4, 9)$, there is little effect. (When $k = 9$ and $\alpha = 0.25$, we see some mild effects.) When we reach $k = 25$, however, contamination begins to hurt our ability to discriminate between π_1 and π_2 using Fisher's LDF with 10% contamination. For $k = 100$, contamination has a definite harmful effect.

The above results are based on single sample scale contamination with $n_1 = n_2$. Similar calculations were made for $n_1 = np$, $n_2 = n(1-p)$, i.e., sample sizes proportional to the prior probabilities of the two populations. In this case, the graph of P^C as a function of p is not symmetric about $p = 0.5$. The contaminated probability of misclassification is closer to the optimal probability when the prior probability of the contaminated population is less than 0.5. Figure 6 illustrates the effect of scale contamination for $\delta = 2$, $k = 100$, and $n_1 = np$, $n_2 = n(1-p)$. This figure corresponds to Figure 5 with equal sample sizes. The general conclusions regarding the performance of the LDF with contaminated initial samples are the same, whether we take equal sample sizes or sample size proportional to the prior probabilities of the two populations.

Similar conclusions are drawn when both samples are contaminated. Figures 7 and 8 illustrate the effect of two-sample scale contamination for $n_1 = n_2$, ($\delta = 3$, $k_1 = k_2 = 9$) and ($\delta = 3$, $k_1 = k_2 = 100$). The four graphs in each figure (from top to bottom) correspond to $(\alpha, \beta) =$ (0.25, 0.25); $(\alpha, \beta) = (0.10, 0.25)$; $(\alpha, \beta) = (0.10, 0.10)$; $(\alpha, \beta) = (0, 0)$. As expected, two sample scale contamination hurts our ability to make correct classifications more than single-sample contamination. Even with $k_1 = k_2 = 9$, our probability of misclassification may be significantly increased (depending upon the value of p, the prior probability of population 1).

In closing this section, we note that the overall probability of misclassification can be greatly affected by location contamination, depending upon the direction of the contamination. The effects of location contamination are illustrated in Figures 9-11. For example, for the second population being contaminated, the contamination has little effect when the contaminating mean lies between the two population means (Figure 10). When the contaminating mean is opposite to the uncontaminated population, the contamination may have some effect, depending upon the size of δ and p (Figure 11). As expected, the greatest effect of location contamination occurs when the contaminating mean is on the opposite side of the uncontaminated population mean (Figure 9).

Figure 1. Probability of misclassification in population I with single sample scale contamination: $\delta=2$; $k_1=100$; $n_1=n_2$; $\alpha=0.0$, 0.10, 0.25.*

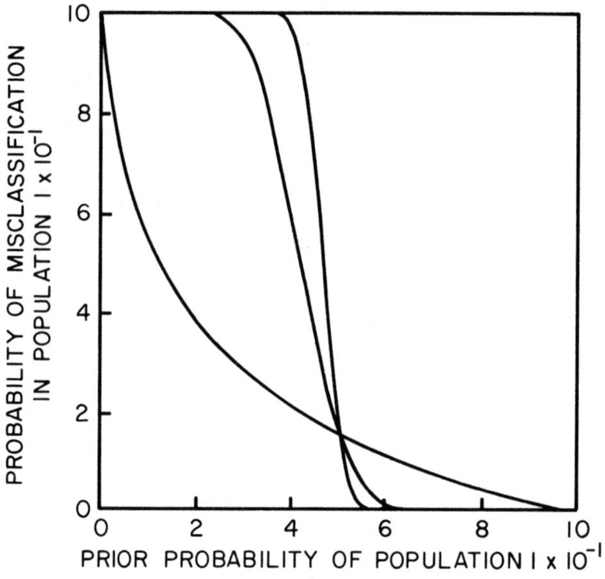

*The three graphs in the figure (from top to bottom at $p<.5$) correspond to $\alpha=0.25, 0.10, 0.0$.

Figure 2. Overall probability of misclassification with single sample scale contamination: $\delta=2$; $k_1=4$; $n_1=n_2$; $\alpha=0.0, 0.10, 0.25$.*

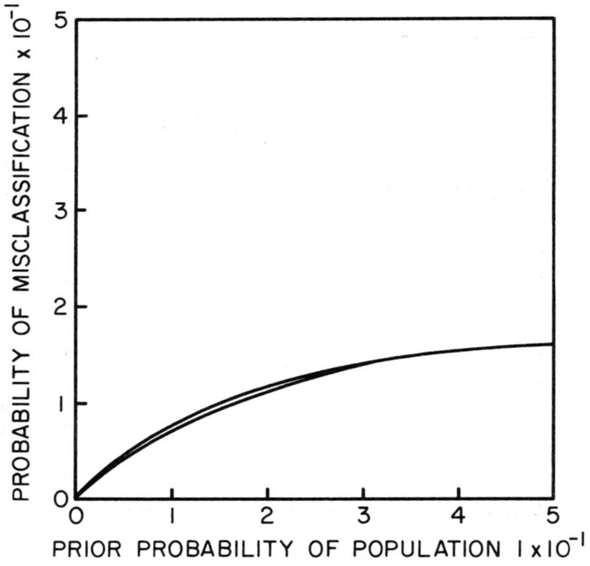

*The three graphs in the figure (from top to bottom) correspond to $\alpha=0.25, 0.10, 0.0$.

Figure 3. Overall probability of misclassification with single sample scale contamination: $\delta=2$, $k_1=9$; $n_1=n_2$; $\alpha=0.0, 0.10, 0.25$.*

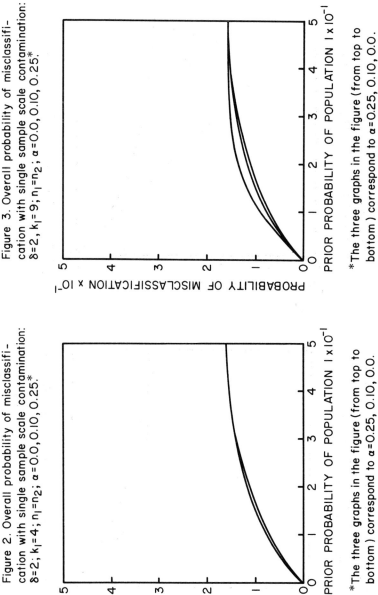

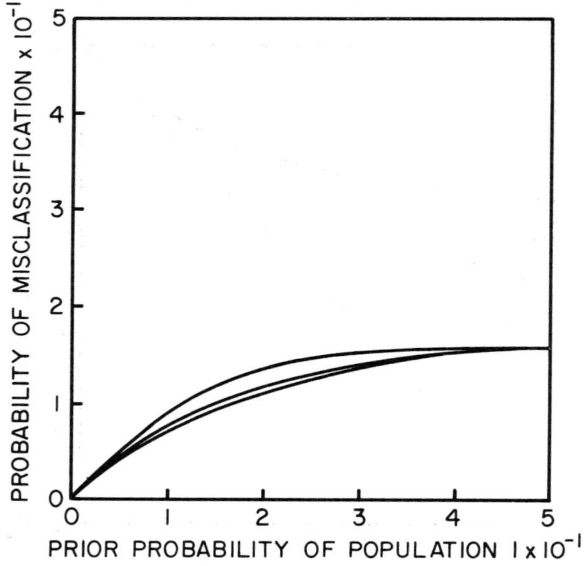

*The three graphs in the figure (from top to bottom) correspond to $\alpha=0.25, 0.10, 0.0$.

Figure 4. Overall probability of misclassification with single sample scale contamination: $\delta=2$; $k_1=25$; $n_1=n_2$; $\alpha=0.0, 0.10, 0.25$.*

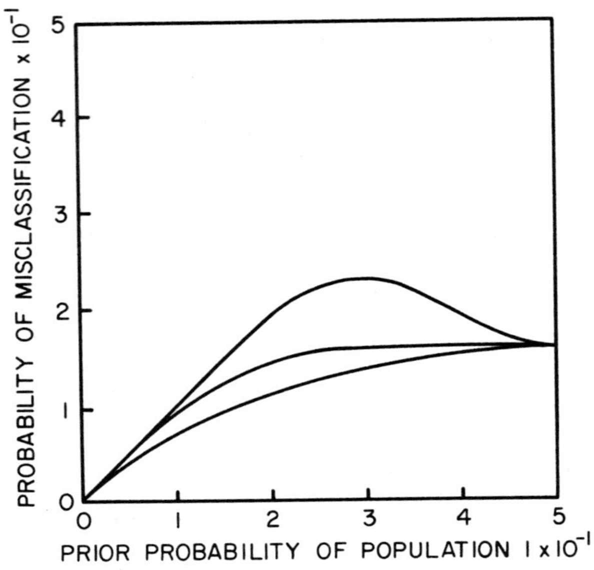

*The three graphs in the figure (from top to bottom) correspond to $\alpha=0.25, 0.10, 0.0$.

Figure 5. Overall probability of misclassification with single sample scale contamination: $\delta=2$; $k_1=100$; $n_1=n_2$; $\alpha=0.0, 0.10, 0.25$.*

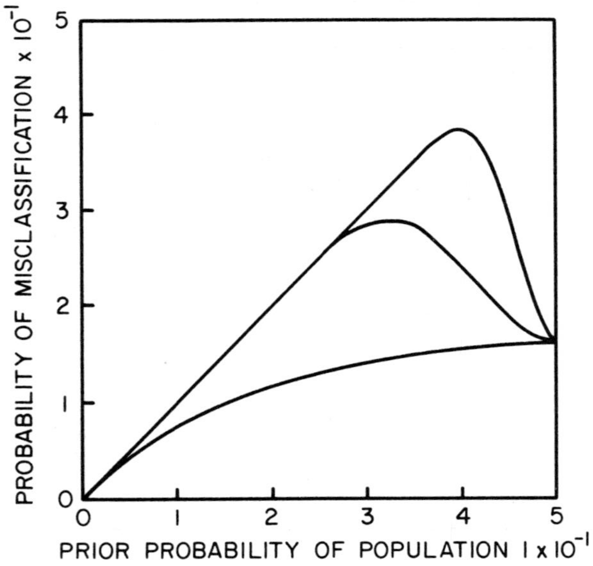

*The three graphs in the figure (from top to bottom) correspond to $\alpha=0.25, 0.10, 0.0$.

Figure 6. Overall probability of misclassification with single sample scale contamination: $\delta = 2$; $k_1 = 100$; $n_1 = np$; $\alpha = 0.0, 0.10, 0.25$.*

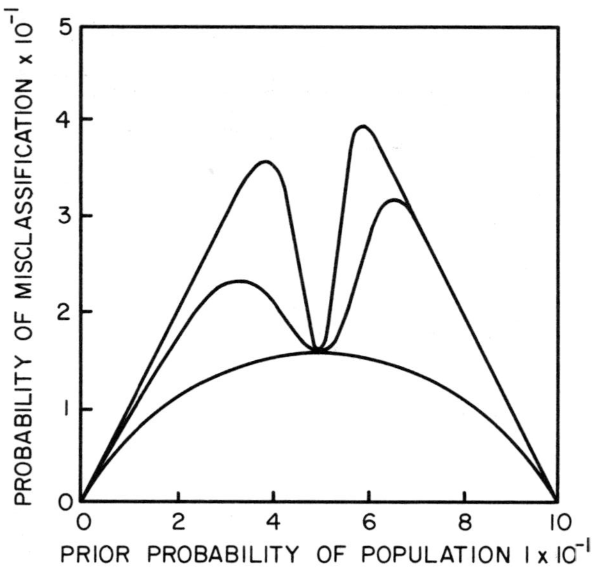

*The three graphs in the figure (from top to bottom) correspond to $\alpha = 0.25, 0.10, 0.0$.

Figure 7. Overall probability of misclassification with two sample scale contamination: $\delta = 3$; $k_1 = 9$, $k_2 = 9$; $n_1 = n_2$; $(\alpha, \beta) = (.25, .25)$, $(.10, .25)$, $(.10, .10)$, $(0.0, 0.0)$.*

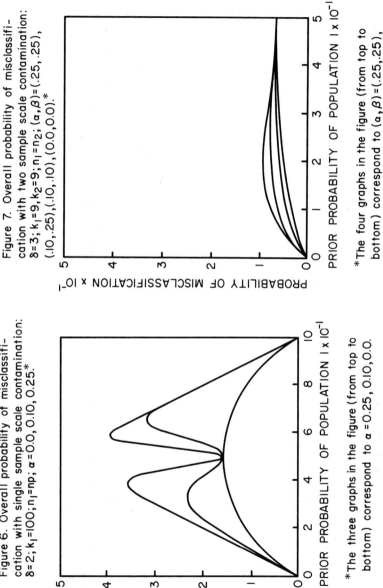

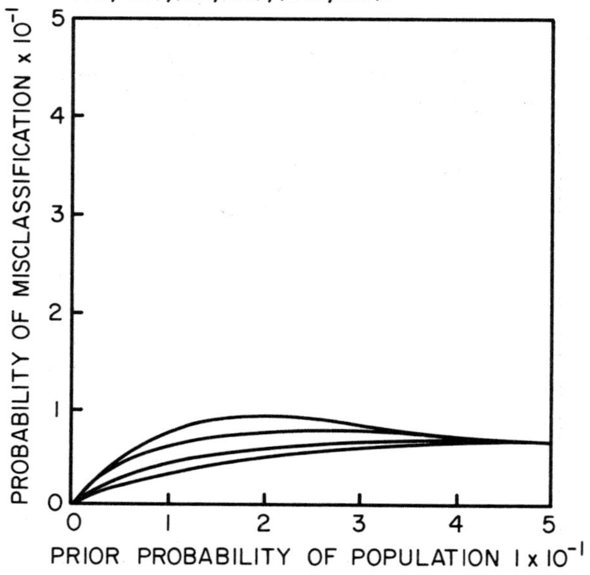

*The four graphs in the figure (from top to bottom) correspond to $(\alpha, \beta) = (.25, .25)$, $(.10, .25)$, $(.10, .10)$, $(0.0, 0.0)$.

Figure 8. Overall probability of misclassification with two sample scale contamination: $\delta=3$; $k_1=100$, $k_2=100$; $n_1=n_2$; $(\alpha,\beta)=(.25,.25)$, $(.10,.25)$, $(.10,.10)$, $(0.0,0.0)$.*

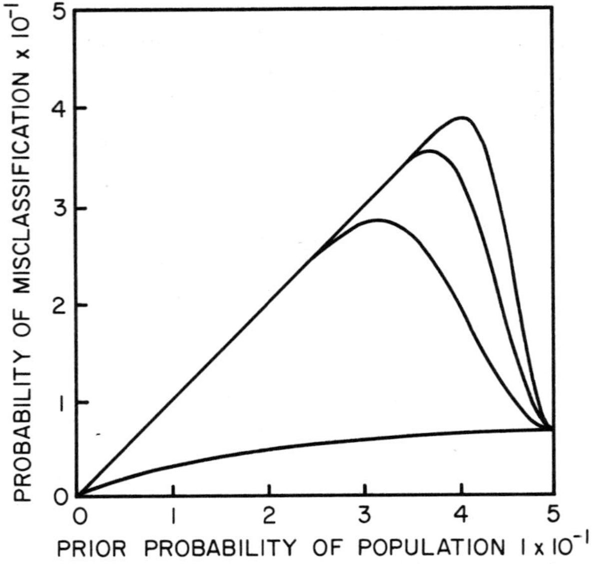

*The four graphs in the figure (from top to bottom) correspond to $(\alpha,\beta)=(.25,.25)$, $(.10,.25)$, $(.10,.10)$, $(0.0,0.0)$.

Figure 9. Overall probability of misclassification with single sample location contamination: $\delta=3$; $c_{-2}=(-1.0,0.0,0.0)$; $n_1=n_2$; $\beta=0.0$, 0.10, 0.25.*

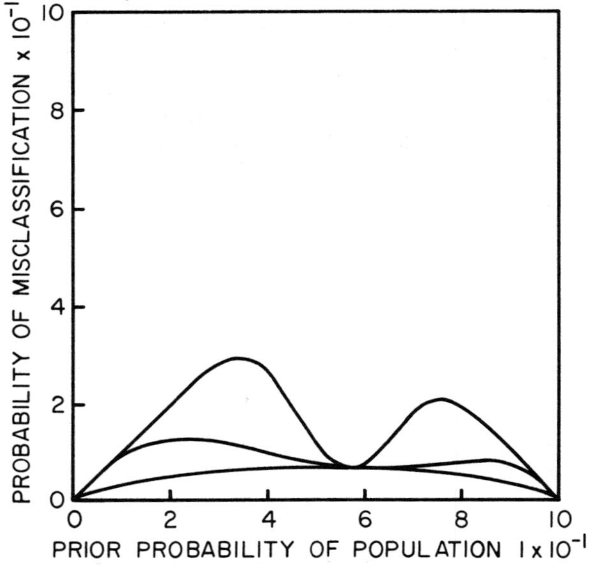

*The three graphs in the figure (from top to bottom) correspond to $\beta=0.25$, 0.10, 0.0.

Figure 10. Overall probability of misclassification with single sample location contamination: $\delta=3$; $c_{-2}=(0.5, 0.0, 0.0)$; $n_1=n_2$; $\beta=0.0, 0.10, 0.25$.*

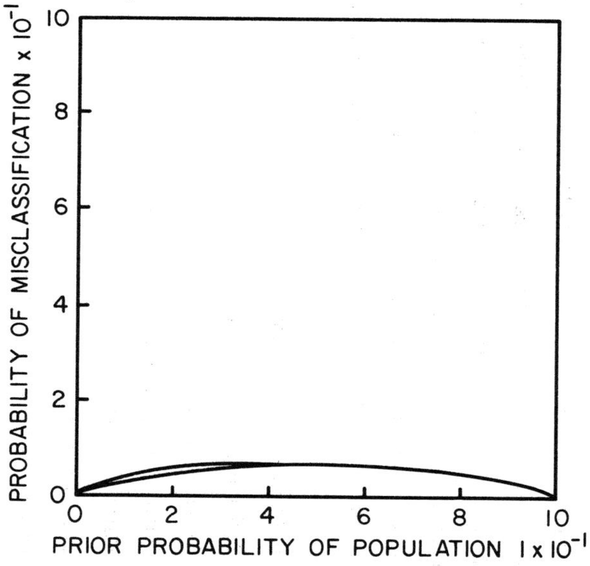

*The three graphs in the figure (from top to bottom) correspond to $\beta=0.25, 0.10, 0.0$.

Figure 11. Overall probability of misclassification with single sample location contamination: $\delta=3$; $c_{-2}=(3.0, 0.0, 0.0)$; $n_1=n_2$; $\beta=0.0, 0.10, 0.25$.*

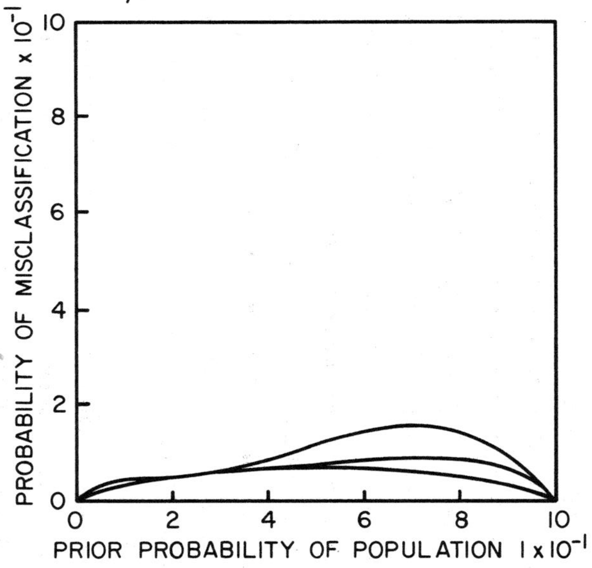

*The three graphs in the figure (from top to bottom) correspond to $\beta=0.25, 0.10, 0.0$.

3. Finite Sample Studies.

It was noted that for large samples scale contamination of the initial samples has no effect on our ability to discriminate between the two populations $\pi_1 : N(\underset{\sim}{0},\underset{\sim}{I})$ and $\pi_2 : N((\delta,0,\ldots,0), \underset{\sim}{I})$ when the prior probabilities of the two populations are equal, i.e., $p = \frac{1}{2}$. We shall now examine the small sample effects of such scale contamination. We wish to answer the following questions:

(1) How well does the usual sample LDF perform with scale-contaminated initial samples when $p = \frac{1}{2}$?

(2) How is the performance of the LDF affected by changing the amount of contamination, α and β, and the type of contaminating distribution?

(3) In investigating questions (1) and (2) we will find that with small samples, scale contamination can have a harmful effect on our ability to discriminate between π_1 and π_2 using Fisher's LDF. Therefore, several "robust linear discriminant functions" will be proposed. Are any of these functions worthwhile alternatives to the usual LDF when the initial samples are scale contaminated? How do these robust LDF's perform when $\alpha = \beta = 0$, i.e., when there is no contamination?

In Ahmed (1), three types of robust linear discriminant functions were considered. The Type I discriminant combined robust location and scale estimates to calculate the estimates of mean and variance for the discriminant function. The location estimators (and their abbreviations) were:

1. 15% Trimmed Mean (TRIM 15)
2. 25% Trimmed Mean (Trim 25)
3. Huber proposal 2 (HUB)
4. Hampel's Huber (HAMHUB)
5. Hampel's estimate (HAMP)
6. Sin estimate (SIN)
7. Gastwirth's estimate (GAST)
8. Trimean (TRI).

All of these were considered by Andrews et. al. (3) and their definitions may be found there or in Ahmed (1).

The robust covariance estimates were the double deletion method of Gnanadesikan and Kettenring (6), using 5% or 10% for the proportion deleted.

The Type II discriminant was based on the use of robust regression coefficients obtained by minimizing p-th power deviations.

The third type of linear discriminant function proposed as an alternative to Fisher's LDF when the initial samples are contaminated was the "a% trimmed LDF." It is constructed in a manner similar to the usual sample LDF. However, those observations which are poor for discrimination are eliminated before forming the LDF. The procedure is outlined below.

(1) Draw a sample of size n_1 from π_1 and a sample of size n_2 from π_2. (The samples actually obtained are from π_1^c and π_2^c.)

(2) For each of the n_1 individuals in sample 1, compute the usual sample LDF: $D_s(\underset{\sim}{x}) = \{\underset{\sim}{x} - \frac{1}{2}(\bar{\underset{\sim}{x}}_1 + \bar{\underset{\sim}{x}}_2)\}'S^{-1}(\bar{\underset{\sim}{x}}_1 - \bar{\underset{\sim}{x}}_2)$.

(3) Rank the $D_s(\underset{\sim}{x})$ computed in (2) and eliminate the observations corresponding to the a% smallest ranks and the observations corresponding to the a% largest ranks.

(4) Repeat steps (2) and (3) for the n_2 individuals in sample 2.

(5) Compute a new linear discriminant function based on those observations not eliminated in steps (3) and (4).

Trimming values of 15% and 25% were used in this study. This procedure is similar to the "Middles" procedure studied by Clarke (4).

Samples were generated from π_1^c and π_2^c for a variety of combinations of parameters. It is possible to obtain the exact probability of misclassification from any sample discriminant in the following manner:

Any of the discriminants has the form

$$D(\underset{\sim}{x}) = \underset{\sim}{b}'\underset{\sim}{x} + c .$$

Since $\underset{\sim}{x}$ is assumed normal, $\underset{\sim}{b}'\underset{\sim}{x} + c$ is normal with mean $\underset{\sim}{b}'\underset{\sim}{u}_i + c$ in the i-th group and variance $\underset{\sim}{b}'\underset{\sim}{\Sigma}\underset{\sim}{b}$. From this the probability that $D(\underset{\sim}{x})$ is less than a given number is easily calculated. Using this method, the goodness of the various robust discriminants were evaluated. In this paper, we present abbreviated results for the univariate case and for the four variate case. Table 1 presents results for one variable, $\delta = 3$, and a variety of sample sizes, contaminants, and methods.

General conclusions from the univariate study are:

(1) When there is no contamination, i.e., $\alpha = \beta = 0$, all methods are equivalent. Thus, in the uncontaminated situation, any of the eleven robust procedures is as good as Fisher's LDF.

(2) For mild single sample contamination, e.g., 10% of $N(\mu,9)$ or $N(\mu,25)$, Fisher's LDF performs as well as the robust procedures. For samples of size 25, Fisher's LDF performs well for any amount of $N(\mu,9)$ contamination, i.e., $(\alpha,\beta) = (0,0.10)$, $(0,0.25)$, $(0.10,0.10)$, $(0.25, 0.25)$.

In general, however, Fisher's linear discriminant function does not perform as well as the robust procedures when the initial samples are scale contaminated.

(3) The Type II robust LDF, the Robust Regression discriminant function does not perform well when the initial samples are scale contaminated. In general, when Fisher's LDF performs poorly, so does the Type II discriminant function.

(4) For mild contaminating distributions, e.g., $N(\mu,9)$, the other robust discriminant functions (Type I and Type III) are equivalent, i.e., they give essentially the same probabilities of misclassification.

(5) For moderate contaminating distributions, e.g., $N(\mu,25)$, most of the Type I and Type III procedures remain similar to each other. The Type I Hampel and Sin LDF's tend to perform better than the other procedures when $n = 25$.

TABLE 1

AVERAGE PROBABILITY OF MISCLASSIFICATION FOR EACH PROCEDURE ($m=1$, $\delta=3$)
(OPTIMAL PROBABILITY = 0.0668)

			Discriminant Procedure											
G(x)	(α,β)	n	I TRIM15	I TRIM25	I HUB	I HAMHUB	I HAMP	I SIN	I GAST	I TRI	II ROBREG	III TLDF15	III TLDF25	FISHER
--	(0,0)	12	0.070	0.070	0.070	0.070	0.070	0.071	0.070	0.070	0.070	0.070	0.070	0.070
		25	0.068	0.068	0.068	0.068	0.068	0.068	0.069	0.069	0.068	0.068	0.068	0.068
N(μ,9)	(0,0.10)	12	0.074	0.075	0.074	0.074	0.073	0.074	0.075	0.074	0.074	0.074	0.075	0.074
		25	0.070	0.070	0.070	0.070	0.070	0.070	0.070	0.070	0.070	0.070	0.070	0.070
N(μ,9)	(0,0.25)	12	0.077	0.076	0.076	0.077	0.076	0.075	0.076	0.077	0.078	0.077	0.076	0.082
		25	0.070	0.071	0.070	0.070	0.071	0.071	0.072	0.072	0.070	0.070	0.071	0.070
N(μ,9)	(0.10,0.10)	12	0.070	0.071	0.070	0.070	0.070	0.071	0.071	0.070	0.071	0.071	0.071	0.071
		25	0.070	0.070	0.070	0.070	0.070	0.070	0.070	0.069	0.069	0.069	0.070	0.069
N(μ,9)	(0.25,0.25)	12	0.076	0.075	0.075	0.075	0.075	0.075	0.075	0.076	0.076	0.076	0.075	0.082
		25	0.070	0.070	0.070	0.070	0.070	0.070	0.070	0.070	0.070	0.070	0.070	0.070
N(μ,9)	--	--	0.072	0.072	0.072	0.072	0.072	0.072	0.073	0.072	0.072	0.072	0.072	0.074
N(μ,25)	(0,0.10)	12	0.070	0.071	0.071	0.071	0.071	0.071	0.072	0.070	0.071	0.071	0.071	0.072
		25	0.071	0.071	0.070	0.070	0.070	0.070	0.072	0.070	0.072	0.071	0.071	0.074
N(μ,25)	(0,0.25)	12	0.078	0.079	0.078	0.079	0.078	0.077	0.079	0.077	0.080	0.080	0.079	0.087
		25	0.071	0.071	0.071	0.072	0.071	0.070	0.072	0.072	0.073	0.071	0.071	0.077
N(μ,25)	(0.10,0.10)	12	0.073	0.072	0.073	0.073	0.072	0.072	0.072	0.072	0.159	0.073	0.072	0.160
		25	0.072	0.073	0.072	0.072	0.072	0.073	0.072	0.073	0.071	0.072	0.073	0.070
N(μ,25)	(0.25,0.25)	12	0.080	0.075	0.080	0.077	0.075	0.075	0.078	0.078	0.166	0.085	0.075	0.180
		25	0.073	0.072	0.072	0.073	0.074	0.072	0.072	0.071	0.073	0.073	0.072	0.079
N(μ,25)	--	--	0.073	0.073	0.073	0.073	0.073	0.073	0.073	0.073	0.096	0.074	0.073	0.100
N(μ,100)	(0,0.10)	12	0.074	0.071	0.071	0.071	0.070	0.070	0.071	0.073	0.074	0.078	0.071	0.103
		25	0.071	0.071	0.070	0.070	0.069	0.069	0.071	0.072	0.072	0.070	0.071	0.085
N(μ,100)	(0,0.25)	12	0.092	0.080	0.085	0.084	0.081	0.077	0.081	0.088	0.089	0.106	0.080	0.113
		25	0.074	0.071	0.073	0.074	0.073	0.071	0.070	0.073	0.076	0.077	0.071	0.104
N(μ,100)	(0.10,0.10)	12	0.072	0.071	0.070	0.070	0.071	0.072	0.070	0.072	0.072	0.074	0.071	0.100
		25	0.069	0.069	0.070	0.070	0.070	0.070	0.069	0.070	0.070	0.070	0.069	0.086
N(μ,100)	(0.25,0.25)	12	0.114	0.090	0.110	0.100	0.081	0.075	0.099	0.105	0.200	0.146	0.090	0.268
		25	0.070	0.072	0.070	0.070	0.072	0.072	0.073	0.072	0.070	0.070	0.071	0.074
N(μ,100)	--	--	0.080	0.074	0.077	0.076	0.073	0.072	0.075	0.078	0.090	0.086	0.074	0.117
Cauchy	(0,0.10)	12	0.071	0.071	0.071	0.070	0.071	0.071	0.071	0.071	0.071	0.070	0.071	0.087
		25	0.068	0.068	0.068	0.068	0.068	0.068	0.068	0.069	0.068	0.068	0.068	0.068
Cauchy	(0,0.25)	12	0.072	0.072	0.072	0.072	0.072	0.072	0.072	0.072	0.156	0.072	0.072	0.118
		25	0.070	0.070	0.070	0.070	0.070	0.070	0.070	0.071	0.073	0.070	0.070	0.122
Cauchy	(0.10,0.10)	12	0.074	0.075	0.074	0.074	0.074	0.075	0.075	0.075	0.074	0.073	0.075	0.126
		25	0.070	0.070	0.070	0.070	0.069	0.069	0.070	0.070	0.155	0.070	0.070	0.133
Cauchy	(0.25,0.25)	12	0.079	0.076	0.076	0.076	0.075	0.075	0.077	0.076	0.084	0.099	0.076	0.122
		25	0.070	0.070	0.070	0.070	0.070	0.069	0.070	0.069	0.157	0.070	0.070	0.151
Cauchy	--	--	0.072	0.072	0.071	0.071	0.071	0.071	0.072	0.071	0.105	0.074	0.072	0.116
--	--	--	0.074	0.073	0.074	0.073	0.072	0.072	0.073	0.074	0.091	0.077	0.073	0.101

(6) For heavier contamination, e.g., $N(\mu,100)$, a few other differences appear, especially for $n = 12$. The 15% Trimmed LDF tends not to perform as well as some of the other robust procedures. The Type I 15% Trimmed Mean LDF and the Type I Trimean LDF also tend not to perform as well as the other procedures. The Type I Sin estimate again tends to perform better than the other methods in many cases.

(7) For Cauchy contamination, the Type I and Type III discriminant functions perform quite similarly to one another for most parameter combinations. Note, however, the following exceptions. For $(\alpha,\beta) = (0,0.25)$, $n = 12$, $\delta = 1$, the Type I Huber, Hampel Huber, Hampel and Sin LDFs perform better than the other procedures. For $(\alpha,\beta) = (0.25,0.25)$ $n = 12$, $\delta = 1$ and 3, the 15% Trimmed LDF does not perform as well as the other procedures.

(8) Differences among the methods decrease as the sample size increases, i.e., there are smaller differences among the methods when $n = 25$ than when $n = 12$. This is to be expected from the asymptotic results presented earlier. When n goes to infinity, Fisher's LDF becomes optimal. With $n = 25$, there are still many situations, however, where Fisher's linear discriminant function does not perform well.

Overall, the Type I and Type III LDFs are very similar to one another. Fisher's linear discriminant function and the Type II Robust Regression discriminant function can perform very poorly when the initial samples are scale contaminated.

Table 2 gives results for the four variate case for $\delta = 3$.

(1) Consider first the uncontaminated situation, i.e., $\alpha = \beta = 0$. When $n = 25$, the twenty procedures are very similar to one another. The robust procedures perform as well as Fisher's LDF. When $n = 12$ and $\delta = 1$, the Type I and Type III discriminant functions give probabilities of misclassification 0.01 to 0.02 higher than the Fisher and Robust Regression LDFs. The Type I Sin LDF gives the highest probability of misclassification (0.375 for the 5% Sin LDF, 0.372 for the 10% Sin LDF), as compared to Fisher's LDF (0.357). The best of the Type I

discriminant functions in this case is the 15% Trimmed Mean LDF (0.363, 0.364). When $n = 12$ and $\delta = 3$, the Fisher LDF and the Type I LDFs give essentially the same probabilities of misclassification. The Type II and Type III LDFs give slightly higher probabilities.

(2) For single sample $N(\underset{\sim}{\mu}, 9\underset{\sim}{I})$ contamination, Fisher's LDF performs as well as the other procedures. In general, however, Fisher's linear discriminant function does not perform as well as the robust procedures when the initial samples are scale contaminated.

(3) As in the univariate case, the Type II Robust Regression discriminant function does not perform well. The Type II discriminant function and Fisher's LDF give very similar results.

(4) For mild contaminating distributions, e.g., $N(\underset{\sim}{\mu}, 9\underset{\sim}{I})$, with both samples contaminated, the Type I LDFs perform better than the others. With 10% contamination, the Type I discriminant functions are very similar. For $\delta = 3$ and 25% contamination, there is little difference among the Type I functions. For $\delta = 1$ and 25% contamination, there is a little more variation among the Type I LDFs. The Sin LDF performs best followed by the 25% Trimmed Mean LDF and Hampel's LDF.

(5) For moderate contaminating distributions, e.g., $N(\underset{\sim}{\mu}, 25\underset{\sim}{I})$, the Type I LDFs again perform better than the other discriminant functions. There is generally a difference of 0.01 to 0.02 in probability of misclassification between the best and worst of the Type I LDFs. The Sin LDF and the Trimmed Mean LDFs tend to perform very well in this case.

In the following special case Fisher's LDF and the Type II and III LDFs performed very well: $(\delta=1, (\alpha,\beta) = (0.25, 0.25), n=25)$ and $(\delta=3, (\alpha,\beta) = (0.25, 0.25), n=12)$.

(6) For heavier contaminating distributions, e.g., $N(\underset{\sim}{\mu}, 100\underset{\sim}{I})$, the situation is very similar to that of moderate contamination. The Type I LDFs perform better than the others with the following exceptions: $(\delta=1, (\alpha,\beta) = (0.25, 0.25), n=12)$, $(\delta=3, (\alpha,\beta) = (0, 0.25), n=12)$, $(\delta=3, (\alpha,\beta) = (0.25, 0.25), n=12)$. There is generally a difference of 0.01 to 0.03 in probability of misclassification between the best and worst of the Type I

LDFs. The Hampel, Sin, and Trimmed Mean LDFs often give the smallest probabilities of misclassification.

(7) Cauchy contamination gave results similar to those for moderate and heavy contamination. A Type I LDF is generally the safest procedure in this situation, with the Hampel and Sin discriminant functions usually performing best. There were several cases, however, where Fisher's LDF or one of the Type III LDFs performed as well as or better than the Type I discriminant functions: for $(\delta=1,(\alpha,\beta) = (0,0.10),n=12)$, the 15% Trimmed LDF performed well; for $(\delta=1,(\alpha,\beta) = (0.10,0.10),n=25)$, the Type III LDFs performed well; for $(\delta=3,(\alpha,\beta) = (0,0.10),n=12)$ Fisher's LDF and the Type II and Type III LDFs gave smaller probabilities of misclassification than the Type I LDFs; for $(\delta=3,(\alpha,\beta) = (0.25,0.25),n=12$ and $n = 25)$, the 15% Trimmed LDF performed as well as or better than the Type I discriminant functions.

(8) As in the univariate study, differences among the methods decrease as the sample size increases, supporting once again our asymtotic results which indicate that for large samples, Fisher's linear discriminant function is optimal even when the initial samples are scale contaminated. With samples of size 25, however, Fisher's LDF may still perform very poorly with contaminated samples.

Overall, the Type I LDFs are very similar to each other. The Type III Trimmed discriminant functions usually do not perform as well as the Type I LDFs but perform considerably better than the Robust Regression LDF or Fisher's LDF.

We have noted that the linear discriminant function is vulnerable to contamination effects even when the a priori probability is $\frac{1}{2}$. A number of alternatives appear promising, these should be studied for the case $p \neq \frac{1}{2}$ and for location contamination robustness. As is usually the case, no one single estimator is uniformly the best, but several seem to perform quite well in a wide variety of situations.

TABLE 2

AVERAGE PROBABILITY OF MISCLASSIFICATION FOR EACH PROCEDURE (m=4, δ=3)
(OPTIMAL PROBABILITY = 0.0668)

			Discriminant Procedure							
G(x)	(α,β)	n	I TRIM15 -05	I TRIM25 -05	I HUB -05	I HAMHUB -05	I HAMP -05	I SIN -05	I GAST -05	I TRI -05
--	(0,0)	12	0.096	0.096	0.096	0.096	0.096	0.097	0.095	0.095
		25	0.083	0.082	0.083	0.083	0.084	0.084	0.082	0.082
N(μ,9I)	(0,0.10)	12	0.089	0.091	0.090	0.090	0.090	0.089	0.091	0.089
		25	0.081	0.081	0.081	0.081	0.081	0.082	0.081	0.082
N(μ,9I)	(0,0.25)	12	0.109	0.106	0.110	0.110	0.109	0.110	0.107	0.108
		25	0.079	0.079	0.078	0.078	0.079	0.079	0.078	0.079
N(μ,9I)	(0.10,0.10)	12	0.098	0.100	0.098	0.098	0.097	0.098	0.098	0.097
		25	0.079	0.080	0.079	0.079	0.079	0.080	0.081	0.080
N(μ,9I)	(0.25,0.25)	12	0.104	0.102	0.102	0.105	0.105	0.104	0.103	0.103
		25	0.088	0.089	0.089	0.089	0.090	0.090	0.090	0.091
N(μ,9I)	--	--	0.091	0.091	0.091	0.091	0.091	0.091	0.091	0.091
N(μ,25I)	(0,0.10)	12	0.090	0.090	0.089	0.090	0.090	0.090	0.090	0.089
		25	0.076	0.076	0.077	0.076	0.076	0.076	0.075	0.075
N(μ,25I)	(0,0.25)	12	0.108	0.111	0.107	0.107	0.108	0.110	0.110	0.110
		25	0.096	0.098	0.096	0.096	0.099	0.098	0.101	0.097
N(μ,25I)	(0.10,0.10)	12	0.109	0.106	0.103	0.102	0.098	0.099	0.105	0.103
		25	0.089	0.089	0.089	0.090	0.088	0.088	0.090	0.090
N(μ,25I)	(0.25,0.25)	12	0.123	0.118	0.124	0.125	0.123	0.123	0.119	0.123
		25	0.119	0.116	0.119	0.122	0.129	0.130	0.116	0.116
N(μ,25I)	--	--	0.101	0.101	0.101	0.101	0.101	0.101	0.101	0.100
N(μ,100I)	(0,0.10)	12	0.118	0.117	0.118	0.119	0.119	0.119	0.117	0.116
		25	0.098	0.098	0.098	0.099	0.101	0.099	0.098	0.098
N(μ,100I)	(0,0.25)	12	0.150	0.149	0.151	0.150	0.157	0.155	0.150	0.151
		25	0.109	0.107	0.107	0.107	0.108	0.107	0.108	0.113
N(μ,100I)	(0.10,0.10)	12	0.116	0.118	0.115	0.115	0.117	0.117	0.118	0.115
		25	0.090	0.090	0.089	0.090	0.089	0.088	0.090	0.091
N(μ,100I)	(0.25,0.25)	12	0.251	0.230	0.249	0.242	0.240	0.230	0.243	0.248
		25	0.129	0.127	0.128	0.128	0.131	0.129	0.126	0.129
N(μ,100I)	--	--	0.133	0.129	0.132	0.131	0.133	0.131	0.131	0.133
Cauchy	(0,0.10)	12	0.101	0.105	0.101	0.100	0.099	0.101	0.103	0.100
		25	0.080	0.080	0.080	0.080	0.080	0.080	0.080	0.081
Cauchy	(0,0.25)	12	0.125	0.126	0.125	0.125	0.124	0.124	0.125	0.125
		25	0.077	0.077	0.077	0.078	0.080	0.078	0.078	0.077
Cauchy	(0.10,0.10)	12	0.096	0.098	0.096	0.095	0.095	0.096	0.099	0.098
		25	0.084	0.085	0.084	0.084	0.085	0.085	0.085	0.085
Cauchy	(0.25,0.25)	12	0.133	0.131	0.136	0.139	0.141	0.140	0.132	0.133
		25	0.086	0.086	0.086	0.085	0.085	0.085	0.087	0.086
Cauchy	--	--	0.098	0.098	0.098	0.098	0.098	0.099	0.099	0.098
--	--	--	0.106	0.105	0.105	0.105	0.106	0.106	0.106	0.106

TABLE 2 (Continued)

			Discriminant Procedure											
$G(\underline{x})$	(α,β)	n	I TRIM15 -10	I TRIM25 -10	I HUB -10	I HAMHUB -10	I HAMP -10	I SIN -10	I GAST -10	I TRI -10	II ROBREG	III TLDF15	III TLDF25	FISHE
--	(0,0)	12	0.094	0.095	0.094	0.096	0.094	0.096	0.093	0.094	0.098	0.099	0.104	0.09
		25	0.082	0.083	0.082	0.082	0.083	0.083	0.083	0.083	0.077	0.080	0.083	0.07
$N(\underline{\mu}, 9I)$	(0,0.10)	12	0.089	0.092	0.089	0.089	0.090	0.091	0.090	0.089	0.101	0.100	0.112	0.10
		25	0.082	0.083	0.082	0.081	0.081	0.081	0.084	0.084	0.083	0.079	0.082	0.08
$N(\underline{\mu}, 9I)$	(0,0.25)	12	0.106	0.108	0.108	0.109	0.109	0.108	0.108	0.107	0.118	0.109	0.110	0.11
		25	0.078	0.078	0.078	0.078	0.078	0.078	0.077	0.077	0.079	0.079	0.081	0.08
$N(\underline{\mu}, 9I)$	(0.10,0.10)	12	0.101	0.101	0.100	0.100	0.099	0.100	0.100	0.101	0.107	0.099	0.106	0.10
		25	0.076	0.077	0.076	0.076	0.076	0.076	0.078	0.079	0.078	0.079	0.084	0.08
$N(\underline{\mu}, 9I)$	(0.25,0.25)	12	0.110	0.111	0.110	0.111	0.110	0.111	0.109	0.109	0.111	0.122	0.117	0.11
		25	0.084	0.084	0.084	0.084	0.084	0.085	0.084	0.086	0.101	0.093	0.098	0.10
$N(\underline{\mu}, 9I)$	--	--	0.091	0.092	0.091	0.091	0.091	0.092	0.091	0.092	0.(97	0.095	0.099	0.09
$N(\underline{\mu}, 25I)$	(0,0.10)	12	0.097	0.097	0.097	0.097	0.097	0.096	0.097	0.099	0.103	0.099	0.112	0.10
		25	0.074	0.074	0.074	0.073	0.074	0.074	0.075	0.075	0.094	0.086	0.093	0.09
$N(\underline{\mu}, 25I)$	(0,0.25)	12	0.105	0.109	0.106	0.106	0.109	0.111	0.107	0.109	0.113	0.121	0.114	0.11
		25	0.091	0.094	0.093	0.097	0.096	0.095	0.096	0.091	0.101	0.094	0.099	0.11
$N(\underline{\mu}, 25I)$	(0.10,0.10)	12	0.107	0.111	0.105	0.105	0.101	0.105	0.110	0.107	0.109	0.125	0.116	0.11
		25	0.081	0.082	0.080	0.081	0.080	0.080	0.084	0.082	0.099	0.095	0.098	0.10
$N(\underline{\mu}, 25I)$	(0.25,0.25)	12	0.120	0.127	0.120	0.120	0.129	0.127	0.129	0.130	0.105	0.103	0.109	0.11
		25	0.102	0.101	0.102	0.104	0.109	0.109	0.100	0.098	0.133	0.125	0.126	0.14
$N(\underline{\mu}, 25I)$	--	--	0.097	0.099	0.097	0.098	0.099	0.100	0.100	0.099	0.107	0.106	0.108	0.11
$N(\underline{\mu}, 100I)$	(0,0.10)	12	0.103	0.095	0.104	0.101	0.098	0.097	0.099	0.100	0.145	0.112	0.111	0.14
		25	0.090	0.090	0.090	0.090	0.090	0.089	0.092	0.088	0.104	0.091	0.103	0.11
$N(\underline{\mu}, 100I)$	(0,0.25)	12	0.139	0.137	0.139	0.138	0.149	0.149	0.140	0.139	0.139	0.121	0.133	0.14
		25	0.117	0.113	0.116	0.117	0.118	0.116	0.114	0.116	0.174	0.119	0.139	0.20
$N(\underline{\mu}, 100I)$	(0.10,0.10)	12	0.119	0.123	0.118	0.118	0.121	0.123	0.119	0.117	0.179	0.155	0.170	0.18
		25	0.100	0.100	0.101	0.101	0.098	0.098	0.103	0.101	0.119	0.104	0.113	0.13
$N(\underline{\mu}, 100I)$	(0.25,0.25)	12	0.234	0.224	0.233	0.226	0.224	0.220	0.232	0.226	0.260	0.239	0.207	0.29
		25	0.126	0.124	0.125	0.125	0.119	0.117	0.125	0.121	0.140	0.133	0.127	0.15
$N(\underline{\mu}, 100I)$	--	--	0.129	0.126	0.128	0.127	0.127	0.126	0.128	0.126	0.158	0.134	0.138	0.17
Cauchy	(0,0.10)	12	0.101	0.105	0.101	0.100	0.099	0.101	0.103	0.100	0.095	0.092	0.093	0.09
		25	0.086	0.087	0.086	0.086	0.086	0.086	0.088	0.086	0.083	0.084	0.085	0.08
Cauchy	(0,0.25)	12	0.088	0.090	0.088	0.087	0.085	0.086	0.089	0.089	0.123	0.096	0.117	0.12
		25	0.079	0.079	0.079	0.079	0.081	0.080	0.081	0.079	0.182	0.086	0.127	0.17
Cauchy	(0.10,0.10)	12	0.092	0.094	0.093	0.092	0.091	0.092	0.095	0.094	0.096	0.097	0.097	0.09
		25	0.087	0.087	0.086	0.087	0.087	0.087	0.086	0.087	0.143	0.133	0.143	0.15
Cauchy	(0.25,0.25)	12	0.101	0.096	0.102	0.105	0.105	0.104	0.098	0.101	0.136	0.089	0.127	0.16
		25	0.083	0.084	0.083	0.083	0.082	0.083	0.084	0.083	0.132	0.083	0.099	0.13
Cauchy	--	--	0.089	0.089	0.089	0.089	0.089	0.089	0.089	0.089	0.124	0.095	0.111	0.12
--	--	--	0.101	0.101	0.101	0.101	0.101	0.101	0.102	0.101	0.121	0.107	0.114	0.12

References

1. Ahmed, S. W. (1975), Discriminant analysis when the initial samples are contaminated. Ph.D. Dissertation, University of North Carolina.

2. Ahmed, S. W. and Lachenbruch, P. A., (1975), Discriminant analysis when one or both of the initial samples is contaminated: Large Sample Results, EDV in Medizin und Biologie, 6, pp.35-42.

3. Andrews, D. F., Bickel, P.J., Hampel, F.R., Huber, P.J., Rogers, W.H., and Tukey, J.W.,(1972), Robust Estimates of Location, Princeton University Press.

4. Clarke, W.R., (1975), Dissertation, University of Iowa.

5. Gilbert, E.S., (1968), On discrimination using qualitative variables, JASA, 63, p. 1399.

6. Gnanadesikan, R. and Kettenring, J., (1972), Robust estimates, residuals, and outlier detection with multiresponse data, Biometrics, 28, pp. 81-124.

7. Lachenbruch, P.A., (1966), Discriminant analysis when the initial samples are misclassified, Technometrics, 8, p. 657.

8. Lachenbruch, P.A., Sneering, D., and Revo, L.T., (1973), Robustness of the linear and quadratic discriminant function to certain types of non-normality, Commun. Stat., 1, pp. 57-59.

9. Revo, L.T., (1970), On classifying with certain types of ordered qualitative data: an evaluation of several procedures. North Carolina Institute of Statistics Mimeo Series 708.

Department of Biostatistics
The University of North Carolina
Chapel Hill, North Carolina 27514

The Statistical Basis of Computerized Diagnosis Using the Electrocardiogram

Jerome Cornfield
Rosalie A. Dunn
Hubert V. Pipberger

Introduction.

This paper describes the computer-based multigroup diagnosis program of the Veterans Administration, a program designed to assign a tracing to one of a set of predesignated diagnostic categories with minimum probability of error, based on the resting electrocardiogram. Unlike most other automatic interpretations of the ECG, it is not intended to simulate a skilled cardiologist, i.e., to approach as closely as possible the interpretation that would have been reached by such a reader. Some cardiologists, perhaps many, would regard the equivalence of these two objectives as axiomatic, but we consider it safer to regard the equivalence as a hypothesis to be tested rather than as an axiom on which to base an algorithm. Algorithms accepting the equivalence usually simulate the cardiologist by using decision trees. Thus, if the Q/R ratio or Q amplitude in lead Z exceed "normal limits," anterior myocardial infraction, pulmonary emphysema or left ventricular hypertrophy would need to be considered. Additional measurements, e.g., R amplitude in lead X, would then be used to distinguish among these possibilities. Evaluation of the performance of such algorithms, obtained by comparing their interpretations in patients whose diagnosis was established by in-pendent means, such as cardiac catheterization, clinical history, or autopsy, is not reassuring. Thus, when a selected group of electrocardiographers were tested in a study by Simonson et al [1], they achieved correct interpretations in only 54 percent of the cases. Similarly, a criterion for the diagnosis of left ventricular hypertrophy, known to achieve 60 percent accuracy in autopsy material [2], led, when embodied in a computer program, to a 98 percent agreement between computer and cardiologist, a concordance rate which was incorrectly interpreted as a measure of the accuracy of the interpretation [3]. Frequently, the decision-tree programs include large numbers of electrocardiogram measurements, leading to overdiagnosis and the labelling of excessive numbers of normal tracings as abnormal [4]. Such misdiagnosis has been called "heart disease of electrocardiographic origin" [5], an ailment with normal life expectation but accompanied by considerable anxiety.

The procedure here described is based on the estimation of the probability that an individual with observational vector $\underline{x}$ falls in a given diagnostic category, since the assignment procedure with minimum classification error is that which assigns an individual to that category for which his probability is greatest [6]. The algorithm only approximates the required probability and so cannot be claimed to have achieved minimum error. Our working hypothesis is that direct efforts to approximate this probability will lead to smaller classification error than efforts to simulate cognitive processes that are at best imperfect, particularly when large numbers of variables are involved and, in any event, poorly understood. Although this working hypothesis can and should be put to independent test, all the evidence to date is in favor of it.

It is convenient to consider separately the methods used to obtain the observational vector $\underline{x}$ from an ECG tracing, and the data and methods used to convert such an $\underline{x}$ to the set of probabilities which constitute the final computer printout.

Wave Recognition and Measurements.

Figure 1 shows schematically an idealized output from one beat for the three orthogonal leads. The voltages in each of the three leads are stored on magnetic tape at a sampling rate of 500/sec. for 10 seconds. What is an appropriate way to represent these data? One possibility is to transform from the time to the frequency domain and to represent the

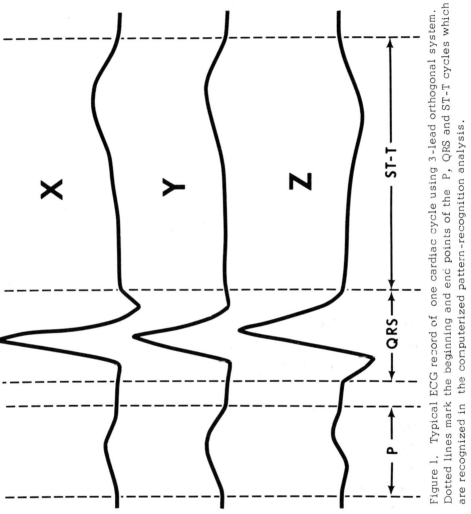

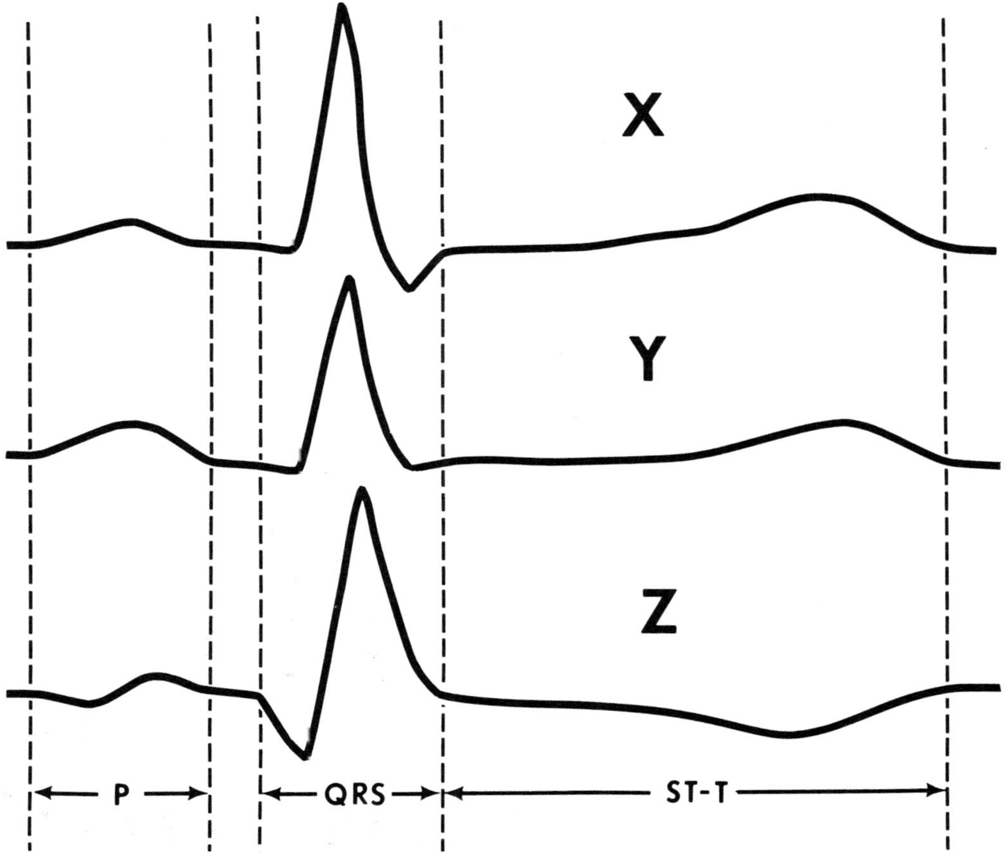

Figure 1. Typical ECG record of one cardiac cycle using 3-lead orthogonal system. Dotted lines mark the beginning and end points of the P, QRS and ST-T cycles which are recognized in the computerized pattern-recognition analysis.

tracing by some set of Fourier coefficients. Another is to regard the problem as one of pattern recognition. Still a third, and the one adopted by us, is to identify and abstract those electrocardiographic variables that the cardiologist has traditionally dealt with. The advantage of the third approach is that familiar variables have a physiologic basis.

As an example of how these variables are defined we consider the onset of QRS, starting with smoothed derivatives of voltage with respect to time for each of the three component leads, $\dot{x}_t$, $\dot{y}_t$, and $\dot{z}_t$, the computation of which for each beat, given the basic digital record, is immediate. Spatial velocity, V_t, defined as $(\dot{x}_t^2 + \dot{y}_t^2 + \dot{z}_t^2)^{\frac{1}{2}}$, is compared with a standard spatial velocity curve, V_T^*, obtained by averaging the V_t for a series of records for which QRS onset had been visually identified by trained readers, with $T = 0$ at QRS onset. Time of onset for a new tracing, t_q, is then defined as that value of t_q which minimizes $\Sigma w_t(V_{t-t_q} - V_T^*)^2$ in the neighborhood of t_q, and w_t is the inverse of the estimated variance around the standard curve. Differences between computer and visually identified QRS onsets fall within a range of about ±12 msec, or about one percent of a beat length. Computations defining the beginning and ends of other traditional waves are also specified.

The definition of the wave onsets and terminations is a first step in defining the phases of each record, this constituting a necessary preliminary to the averaging of beats or of different records. Averaging out-of-phase records will produce a gently undulating function in which the characteristic features of each individual function, such as the sharp QRS peak, are obliterated. The second step preliminary to averaging is time normalization. Thus the interval from QRS onset to termination is divided into eight equal segments, and the amplitude of the projections onto the x, y, and z axis at each eighth are averaged for each beat in the record (Figure 2).

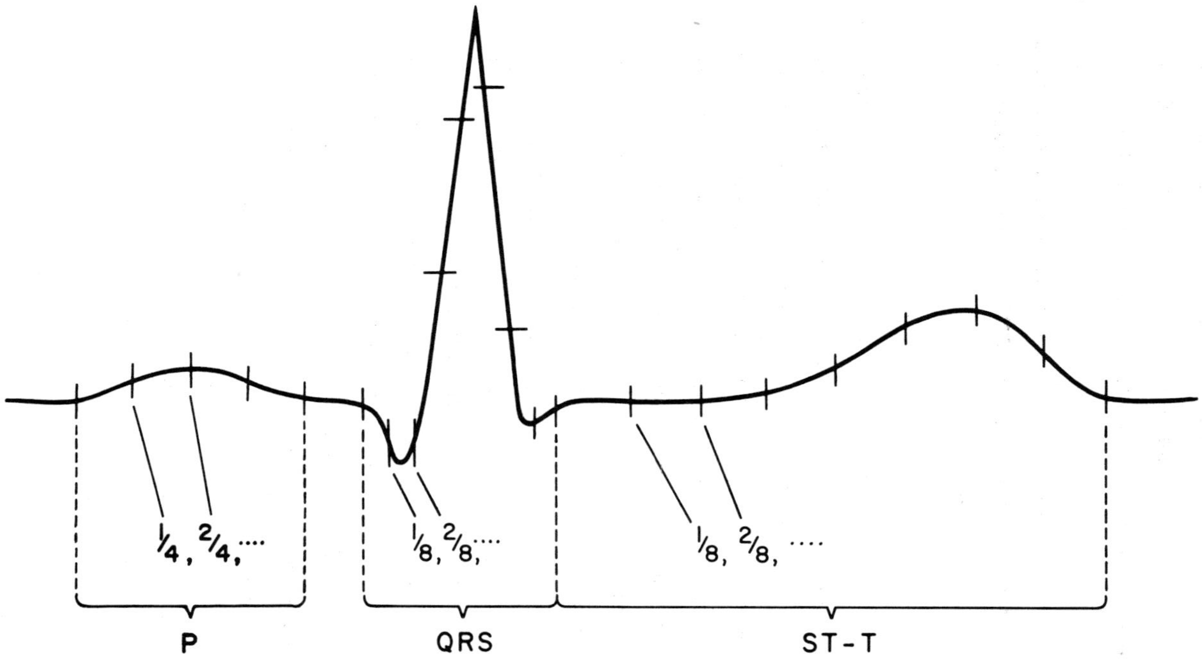

Figure 2. Hypothetical ECG record of one cardiac cycle of one scalar lead from a 3-lead orthogonal system. Dotted lines mark the beginning and end points of the P, QRS, and ST-T cycles which are recognized in the computerized pattern-recognition analysis. The P interval is divided into fourths and the QRS and ST-T complexes into eighths, so that voltages recorded at these points are considered time-normalized.

Procedures of this type reduce the original tracing to about 300 measurements, which in the aggregate include virtually every ECG variable that has been suggested in the electrocardiographic literature. These variables are of six types: voltage amplitudes on time-normalized scale, maximum aplitudes for each wave, angular variables, wave durations, time of peak amplitudes for each wave, and areas, expressed in units of millivolts x milliseconds.

Data Base.

Tracings of this type and processed in this way have been collected for a very large number of subjects of known diagnostic status. We shall consider here the 2336 records collected on male Veterans Administration patients over a wide range of ages. Similar data bases are being established for women and for children, as well as for exercise electrocardiograms on patients with atherosclerosis as indicated by coronary angiograms, and for young adults in whom no or minimal atherosclerosis may be presumed. The diagnosis is determined on the basis of non-electrocardiographic criteria. Thus the sample of 375 patients with left ventricular hypertrophy consisted of 212 cases with aortic valvular disease, 37 cases with mitral regurgitation (both evidenced by left heart catheterization), 78 cases with a history of sustained hypertension, and 68 cases identified at autopsy. But despite the use of a wide range of diagnostic criteria, the diagnostic status of each patient in the data base cannot be considered as known without error, particularly for the normals. Despite the absence of present or past signs or symptoms of cardiovascular disease, non-trivial numbers of them probably had atherosclerosis which would have been disclosed if non-invasive methods of diagnosis had been available. The diagnostic status of any patient in the data base should not therefore be considered as anything absolute, but rather as a state operationally defined by the diagnostic methods used to establish it.

Calculation of Posterior Probabilities.

Let $\underline{x}$ denote a k-dimensional vector of measurements for a patient, m the number of diagnostic classes, and i a particular diagnostic class from the m possibilities. (In what follows, k = 66 and m = 7.) If $f(\underline{x}|i)$ denotes the conditional generalized probability density function of $\underline{x}$ for those in disease category i, and g_i the unconditional (prior) probability of membership in category i, then the posterior probability of membership in category i, given the measurement vector $\underline{x}$, is, by Bayes theorem,

(1) $$P(i|\underline{x}) = f(\underline{x}|i)g_i / \sum_{j=1}^{m} f(\underline{x}|j)g_j \; .$$

The density function, $f(\underline{x}|i)$, has been taken as multivariate normal with mean vector $\underline{\mu}_i$ and common covariance matrix $\underline{\Sigma}$. The function of the data base is then to supply estimates of the likelihood ratios $f(\underline{x}|j)/f(\underline{x}|i)$, where

(2) $$\log f(\underline{x}|j)/f(\underline{x}|i) = [\underline{x} - (\underline{\mu}_i + \underline{\mu}_j)/2]'\underline{\Lambda} \; ,$$

and $\underline{\Lambda}$ is the vector of linear discriminant coefficients, $\underline{\Sigma}^{-1}(\underline{\mu}_j - \underline{\mu}_i)$. The unknown parameters are estimated using sample means, variances and covariances for each diagnostic group in the data base.

A special problem arises in the treatment of angular variables. These arise as follows: The voltages observed on the x, y, and z leads at time t define a point in 3-space, while the ray from the origin to that point defines two angles, which contain diagnostic information. But these angles obey the addition rule that $360° + \alpha° = \alpha°$ so that averaging, let alone treatment as multivariate normal, is inappropriate. What can be averaged, however, are the x, y, and z coordinates, so that the average angle, $\bar{\theta}$, can be defined as the angle corresponding to these average coordinates [7]. This procedure is appropriate if the univariate p.d.f. of the angles in the ith diagnostic category is of the form

(3) $$f(\theta|\bar{\theta}_i) = C\, e^{\kappa \cos(\theta - \bar{\theta}_i)} \; ,$$

where κ is a measure of dispersion and C a normalizing constant depending only on κ. For two diagnostic categories with the same κ, θ affects the log likelihood ratio only through the quantity $\cos(\theta - \bar{\theta}_i) - \cos(\theta - \bar{\theta}_j)$. We have taken this function of θ as the one for which computation of arithmetic means, variances and covariances is appropriate.

Results.

The assumption of multivariate normality with equal covariance matrices is far from satisfied, and one must inquire what effect departures from this assumption have on the calculated probabilities. The result of the preceding analysis is to substitute for each tracing a vector of probabilities $(P_1, P_2, \ldots, P_m)$. If these probabilities actually describe the population, then for all individuals for whom P_i has a given value, the proportion who are actually in diagnostic category i should be P_i. Such a comparison is given in Tables 1 and 2 for two of the diagnostic categories, the normals and those with anterior myocardial infarcts. In each table the 2336 individuals in the data base are broken into ten equal groups of 233 or 234 individuals each, the first group having the smallest values of P_i and the tenth having the largest values. The tables compare the actual number out of 233 falling in each group with the sum of the P_i for all 233 individuals in that group. The posterior probabilities for normals are consistently overestimated by about 10 percent - a systematic error which is considered acceptable.

It will be noted that only about one third of the normals were assigned probabilities of .9 or greater of being normal and that about 15% were assigned probabilities of .5 or less. This, however, describes less the appropriateness or inappropriateness of the methods used to estimate the probabilities than the amount of information contained in the data. Another way of looking at this is provided in Table 3, which shows the number of standard deviations by which the pairwise means of the linear compound (equation 2) differ. A third way of viewing this is in terms of misclassification probabilities, with each individual assigned

Table 1

Expected and Observed Number of Normals by Decile of Posterior Probability of Being Normal, P_1

Decile of probability	P_1	Number of Normals Expected	Observed
1	0.000 - 0.000	0.0	0
2	0.000 - 0.002	0.2	0
3	0.002 - 0.007	0.9	0
4	0.007 - 0.019	2.8	1
5	0.020 - 0.060	8.4	4
6	0.060 - 0.161	23.0	18
7	0.161 - 0.437	67.6	61
8	0.437 - 0.746	141.1	128
9	0.746 - 0.914	197.0	173
10	0.915 - 1.000	225.2	211
Total		666.2	596

Table 2

Expected and Observed Number of AMI by Decile of Posterior Probability of Being AMI, P_2

Decile of probability	P_2	Number of AMI Expected	Observed
1	0.000 - 0.000	0.0	2
2	0.000 - 0.001	0.2	4
3	0.001 - 0.004	0.6	1
4	0.004 - 0.007	1.3	6
5	0.007 - 0.013	2.3	5
6	0.013 - 0.026	4.5	6
7	0.027 - 0.064	9.9	20
8	0.065 - 0.204	26.7	42
9	0.204 - 0.691	94.8	83
10	0.694 - 0.998	209.4	175
Total		349.7	344

Table 3

Number of Standard Deviations by Which Each Pairwise Comparison Differs, Seven Group Comparison

	AMI	PMI	LMI	LVH	RVH	COPD
Normal	3.46	3.55	7.45	2.11	2.34	2.76
AMI	--	3.55	3.80	1.91	2.90	2.26
PMI	--	--	5.36	2.77	3.23	2.82
LMI	--	--	--	4.55	5.62	4.82
LVH	--	--	--	--	2.10	2.19
RVH	--	--	--	--	--	2.17

AMI = anterior myocardial infarction

PMI = posterior myocardial infarction

LMI = lateral myocardial infarction

LVH = left ventricular hypertrophy

RVH = right ventricular hypertrophy

COPD = chronic obstructive pulmonary disease

to the diagnostic category for which his probability is greatest. (Minimization of misclassification cost provides a more general assignment procedure, but one we have so far not employed.) Because the electrocardiogram is not an infallible diagnostic tool, the probability and the assignment for many patients will be importantly influenced by the prior probabilities used. We therefore show the misclassification probabilities for three different sets of prior probabilities (Tables 4, 5, and 6). The probability of a correct diagnosis depends on the diagnostic category into which a patient falls, being highest for the normals and lowest for those with right ventricular hypertrophy.

Table 4

Misclassification Matrix Using Priors from the Well-Diagnosed File (2336 Patients)[†]

Clinical		Computer (1)	(2)	(3)	(4)	(5)	(6)	(7)	Prior Probabilities
N	(1)	88.4	1.3	1.5	0.0	3.4	1.2	4.2	0.255
AMI	(2)	7.6	63.7	3.8	2.9	12.8	0.6	8.7	0.147
PMI	(3)	4.2	4.2	81.1	0.9	3.6	1.8	4.2	0.192
LMI	(4)	1.2	16.5	2.4	69.4	2.4	1.2	7.1	0.036
LVH	(5)	22.8	11.2	6.9	0.8	50.0	2.3	6.1	0.169
RVH	(6)	32.2	5.0	4.1	0.0	4.1	38.0	16.5	0.052
COPD	(7)	13.3	8.6	9.5	0.9	4.0	3.5	60.2	0.149

[†]Total percentage correctly classified $= 88.4 \times 0.255 + 63.7 \times 0.147 + \ldots$
$= 69.4$.

Table 5

Misclassification Matrix Using Priors of the Admitting Office of VA Hospital, Washington, D. C. (2336 Patients)[†]

Clinical		Computer (1)	(2)	(3)	(4)	(5)	(6)	(7)	Prior Probabilities
N	(1)	98.0[‡]	0.3	0.7	0.0	0.5	0.0	0.5	0.60
AMI	(2)	14.5	60.2	3.5	2.6	12.5	0.6	6.1	0.07
PMI	(3)	11.6	4.0	77.3	0.9	3.1	0.7	2.4	0.10
LMI	(4)	8.2	14.1	2.4	69.4	2.4	0.0	3.5	0.02
LVH	(5)	34.8	9.1	5.6	0.8	44.7	2.0	3.0	0.13
RVH	(6)	48.8	4.1	2.5	0.0	4.1	28.9	11.6	0.02
COPD	(7)	27.7	8.4	9.2	0.9	3.4	2.6	47.8	0.06

[†]Total percentage correctly classified $= 98.0 \times 0.60 + 60.2 \times 0.07 + \ldots$
$= 81.4$

[‡]Note that only 2% of the normals are misclassified with a concomitant increase in misclassifications of abnormal records.

Table 6

Misclassification Matrix Using Priors of the Cardiology Service of the VA Hospital, West Roxbury, Ma. (2336 Patients)†

Clinical		(1)	(2)	(3)	(4)	(5)	(6)	(7)	Prior probabilities
N	(1)	76.3‡	1.8	3.2	0.0	15.4	1.0	2.2	0.12
AMI	(2)	3.5	68.0	3.8	3.2	19.2	0.3	2.0	0.19
PMI	(3)	1.3	4.7	84.2	0.9	6.5	1.1	1.3	0.24
LMI	(4)	1.2	17.6	1.2	71.8	7.1	0.0	1.2	0.06
LVH	(5)	12.7	10.9	6.1	0.8	64.7	1.8	3.0	0.30
RVH	(6)	21.5	9.1	8.3	0.0	14.9	35.5	10.7	0.03
PE	(7)	9.8	15.0	12.1	0.9	11.0	3.5	47.8	0.06

†Total percentage correctly classified $= 76.3 \times 0.12 + 68.0 \times 0.19 + \ldots = 69.9$.

‡Note that 24% of the normals were misclassified with an increase in correct classifications of abnormals.

Problems.

The method described provides a way of diagnosis using the resting electrocardiogram, but clearly one capable of improvement - whether slight or substantial remaining for future work to determine. We here mention a number of the more important problems.

Lack of mutually exclusive categories. It is possible for a patient to have two or more conditions, say, both an infarct and ventricular hypertrophy. The present algorithm assumes, however, that he has one or the other, but not both. This introduces a distortion into the probabilities that could be remedied only by collecting or constructing a data base with an expanded number of diagnostic categories. We hope to address ourselves to this task.

Lack of exhaustive categories. If an electrocardiogram is taken on a patient who falls in none of the prespecified categories, the algorithm will in desperation spread the probabilities over those assigned. It seems more appropriate, however, to perform a preliminary test of significance to help decide whether a patient does fall in one of the prespecified categories, and to compute the probabilities only if the decision is that he does, printing out "non-specific abnormality" if he does not. We have used the multivariate distance $(\underline{x} - \underline{\mu}_i)'\underline{\Sigma}^{-1}(\underline{x} - \underline{\mu}_i)$ as the test statistic. Given multivariate normality this has the chi-square distribution with k degrees of freedom when the hypothesis is true. But the actual histogram of this quantity departs markedly from that of χ_k^2, despite Tables 1 and 2, and we have therefore taken our percentage points from the empirical rather than theoretical distributions.

Priors. The required priors are given in principle by the relative frequency of the diagnostic categories in the population being classified. Thus in Table 2, as in Table 4, they are given by the frequencies of the different categories in the basic data base of 2336 patients; in Table 5 by the frequency of admission to the VA hospital in Washington, which includes a high proportion of normals; in Table 6 by those to the cardiology service of the VA hospital in West Roxbury, Massachusetts, which includes a high proportion of infarctions and ventricular hypertrophy. More recently we have adopted a system in which the physician requesting the electrocardiogram is asked to check off the tentative diagnosis and have used different priors for each such diagnosis. For those who wish a calculation minimally dependent on the choice of prior, there is a program option which will print out the ratio of average likelihoods, or Bayes factor [8], rather than the diagnosis probability. Thus, rather than the $P(i|\underline{x})$ of equation (1), we can supply

$$\text{Bayes factor} = \frac{P(i|\underline{x})}{1 - P(i|\underline{x})} \Bigg/ \frac{g_i}{1 - g_i}$$

$$= f(\underline{x}|i) \Big/ \frac{1}{1-g_i} \sum_{j \neq i} g_j f(\underline{x}|j) \ .$$

The priors are involved here only to the extent of defining the composite hypothesis, alternative to hypothesis i. Even this mild dependence could be avoided by computing only the $m(m-1)/2$ pairwise likelihood ratios, but we have found none so ideologically pure as to insist on this.

Multivariate normality. The data tails are heavier than is implied by the assumption of multivariate normality, and the diagnoses are overly influenced by extreme values in the data base. It would be desirable to investigate the use of alternative models, of which log multivariate normality, modified by the addition of a constant, seems the simplest.

Other demographic variables. Weight, skin color, age and sex have non-trivial influences on the electrocardiographic tracing and a systematic expansion of the data base to take account of them is desirable.

Comparative performance. There are other computer interpretations of the electrocardiogram available to the interested user. It would be of great interest and some importance to mount a comparative study in which these interpretations were compared with each other and with the interpretations of a panel of cardiologists on a set of tracings from patients of known diagnostic status. The complexity and cost of such a collaborative undertaking are substantial, and it has not yet been undertaken.

References

1. Simonson, E., Tuna, N., Okamoto, N. et al, Diagnostic Accuracy of the Vectorcardiogram and Electrocardiogram. A Cooperative Study, Am. J. Cardiol. 17 : 829-878, 1966.
2. Romhilt, D.W., Estes, E.H., Jr., A Point-Score System for the ECG Diagnosis of Left Ventricular Hypertrophy, Am. Hear J., 75: 752-758, 1968.
3. Crevasse, L., Ariel, M., A New Scalar Electrocardiographic Computer Program. Clinical Evaluation", J.A.M.A., 226: 1089-1093, 1973.
4. Neufeld, H.N., Sive, P.H., Rise, E., et al , The Use of a Computerized ECG Interpretation System in an Epidemiologic Study", Methods Inf. Med., 10: 85-90, 1971.

5. Prinzmetal, M., Goldman, A., Massumi, R.A., et al, Clinical Implication of Errors in Electrocardiographic Interpretation. Heart Disease of Electrocardiographic Origin, J.A.M.A., 161: 138-143.

6. Anderson, T.W., An Introduction to Multivariate Statistical Analysis, John Wiley and Sons, Inc., New York, 1958.

7. Batschelet, E., Statistical Methods for the Analysis of Problems in Animal Orientation and Certain Biological Rhythms, The American Institute of Biological Sciences, Washington, D. C., 1965.

8. Good, I.J., A Bayesian Significance Test for Multinomial Distributions, JR SS, B, 29: 399-431, 1967.

Supported in part by research grants HL 15191 and HL15047 from the National Heart and Lung Institute, National Institutes of Health, United States Public Health Service.

Departments of Statistics, Clinical Engineering and Medicine,
The George Washington University
and
The Veterans Administration Research Center for Cardiovascular Data Processing,
VA Hospital
Washington, D. C. 20422

Linear Discrimination and Some Further Results on Best Lower Dimensional Representations

Raul Hudlet
Richard Johnson

Introduction

When discriminating among k p-variate normal distributions with common covariance matrix Σ, it is well known that the optimal discrimination procedure is based on the score functions in (1.1) below. For many practical reasons, however, it proves useful to have one- two- or three-dimensional representations of the data. Plotting the transformed observations in these lower dimensional spaces can lead to better understanding of the relationships between populations as well as the detection of outliers. Such a representation should be especially helpful when p is large compared to k or when the means almost lie in a low dimensional hyperplane. Lachenbruch (1975), Chapter 5, contains a discussion of the between - within method, due to Fisher (1938), for reducing dimension. Examples are widespread (c.f. Cooley and Lohnes (1971) Section 9.3) and many program packages, including the BMD series, have an associated plotting option.

Our goal is to find low dimensional hyperplanes which in some sense best represent the populations. The Fisher method is not, however, unique in providing representations and we propose a class of procedures for selecting a low dimensional representation consisting of a few linear combinations of the observations. The population version of Fisher's method is a special case of our Theorem 4.2 and Theorem 4.3. Consequently, our results imply new optimality properties for Fisher's between-within procedure. Our criteria is a weighted sum of loss of distance and it shows explicitly the manner in which the prior enters. This has the advantage of leading to suggestions on how to select the relative sample sizes when forming a learning set and how to modify the Fisher method if the proportions in the learning set differ radically from the prior, subjects which do not seem to have attracted attention thus far. A short discussion appears in Section 6 after Corollary 6.4.

Interestingly, we show in Section 5 that the proposed class of p-dimensional representations, including Fisher's between-within method, can be interpreted as based on orthogonal transformations of the standardized principal components based on the common Σ. Each provides a decomposition of the scores as a sum of squares. Sample interpretations of these results, related to learning sets, are given in Section 6 where Lemma 6.1 provides the key to this extension. Theorems 6.2 and 6.3 take account of the sample scatter about the mean in the construction of a best approximation to the learning set. One further property of Fisher's method, as a maximization of the spread of projected means, is given in the final section.

1. Review of Structure for Linear Distrimination.

In the general classification problem there are k populations $\Pi_1, \ldots, \Pi_k$ having probability densities $f_1(x), \ldots, f_k(x)$ (w.r.t. some σ-finite measure) and relative frequencies of occurrence $p_1, \ldots, p_k$. An observation x becomes available and we wish to determine the population to which it belongs. That is, one needs a rule for associating x with one of the k populations. It seems reasonable to associate the observation with that population Π_i for which the posterior probability is highest, or equivalently to that population Π_i for which the score $S_i'(x) = p_i f_i(x)$ is highest. In fact, Anderson (1958) has shown that this rule is admissible and that it is Bayes when the costs of misclassification are equal.

In the case where $f_i(x)$ is a p-variate normal with mean μ_i and covariance matrix Σ of full rank, the score can be written

$$S_i'(x) = p_i(2\pi)^{-p/2}|\Sigma|^{-\frac{1}{2}} \exp[-\tfrac{1}{2}(x-\mu_i)'\Sigma^{-1}(x-\mu_i)].$$

Removing the common constant $(2\pi)^{-p/2}|\Sigma|^{-\frac{1}{2}}$, taking logarithms and multiplying by two, one obtains the equivalent score

(1.1) $$S_i = -(\underset{\sim}{x}-\underset{\sim}{\mu}_i)'\,\Sigma^{-1}(\underset{\sim}{x}-\underset{\sim}{\mu}_i) + 2\,\ell n\, p_i$$

and the rule is:

> assign $\underset{\sim}{x}$ to that population for which the score S_i, $i = 1,\ldots,k$ is highest.

Letting $D(\underset{\sim}{x},\underset{\sim}{\mu}_i) = (\underset{\sim}{x}-\underset{\sim}{\mu}_i)'\Sigma^{-1}(\underset{\sim}{x}-\underset{\sim}{\mu}_i)$ be the squared Mahalanobis distance from $\underset{\sim}{x}$ to $\underset{\sim}{\mu}_i$, the score $S_i(\underset{\sim}{x})$ may also be written as

(1.2) $$-S_i(\underset{\sim}{x}) = D(\underset{\sim}{x},\underset{\sim}{\mu}_i) - 2\,\ell n\, p_i \quad .$$

The rule may now be interpreted to say:

> assign $\underset{\sim}{x}$ to the nearest population after penalizing each $D(\underset{\sim}{x},\underset{\sim}{\mu}_i)$ by the corresponding non-negative quantity $-2\,\ell n\, p_i$.

The case of two populations. With two populations the rule is assign $\underset{\sim}{x}$ to Π_1 if $(\underset{\sim}{x}-\underset{\sim}{\mu}_1)'\Sigma^{-1}(\underset{\sim}{x}-\underset{\sim}{\mu}_1) - 2\ell n\, p_1 \leq (\underset{\sim}{x}-\underset{\sim}{\mu}_2)'\,\Sigma^{-1}(\underset{\sim}{x}-\underset{\sim}{\mu}_2) - 2\,\ell n\, p_2$, otherwise assign $\underset{\sim}{x}$ to Π_2. Let $\bar{\underset{\sim}{\mu}} = \frac{1}{2}(\underset{\sim}{\mu}_1+\underset{\sim}{\mu}_2)$ be the midpoint between $\underset{\sim}{\mu}_1$ and $\underset{\sim}{\mu}_2$ and $\underset{\sim}{\delta} = \underset{\sim}{\mu}_1-\underset{\sim}{\mu}_2$ be the vectors from $\underset{\sim}{\mu}_2$ to $\underset{\sim}{\mu}_1$. The rule then takes the form, assign $\underset{\sim}{x}$ to Π_1 if

(1.3) $$(\underset{\sim}{x}-\bar{\underset{\sim}{\mu}})'\,\Sigma^{-1}\,\underset{\sim}{\delta} + \ell n(p_1/p_2) \geq 0$$

otherwise assign $\underset{\sim}{x}$ to Π_2 . The set

$$\{\,\underset{\sim}{x} : (\underset{\sim}{x}-\bar{\underset{\sim}{\mu}})'\,\Sigma^{-1}\underset{\sim}{\delta} + \ell n(p_1/p_2) = 0\,\}$$

is a hyperplane, passing through the point $\bar{\underset{\sim}{\mu}} - \dfrac{\ell n(p_1/p_2)}{\underset{\sim}{\delta}'\,\Sigma^{-1}\,\underset{\sim}{\delta}}\,\underset{\sim}{\delta}$ and with normal direction $\Sigma^{-1}\underset{\sim}{\delta}$. The relation $(\underset{\sim}{x}-\bar{\underset{\sim}{\mu}})'\Sigma^{-1}\underset{\sim}{\delta} + \ell n(p_1/p_2) \geq 0$ holds only for those $\underset{\sim}{x}$'s on one side of the plane and is < 0 for the $\underset{\sim}{x}$'s on the other side of the plane.

When $p_1 = p_2$, the separating hyperplane passes through the middle point $\bar{\underset{\sim}{\mu}}$, whereas if $p_1 \neq p_2$ the plane is pushed towards the least likely population thus expanding the region corresponding to the population most likely apriori.

With only two populations, one can easily calculate the probabilities of correct classification. Since $\underset{\sim}{\delta}' \Sigma^{-1}(\underset{\sim}{x}-\underset{\sim}{\mu})$ is a linear combination of the components of $\underset{\sim}{x}$, $\underset{\sim}{\delta}' \Sigma^{-1}(\underset{\sim}{x}-\bar{\underset{\sim}{\mu}}) + \ell n(p_1/p_2)$ has a normal distribution with variance $\Delta^2 = \underset{\sim}{\delta}'\Sigma^{-1}\underset{\sim}{\delta}$ and mean

$$(1.4) \quad \begin{cases} \frac{1}{2}\Delta^2 + \ell n(p_1/p_2) & x \epsilon \Pi_1 \\ -\frac{1}{2}\Delta^2 - \ell n(p_2/p_1) & x \epsilon \Pi_2 \end{cases}$$

where $\Delta^2 = \underset{\sim}{\delta}'\Sigma^{-1}\underset{\sim}{\delta} = (\underset{\sim}{\mu}_1-\underset{\sim}{\mu}_2)'\Sigma^{-1}(\underset{\sim}{\mu}_1-\underset{\sim}{\mu}_2)$ is the squared Mahalanobis distance from $\underset{\sim}{\mu}_1$ to $\underset{\sim}{\mu}_2$. Consequently, the probabilities of correct classification, P [classifying $\underset{\sim}{x}$ as from Π_1 given that $\underset{\sim}{x}$ from Π_1] = $P(1|1)$ and $P(2|2)$ for the second population, are expressible in terms of the standard normal variable Z as

$$(1.5) \quad \begin{aligned} P(1|1) &= P(Z < \tfrac{1}{2}\Delta + \frac{\ell n(p_1/p_2)}{\Delta}) \\ P(2|2) &= P(Z < \tfrac{1}{2}\Delta + \frac{\ell n(p_2/p_1)}{\Delta}) \end{aligned}$$

The observations may also be misclassified, the corresponding probabilities are

$$P(\text{classify as } \Pi_1|\Pi_2) = P(1|2) = 1 - P(2|2)$$

$$P(\text{classify as } \Pi_2|\Pi_1) = P(2|1) = 1 - P(1|1)\,.$$

It is well known that the procedure (1. 3) minimizes $p_1 P(2|1) + p_2 P(1|2)$.

Three or more populations. Similar reasoning applies when there are more than two populations. For instance in the case of three populations there will be three scores S_1, S_2 and S_3. The region assigned to Π_1 is that region where $S_1(\underset{\sim}{x}) > S_2(\underset{\sim}{x}) > S_3(\underset{\sim}{x})$ which is that region on the side of $\underset{\sim}{\mu}_1$ of the planes separating Π_1 from Π_2 and Π_1 from Π_3. The probabilities of correct classification are

$$(1.6) \quad \begin{aligned} P(1|1) &= P(Z < \tfrac{1}{2}\Delta_{1j} + (\Delta_{1j})^{-1}\ell n(p_1/p_j);\ j = 2,3) \\ P(2|2) &= P(Z < \tfrac{1}{2}\Delta_{2j} + (\Delta_{2j})^{-1}\ell n(p_2/p_j);\ j = 1,3) \\ P(3|3) &= P(Z < \tfrac{1}{2}\Delta_{3j} + (\Delta_{3j})^{-1}\ell n(p_3/p_j);\ j = 1,2) \end{aligned}$$

where $\underset{\sim}{\delta}_{ij} = \underset{\sim}{\mu}_i - \underset{\sim}{\mu}_j$ and $\Delta_{ij} = (\underset{\sim}{\delta}'_{ij}\, \Sigma^{-1}\, \underset{\sim}{\delta}_{ij})^{\frac{1}{2}}$ is the Mahalanobis distance from $\underset{\sim}{\mu}_i$ to $\underset{\sim}{\mu}_j$ and Z is a N(0,1) variable. The form of (1.6) extends in an obvious manner to the k population setting.

2. Working in the Standardized Principal Components Space.

Some clarity is gained by recasting the above formulation in terms of principal components derived from the common Σ. First, let $\underset{\sim}{y} = T\underset{\sim}{x}$ where T is $p \times p$ nonsingular. If $\underset{\sim}{x} \in \Pi_i$, $\underset{\sim}{y} \sim N(T\, \underset{\sim}{\mu}_i,\ T\,\Sigma\, T')$. Working in the $\underset{\sim}{y}$ space and proceeding to obtain scores as in (1.2), the rule is to assign $\underset{\sim}{y}$ to that population for which,

$$-S_i(\underset{\sim}{y}) = (\underset{\sim}{y} - \underset{\sim}{\eta}_i)'\, \Sigma_y^{-1} (\underset{\sim}{y} - \underset{\sim}{\eta}_i) - 2\ell n\, p_i$$

is a minimum. Here $\underset{\sim}{\eta}_i = T\underset{\sim}{\mu}_i$ and $\Sigma_y = T\,\Sigma\, T'$. But

$$(2.1) \qquad -S_i(\underset{\sim}{y}) = (T(\underset{\sim}{x} - \underset{\sim}{\mu}_i))'(T\Sigma T')^{-1}(T(\underset{\sim}{x} - \underset{\sim}{\mu}_i)) - 2\ell n\, p_i = (\underset{\sim}{x} - \underset{\sim}{\mu}_i)'\Sigma^{-1}(\underset{\sim}{x} - \underset{\sim}{\mu}_i) - 2\ell n p_i = -S_i(\underset{\sim}{x})$$

showing that the distance and hence the rule is invariant under nonsingular transformations.

Consider the spectral decomposition of Σ, $P\,\Lambda\, P'$ with P orthogonal and Λ diagonal. By selecting $T = \Lambda^{-\frac{1}{2}} P'$, $\underset{\sim}{y}$ becomes $\underset{\sim}{y} = \Lambda^{-\frac{1}{2}} P'\underset{\sim}{x}$. That is, y is a realization of the principal components standardized to have variance one. We say that y are the <u>standardized principal components.</u>

From (2.1), we find

$$(2.2) \qquad -S_i(\underset{\sim}{y}) = (\underset{\sim}{y} - \underset{\sim}{\eta}_i)'(\underset{\sim}{y} - \underset{\sim}{\eta}_i) - 2\ell n\, p_i$$

and the rule is:

> Assign $\underset{\sim}{x}$ to that population Π_i for which $\underset{\sim}{y}$ is closest to $\underset{\sim}{\eta}_i$ in the usual Euclidean sense after penalizing each distance by the corresponding term $-2\ell n\, p_i$.

In the space of standardized principal components, the case of two populations is easily visualized geometrically. The set of equidistant $\underset{\sim}{y}$'s is the set $\{\underset{\sim}{y} \mid (\underset{\sim}{y} - \underset{\sim}{\eta}_1)'(\underset{\sim}{y} - \underset{\sim}{\eta}_1) - 2\ell n\, p_1 = (\underset{\sim}{y} - \underset{\sim}{\eta}_2)'(\underset{\sim}{y} - \underset{\sim}{\eta}_2) - 2\ell n p_2\}$

which is a hyperplane whose normal is the segment from η_2 to η_1 and passing through the point $\frac{1}{2}(\eta_1 + \eta_2) - (\eta_1 - \eta_2)\frac{\ell n(p_1/p_2)}{(\eta_1 - \eta_2)'(\eta_1 - \eta_2)}$.

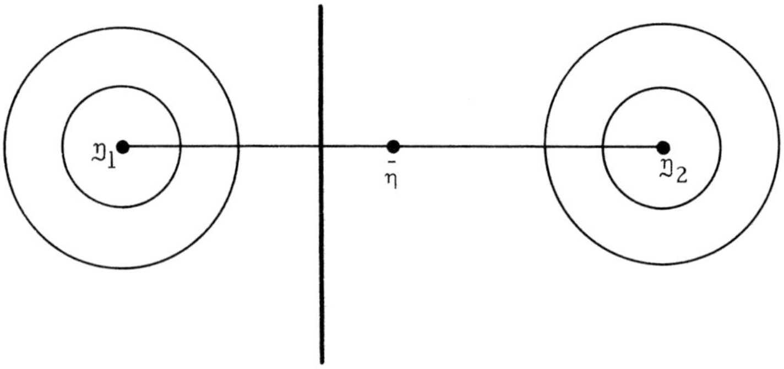

Figure 2.1. Separating hyperplane for two populations in the standardized principal component space ($p_1 > p_2$).

With several populations one deals only with the spherical or hyper-spherical contours of the usual Euclidean distance.

3. Projecting to Minimize the Total Misclassification Probability.

If a set of observations from each of the k populations has been observed, it is a good idea to plot these points both to see the overall pattern and to detect possible outliers. However, even for $p = 3$ characteristics this may be cumbersome and for larger p alternative representations of the data must be employed. To alleviate this problem, each observed vector $\underset{\sim}{x}$ may be condensed into a univariate observation by means of a linear combination $\underset{\sim}{\ell}_1'\underset{\sim}{x}$ and then these univariate observations may be plotted. A second $\underset{\sim}{\ell}_2'\underset{\sim}{x}$ uncorrelated with the first, may then be formed and the points $(\underset{\sim}{\ell}_1'\underset{\sim}{x}, \underset{\sim}{\ell}_2'\underset{\sim}{x})$ plotted. Where a 3-dimensional visual display is available, this can be carried one step further to obtain $\underset{\sim}{\ell}_3'\underset{\sim}{x}$. One choice of a criterion for selecting $\underset{\sim}{\ell}_1, \underset{\sim}{\ell}_2, \ldots$ is to minimize the total misclassification probability as in Welch (1939).

For an arbitrary $\underset{\sim}{\ell}$ if $\underset{\sim}{x} \in \Pi_i$, the variable $\underset{\sim}{\ell}'\underset{\sim}{x}$ is $\sim N(\underset{\sim}{\ell}'\underset{\sim}{\mu}_i, \underset{\sim}{\ell}'\Sigma\underset{\sim}{\ell})$ so the scoring method applied to this variable assigns $\underset{\sim}{\ell}'\underset{\sim}{x}$ to that population for which the score

$$-S_i(\underset{\sim}{\ell}'\underset{\sim}{x}) = \underset{\sim}{\ell}'(\underset{\sim}{x}-\underset{\sim}{\mu}_i)(\underset{\sim}{\ell}'\Sigma\underset{\sim}{\ell})^{-1}\underset{\sim}{\ell}'(\underset{\sim}{x}-\underset{\sim}{\mu}_i) - 2\,\ell n\, p_i$$

is a minimum. In the case of two populations the total probability of misclassification is

$$(3.1) \qquad p_1\Phi\left(-\frac{\Delta}{2} - \frac{\ell n(p_1/p_2)}{\Delta}\right) + p_2\Phi\left(-\frac{\Delta}{2} - \frac{\ell n(p_2/p_1)}{\Delta}\right)$$

where here $\Delta = |\underset{\sim}{\ell}'(\underset{\sim}{\mu}_1-\underset{\sim}{\mu}_2)|(\underset{\sim}{\ell}'\Sigma\underset{\sim}{\ell})^{-\frac{1}{2}}$ and Φ is the standard normal cumulative distribution. Letting $k = \ell n\, p_1/p_2$ and differentiating (3.1) with respect to Δ gives $-p_1 \exp[-\frac{\Delta^2}{8} - \frac{k^2}{2\Delta} - \frac{k}{2}] < 0$, so that total misclassification probability is decreasing in Δ. Thus, we must take Δ as large as possible or $\Delta = [(\underset{\sim}{\mu}_2-\underset{\sim}{\mu}_1)'\Sigma^{-1}(\underset{\sim}{\mu}_2-\underset{\sim}{\mu}_1)]^{\frac{1}{2}}$ which is attained when $\underset{\sim}{\ell} \propto \Sigma^{-1}(\underset{\sim}{\mu}_1-\underset{\sim}{\mu}_2)$. This was to be expected since the Bayes procedure depends on a single linear combination and minimizes the probability of misclassification.

For more than two populations the situation becomes rather complicated and no workable form is known (c.f. Gusemann, Peters, Walker (1975)). The main reasons for projecting seem to be the added simplicity and the visual display but simplicity is lost when the solution becomes untractable. In the next section, we take an alternative approach.

4. Linear Discrimination and Best Lower Dimensional Representations.

In this section, we obtain some new results concerning best representations of populations. Our goal here is to select $\underset{\sim}{\ell}_1, \underset{\sim}{\ell}_2, \ldots, \underset{\sim}{\ell}_p$ so that, successively, the increasing sets $(\underset{\sim}{\ell}_1'\underset{\sim}{x})$, $(\underset{\sim}{\ell}_1'\underset{\sim}{x}, \underset{\sim}{\ell}_2'\underset{\sim}{x})$, $(\underset{\sim}{\ell}_1'\underset{\sim}{x}, \underset{\sim}{\ell}_2'\underset{\sim}{x}, \underset{\sim}{\ell}_3'\underset{\sim}{x}), \ldots, (\underset{\sim}{\ell}_1'\underset{\sim}{x}, \ldots, \underset{\sim}{\ell}_p'\underset{\sim}{x})$ 'best' reproduce all of the discriminant scores $S_i(\underset{\sim}{x})$ or the classification procedure. The objective is to select the first few for plotting and visual inspection of the learning sets whereas the optimal normal theory classification procedure, based on the $S_i(\underset{\sim}{x})$, necessarily employs all of the $\underset{\sim}{\ell}_i'\underset{\sim}{x}$.

Whatever criterion we employ for selecting $\underset{\sim}{\ell}_1$, we are going to demand at a second stage that, given the population to which $\underset{\sim}{x}$ belongs, $\underset{\sim}{\ell}_2'\underset{\sim}{x}$ be uncorrelated with $\underset{\sim}{\ell}_1'\underset{\sim}{x}$ and so on. This requirement however does not produce uncorrelated linear transformations since we are only asking for conditional uncorrelatedness. Thus, in terms of $\underset{\sim}{\ell}_1'\underset{\sim}{x},\ldots,\underset{\sim}{\ell}_p'\underset{\sim}{x}$, we are interested in $p\times p$ matrices $L' = (\underset{\sim}{\ell}_1,\ldots,\underset{\sim}{\ell}_p)$ such that $L\Sigma L'$ is diagonal and nonsingular. It will prove convenient in our derivation to employ the symmetric square root $\Sigma^{-\frac{1}{2}} = P\Delta^{-\frac{1}{2}}P'$ where $\Sigma = P\Delta P'$ is the spectral decomposition of Σ. We express any L as $A\Sigma^{-\frac{1}{2}}$ where $A' = (\underset{\sim}{a}_1,\ldots,\underset{\sim}{a}_p)$ and AA' is diagonal. Denoting the set of linear combinations by $\underset{\sim}{y} = A\Sigma^{-\frac{1}{2}}\underset{\sim}{x}$, from the invariance of $S_i(\underset{\sim}{x})$ under non-singular transformation we obtain

$$-S_i(\underset{\sim}{x}) = -S_i(\underset{\sim}{y}) = (A\Sigma^{-\frac{1}{2}}(\underset{\sim}{x}-\underset{\sim}{\mu}_i))'(A\Sigma^{-\frac{1}{2}}\Sigma\Sigma^{-\frac{1}{2}}A')^{-1}(A\Sigma^{-\frac{1}{2}}(\underset{\sim}{x}-\underset{\sim}{\mu}_i))-2\ln p_i$$

$$= \frac{(\underset{\sim}{a}_1'\Sigma^{-\frac{1}{2}}(\underset{\sim}{x}-\underset{\sim}{\mu}_i))^2}{\underset{\sim}{a}_1'\underset{\sim}{a}_1} + \ldots + \frac{(\underset{\sim}{a}_p'\Sigma^{-\frac{1}{2}}(\underset{\sim}{x}-\underset{\sim}{\mu}_i))^2}{\underset{\sim}{a}_p'\underset{\sim}{a}_p} - 2\ln p_i \tag{4.1}$$

Notice that $(\underset{\sim}{a}_j'\underset{\sim}{a}_j)^{-1}(\underset{\sim}{a}_j'\Sigma^{-\frac{1}{2}}(\underset{\sim}{x}-\underset{\sim}{\mu}_i))^2$ is invariant to changes in the modulus of $\underset{\sim}{a}_j$ so, without loss of generality, we assume that $\underset{\sim}{a}_1,\ldots,\underset{\sim}{a}_p$ have norm one. In other words we may assume that A is orthogonal. Then (4.1) takes the form

$$-S_i(\underset{\sim}{x}) = \sum_{j=1}^{p}(\underset{\sim}{a}_j'\Sigma^{-\frac{1}{2}}(\underset{\sim}{x}-\underset{\sim}{\mu}_i))^2 - 2\ln p_i \tag{4.2}$$

Relation (4.2) is important in that it provides a total decomposition of the distance part of $S_i(\underset{\sim}{x})$ into a sum of squares.

On the other hand, suppose we decide to work with only the first r linear combinations $\underset{\sim}{a}_1'\Sigma^{-\frac{1}{2}}\underset{\sim}{x},\ldots,\underset{\sim}{a}_r'\Sigma^{-\frac{1}{2}}\underset{\sim}{x}(1\le r\le p)$. Then, the vector $A_r\Sigma^{-\frac{1}{2}}\underset{\sim}{x}$ where $A_r' = (\underset{\sim}{a}_1,\ldots,\underset{\sim}{a}_r)$ is distributed as $N(A_r\Sigma^{-\frac{1}{2}}\underset{\sim}{\mu}_i, I_r)$ when $\underset{\sim}{x}\in\Pi_i$. The scoring method applied to this variable assigns it to that population for which the score

$$-S_i(A_r\Sigma^{-\frac{1}{2}}\underset{\sim}{x}) = \sum_{j=1}^{r}(\underset{\sim}{a}_j'\Sigma^{-\frac{1}{2}}(\underset{\sim}{x}-\underset{\sim}{\mu}_i))^2 - 2\ln p_i$$

is minimum.

The distance part of the scores for $A_r \Sigma^{-\frac{1}{2}} \underset{\sim}{x}$ is always smaller than the distance part of the scores for $\underset{\sim}{x}$ since

$$(4.3) \qquad -S_i(\underset{\sim}{x}) + S_i(A_r \Sigma^{-\frac{1}{2}} \underset{\sim}{x}) = \sum_{j=r+1}^{p} (\underset{\sim}{a}_j' \Sigma^{-\frac{1}{2}} (\underset{\sim}{x} - \underset{\sim}{\mu}_i))^2 .$$

This difference represents how much distance we lose by considering only $A_r \Sigma^{-\frac{1}{2}} \underset{\sim}{x}$ when the full matrix A is needed for the total decomposition of the p dimensional distance from $\underset{\sim}{x}$ to each $\underset{\sim}{\mu}_i$.

We propose two criterions for selecting A. The first selects A so that the expected value of the

$$(4.4) \qquad \begin{array}{l}\text{weighted average}\\ \text{loss of distance}\end{array} = \mathcal{L}_r(p_1, \ldots, p_k) = \sum_{i=1}^{k} p_i(-S_i(\underset{\sim}{x}) + S_i(A_r \Sigma^{-\frac{1}{2}} \underset{\sim}{x}))$$

over $\underset{\sim}{x}$ and the prior, is a minimum. The second selects A so that the expected value of the

$$(4.5) \qquad \text{average loss of distance} = \mathcal{L}_r(1/k, \ldots, 1/k) = \frac{1}{k} \sum_{i=1}^{k} (-S_i(\underset{\sim}{x}) + S_i(A_r \Sigma^{-\frac{1}{2}} \underset{\sim}{x}))$$

is a minimum. As we shall see the first method reduces to Fisher's between-within method, a fact seemingly unnoticed thus far.

We begin by obtaining an expression for the expected value of the more general weighted loss

$$(4.6) \qquad \mathcal{L}_r(q_1, \ldots, q_k) = \sum_{i=1}^{k} q_i(-S_i(\underset{\sim}{x}) + S_i(A_r \Sigma^{-\frac{1}{2}} \underset{\sim}{x}))$$

where $q_1, \ldots, q_k$ are non-negative weights that add to one and A_r is $r \times p$ with orthonormal rows $\underset{\sim}{a}_1', \ldots, \underset{\sim}{a}_r'$. First, we introduce some notation.

Let

$$(4.7) \qquad \begin{aligned} \bar{\underset{\sim}{\mu}} &= p_1 \underset{\sim}{\mu}_1 + \ldots + p_k \underset{\sim}{\mu}_k , \\ B = B(p_1, \ldots, p_k) &= \sum_{i=1}^{k} p_i (\underset{\sim}{\mu}_i - \bar{\underset{\sim}{\mu}})(\underset{\sim}{\mu}_i - \bar{\underset{\sim}{\mu}})' \end{aligned}$$

then $\bar{\underset{\sim}{\mu}}$ is the 'total' population mean and we may think of B as the populations between groups cross products matrix.

Lemma 4.1.

Let $\mathcal{L}_r(q_1,\ldots,q_k)$ be as in (4.6) and $D = \sum_{i=1}^{k} (p_i + q_i)(\underset{\sim}{\mu}_i - \bar{\underset{\sim}{\mu}})(\underset{\sim}{\mu}_i - \bar{\underset{\sim}{\mu}})'$ then, for $1 \le r \le p$,

$$E(\mathcal{L}_r(q_1,\ldots,q_k)) = p - r + \operatorname{tr}\Sigma^{-\frac{1}{2}} C \Sigma^{-\frac{1}{2}} - \operatorname{tr} A_r \Sigma^{-\frac{1}{2}} D \Sigma^{-\frac{1}{2}} A_r' . \tag{4.8}$$

Proof:

We will use the fact that if $\underset{\sim}{Z}$ is a $t\times 1$ random vector with $E\underset{\sim}{Z} = \underset{\sim}{\eta}$ and covariance matrix $M(|M| \neq 0)$ then $E(\underset{\sim}{Z}-\underset{\sim}{a})'M^{-1}(\underset{\sim}{Z}-\underset{\sim}{a}) = t + \operatorname{tr} M^{-1}(\underset{\sim}{\eta}-\underset{\sim}{a})(\underset{\sim}{\eta}-\underset{\sim}{a})'$. We have

$$E(\mathcal{L}_r) = E \sum_{j=1}^{k} q_j(-S_j(\underset{\sim}{X}) + S_j(A_r \Sigma^{-\frac{1}{2}}\underset{\sim}{X})) \tag{4.9}$$

$$= \sum_{i=1}^{k} p_i \sum_{j=1}^{k} q_j\, E[-S_j(X) + S_j(A_r\Sigma^{-\frac{1}{2}}\underset{\sim}{X}) \mid \underset{\sim}{X} \in \Pi_i]$$

$$= \sum_{i=1}^{k} p_i \sum_{j=1}^{k} q_j\, E[(\underset{\sim}{X}-\underset{\sim}{\mu}_j)'\Sigma^{-1}(\underset{\sim}{X}-\underset{\sim}{\mu}_j) - (\underset{\sim}{X}-\underset{\sim}{\mu}_j)'\Sigma^{-\frac{1}{2}}A_r'A_r\Sigma^{-\frac{1}{2}}(\underset{\sim}{X}-\underset{\sim}{\mu}_j) \mid \underset{\sim}{X} \in \Pi_i]$$

$$= \sum_{i=1}^{k} p_i \sum_{j=1}^{k} q_j[p + \operatorname{tr}\Sigma^{-1}(\underset{\sim}{\mu}_i-\underset{\sim}{\mu}_j)(\underset{\sim}{\mu}_i-\underset{\sim}{\mu}_j)' - r - \operatorname{tr}A_r\Sigma^{-\frac{1}{2}}(\underset{\sim}{\mu}_i-\underset{\sim}{\mu}_j)(\underset{\sim}{\mu}_i-\underset{\sim}{\mu}_j)'\Sigma^{-\frac{1}{2}}A_r']$$

$$= p - r + \operatorname{tr}\Sigma^{-1}\Big[\sum_{i=1}^{k}\sum_{j=1}^{k} p_i q_j(\underset{\sim}{\mu}_i-\underset{\sim}{\mu}_j)(\underset{\sim}{\mu}_i-\underset{\sim}{\mu}_j)'\Big] - \operatorname{tr}A_r\Sigma^{-\frac{1}{2}}\Big[\sum_{i=1}^{k}\sum_{j=1}^{k} p_i q_j(\underset{\sim}{\mu}_i-\underset{\sim}{\mu}_j)(\underset{\sim}{\mu}_i-\underset{\sim}{\mu}_j)'\Big]\Sigma^{-\frac{1}{2}}$$

But

$$\sum_{i,j} p_i q_j(\underset{\sim}{\mu}_i-\underset{\sim}{\mu}_j)(\underset{\sim}{\mu}_i-\underset{\sim}{\mu}_j)' = \sum_{i,j} p_i q_j(\underset{\sim}{\mu}_i-\bar{\underset{\sim}{\mu}})(\underset{\sim}{\mu}_i-\bar{\underset{\sim}{\mu}})' +$$

$$\sum_{i,j} p_i q_j(\underset{\sim}{\mu}_j-\bar{\underset{\sim}{\mu}})(\underset{\sim}{\mu}_j-\bar{\underset{\sim}{\mu}})' + \sum_{i,j} p_i q_j(\underset{\sim}{\mu}_i-\bar{\underset{\sim}{\mu}})(\bar{\underset{\sim}{\mu}}-\underset{\sim}{\mu}_j)' + \sum_{i,j} p_i q_j(\bar{\underset{\sim}{\mu}}-\underset{\sim}{\mu}_j)(\underset{\sim}{\mu}_i-\bar{\underset{\sim}{\mu}})'$$

$$= \sum_i p_i(\underset{\sim}{\mu}_i-\bar{\underset{\sim}{\mu}})(\underset{\sim}{\mu}_i-\bar{\underset{\sim}{\mu}})' + \sum_j q_j(\underset{\sim}{\mu}_j-\bar{\underset{\sim}{\mu}})(\underset{\sim}{\mu}_j-\bar{\underset{\sim}{\mu}})' + 0 + 0 = D . \tag{4.10}$$

Substituting in (4.9), we obtain the result. ∎

Theorem 4.2.

Under the conditions of Lemma 4.1, $E(\mathcal{L}_r)$ is minimized when $\underset{\sim}{a}_1,\ldots,\underset{\sim}{a}_r$ constitute a set of orthonormal eigenvectors of the matrix $\Sigma^{-\frac{1}{2}} D \Sigma^{-\frac{1}{2}}$ corresponding to the r largest eigenvalues. Then

$$E(\mathcal{L}_r) = p - r + (\delta_{r+1} + \ldots + \delta_p) \tag{4.11}$$

where $\delta_1 \geq \ldots \geq \delta_p$ are the eigenvalues of the matrix $\Sigma^{-\frac{1}{2}} D \Sigma^{-\frac{1}{2}}$.

Proof: From Lemma 4.1

$$
\begin{aligned}
E(\mathcal{L}_r) &= p-r-\text{tr}\, \Sigma^{-\frac{1}{2}} D \Sigma^{-\frac{1}{2}} - \text{tr}\, A_r \Sigma^{-\frac{1}{2}} D \Sigma^{-\frac{1}{2}} A_r' \\
&= p-r-\text{tr}\, \Sigma^{-\frac{1}{2}} D \Sigma^{-\frac{1}{2}} - \sum_{i=1}^{r} \underset{\sim}{a}_i' \Sigma^{-\frac{1}{2}} D \Sigma^{-\frac{1}{2}} \underset{\sim}{a}_i .
\end{aligned} \tag{4.12}
$$

Equation (4.12) shows that it is enough to maximize the term $\sum_{i=1}^{r} \underset{\sim}{a}_i' \Sigma^{-\frac{1}{2}} D \Sigma^{-\frac{1}{2}} \underset{\sim}{a}_i$ which is achieved when $\underset{\sim}{a}_1, \ldots, \underset{\sim}{a}_r$ constitute a set of r orthonormal eigenvectors of the matrix $\Sigma^{-\frac{1}{2}} D \Sigma^{-\frac{1}{2}}$ corresponding to the r largest eigenvalues (c.f. Rao (1973) pg. 63).

With this choice of $\underset{\sim}{a}_1, \ldots, \underset{\sim}{a}_r$ we obtain

$$E(\mathcal{L}_r) = p-r+(\delta_{r+1} + \ldots + \delta_p) . \qquad \blacksquare$$

The linear combinations $a_i' \Sigma^{-\frac{1}{2}} x$ may also be selected sequentially.

Theorem 4.3.

Let $q_i \geq 0$, $i = 1, \ldots, k$, $\sum_{i=1}^{k} q_i = 1$ and let $D = \Sigma(p_i + q_i)(\underset{\sim}{\mu}_i - \bar{\underset{\sim}{\mu}})(\underset{\sim}{\mu}_i - \bar{\underset{\sim}{\mu}})'$. For each $i = 1, \ldots, p$, the linear combination $\underset{\sim}{a}_i' \Sigma^{-\frac{1}{2}} \underset{\sim}{x}$ that minimizes $E\,\mathcal{L}_1(q_1, \ldots, q_k)$, subject to $\underset{\sim}{a}_i'\underset{\sim}{a}_j = 0$ $j = 1, \ldots, i-1$ and satisfying $\underset{\sim}{a}_i'\underset{\sim}{a}_i = 1$, is obtained when and only when $\underset{\sim}{a}_i$ is a normalized eigenvector of the matrix $\Sigma^{-\frac{1}{2}} D \Sigma^{-\frac{1}{2}}$ corresponding to the ith largest eigenvalue and is perpendicular to $\underset{\sim}{a}_1, \ldots, \underset{\sim}{a}_{i-1}$.

Proof:

First notice that the requirement $\underset{\sim}{a}_i'\underset{\sim}{a}_j = 0$ $j = 1, \ldots, i-1$ is equivlent to the requirement that Corr. $(\underset{\sim}{a}_i' \Sigma^{-\frac{1}{2}} X, \underset{\sim}{a}_j' \Sigma^{-\frac{1}{2}} X \mid X \in \Pi) = 0$ for $\ell = 1, \ldots, k$. From (4.8) it is enough to find $\underset{\sim}{a}_1, \ldots, \underset{\sim}{a}_p$ so that for $i = 1, \ldots, p$ $\underset{\sim}{a}_i' \Sigma^{-\frac{1}{2}} D \Sigma^{-\frac{1}{2}} \underset{\sim}{a}_i$ is a maximum subject to $\underset{\sim}{a}_i \perp \underset{\sim}{a}_1, \ldots, \underset{\sim}{a}_{i-1}$ and $\underset{\sim}{a}_i'\underset{\sim}{a}_i = 1$. Equivalently, since $1 = \underset{\sim}{a}_i'\underset{\sim}{a}_i = \underset{\sim}{a}_i' \Sigma^{-\frac{1}{2}} \Sigma \Sigma^{-\frac{1}{2}} \underset{\sim}{a}_i$ we maximize

$$\frac{\underset{\sim}{a}_i' \Sigma^{-\frac{1}{2}} D \Sigma^{-\frac{1}{2}} \underset{\sim}{a}_i}{\underset{\sim}{a}_i' \Sigma^{-\frac{1}{2}} \Sigma \Sigma^{-\frac{1}{2}} \underset{\sim}{a}_i} \tag{4.13}$$

The Courant-Fisher minimax theorem (c.f. Bellman (1970) pg. 115) then establishes that $\underset{\sim}{a}_1, \ldots, \underset{\sim}{a}_p$ must constitute a set of orthonormal eigenvectors of the matrix $\Sigma^{-\frac{1}{2}} D \Sigma^{-\frac{1}{2}}$ with $\underset{\sim}{a}_i$ corresponding to the ith largest eigenvalue, $(i = 1, \ldots, p)$.

Theorem 4.3 gives some insight into the manner in which the selection of linear combinations is influenced by the choice of weights $q_1, \ldots, q_k$. When $q_i = p_i$, $i = 1, \ldots, p$, the matrix D takes the form $D = 2 \sum_{i=1}^{k} p_i(\underset{\sim}{\mu}_i - \bar{\underset{\sim}{\mu}})(\underset{\sim}{\mu}_i - \bar{\underset{\sim}{\mu}})' = 2B$. In this case Theorem 4.3 shows that the sequential application of the $\mathcal{I}_1(p_1, \ldots, p_k)$ criterion yields, in p stages, the p canonical variables $a_1' \Sigma^{-\frac{1}{2}} x, \ldots, a_p' \Sigma^{-\frac{1}{2}} x$ produced by the between-within method of Fisher. The population version of his method leads to maximizing $\underset{\sim}{\ell}' B \underset{\sim}{\ell} / \underset{\sim}{\ell}' \Sigma \underset{\sim}{\ell}$. Our expression (4.7) shows explicitly the role played by the prior, a point that has been overlooked by some writers (c.f. Lachenbruch (1975) page 66).

When one selects equal weights, $q_i = 1/k$, $i = 1, \ldots, k$, the matrix D becomes $D = \sum_{i=1}^{k} (p_i + 1/k)(\mu_i - \mu)(\mu_i - \mu)'$. The answer will, in general, be different from Fisher's method unless $k = 2$ or $p_1 = \ldots = p_k$.

5. Relation to Principal Components.

In this section we relate the above results to the standardized principal components described in Section 2. Whatever the choice of weights q_i, Theorem 4.3 yields, in p stages, the p variables $\underset{\sim}{a}_1' \Sigma^{-\frac{1}{2}} \underset{\sim}{x}, \ldots, \underset{\sim}{a}_p' \Sigma^{-\frac{1}{2}} \underset{\sim}{x}$ which can be thought of as a nonsingular linear transformation $A \Sigma^{-\frac{1}{2}} \underline{x}$ of $\underline{x}$ with $A' = (\underset{\sim}{a}_1, \ldots, \underset{\sim}{a}_p)$ orthogonal. The matrix $A\Sigma^{-\frac{1}{2}}$ of the transformation is determined by A since Σ is fixed. Of course A changes when the weights $(q_1, \ldots, q_k)$ are varied.

Lemma 5.1.

Whatever the choice of the weights q_i, the p canonical variables derived in Theorem 4.3 constitute an orthogonal transformation of the standardized principal components.

Proof:

Let $A \Sigma^{-\frac{1}{2}} \underset{\sim}{x}$ be the p canonical variables, so that A is orthogonal. Also let $\Sigma = P \Lambda P'$ be the spectral decomposition of Σ with P orthogonal and Λ diagonal. Then $\Sigma^{-\frac{1}{2}} = P \Lambda^{-\frac{1}{2}} P'$ and for any A

$$A \Sigma^{-\frac{1}{2}} \underset{\sim}{x} = A P \Lambda^{-\frac{1}{2}} P' \underset{\sim}{x} = A P(\Lambda^{-\frac{1}{2}} P' \underset{\sim}{x})$$

where $\Lambda^{-\frac{1}{2}} P'x$ are the standardized principal components. □

Corollary 5.2.

Whatever the choice of the q_i, the p canonical variables given by Theorem 4.3 provide a total decomposition of each score $S_i(\underset{\sim}{x})$, $i = 1, 2, \ldots, p$ as

$$(5.11) \qquad -S_j(\underset{\sim}{x}) = \sum_{i=1}^{p} [\underset{\sim}{a}_i' \Sigma^{-\frac{1}{2}}(\underset{\sim}{x} - \underset{\sim}{\mu}_j)]^2 - 2 \ln p_j \quad .$$

Proof:

This is a direct application of (4.2). □

For an independent derivation of the decomposition (5.11) in the special case of Fisher's between-within method, see Kshirsagar and Arseven (1975). Our development not only shows this as a special case but permits us to take weights $q_1, \ldots, q_k$ different from $p_1, \ldots, p_k$.

6. Interpretation of the Sample Values.

In general the population parameters are not known and instead a learning set is available consisting of n_i observations from population Π_i, $i = 1, \ldots, k$. We may then use the estimated population values in an attempt to implement the procedures of Section 4. Alternatively, we may treat these $n = \sum_{i=1}^{k} n_i$ points in R^p as our total population, each of the n points considered to be equally likely. In this framework we still have k subpopulations, the ith one being the set $\{\underset{\sim}{x}_{i1}, \ldots, \underset{\sim}{x}_{in_i}\}$ $i = 1, \ldots, k$. Moreover, with this convention we have, in correspondence with our previous notation,

$$p_i \longleftrightarrow n_i/n, \quad \underset{\sim}{\mu}_i \longleftrightarrow \bar{\underset{\sim}{x}}_{i\cdot} = \frac{1}{n_i} \sum_{j=1}^{n_i} \underset{\sim}{x}_{ij} \qquad i = 1,\ldots,k$$

(6.1)
$$\bar{\underset{\sim}{\mu}} = \sum_{i=1}^{k} p_i \underset{\sim}{\mu}_i \longleftrightarrow \sum_{i=1}^{k} \frac{n_i}{n} \bar{\underset{\sim}{x}}_{i\cdot} = \bar{\underset{\sim}{x}}$$

$$B = \sum_{i=1}^{k} p_i(\mu_i - \bar{\mu})(\mu_i - \bar{\mu})' \longleftrightarrow \sum_{i=1}^{k} \frac{n_i}{n} (\bar{x}_{i\cdot} - \bar{x})(\bar{x}_{i\cdot} - \bar{x})' \ .$$

The k populations, however, do not have a common covariance matrix Σ since the ith population has covariance

$$W_i = \frac{1}{n_i} \sum_{j=1}^{n_i} (\underset{\sim}{x}_{ij} - \bar{\underset{\sim}{x}}_{i\cdot})(\underset{\sim}{x}_{ij} - \bar{\underset{\sim}{x}}_{i\cdot})' \ .$$

Consequently we decide, as with maximum likelihood estimation, to use the pooled covariance matrix

(6.2)
$$W = \sum_{i=1}^{k} \frac{n_i}{n} W_i$$

in the role of the common covariance matrix Σ.

If $\underset{\sim}{x}$ is an "observation" from this finite population then

$$E \underset{\sim}{x} = \frac{1}{n} \sum_{i=1}^{k} \sum_{j=1}^{n_i} \underset{\sim}{x}_{ij} = \bar{\underset{\sim}{x}}$$

and the covariance matrix T,

(6.3)
$$T = \mathrm{Var}(\underset{\sim}{x}) = \frac{1}{n} \sum_{i=1}^{k} \sum_{j=1}^{n_i} (\underset{\sim}{x}_{ij} - \bar{\underset{\sim}{x}})(\underset{\sim}{x}_{ij} - \bar{\underset{\sim}{x}})'$$

satisfies

(6.4)
$$T = B + W \ .$$

Notice that W_i is the variance of $\underset{\sim}{x}$ given that $\underset{\sim}{x}$ belongs to the ith population, so that $W = \sum_{i=1}^{k} \frac{n_i}{n} W_i$ is the expected value of the conditional variance. Also $\bar{\underset{\sim}{x}}_{i\cdot}$ is the expected value of $\underset{\sim}{x}$ given that $\underset{\sim}{x}$ belongs to the ith population, so $B = \sum_{i=1}^{k} \frac{n_i}{n} (\bar{\underset{\sim}{x}}_{i\cdot} - \bar{\underset{\sim}{x}})(\bar{\underset{\sim}{x}}_{i\cdot} - \bar{\underset{\sim}{x}})'$ is the variance of the conditional expectation. That is, (6.4) is but another special case of the well known result of expressing the variance in terms of conditional moments.

The scores for discriminating in the learning set are

$$-\hat{S}_j(\underset{\sim}{x}) = (\underset{\sim}{x}-\underset{\sim}{\bar{x}}_{j\cdot})' W^{-1}(\underset{\sim}{x}-\underset{\sim}{\bar{x}}_{j\cdot}) - 2\,\ell n \frac{n_j}{n} \qquad j = 1,\ldots,k$$

and the rule is to assign an observation $\underset{\sim}{x}$ to that population where it scores highest. These scores are no longer optimal since parameters have been estimated. What we are able to show is that the results of Section 4 may be extended to include approximating the scores $\hat{S}_j(\underset{\sim}{x})$ from the finite population.

Suppose we consider a set of $r\ (1 \leq r \leq p)$ linear combinations

$$(6.5) \qquad \begin{bmatrix} \underset{\sim}{a}_1' \\ \vdots \\ \underset{\sim}{a}_r' \end{bmatrix} W^{-\frac{1}{2}}\underset{\sim}{x} = A_r W^{-\frac{1}{2}}\underset{\sim}{x}$$

with

$$(6.6) \qquad \underset{\sim}{a}_i'\underset{\sim}{a}_j = 0 \quad i \neq j\,, \quad \underset{\sim}{a}_i'\underset{\sim}{a}_i = 1\,.$$

Then, the scores for these new variables are,

$$(6.7) \qquad \hat{S}_j(A_r W^{-\frac{1}{2}}\underset{\sim}{x}) = (A_r W^{-\frac{1}{2}}(\underset{\sim}{x}-\underset{\sim}{\bar{x}}_{j\cdot}))'(A_r W^{-\frac{1}{2}}(\underset{\sim}{x}-\underset{\sim}{x}_{j\cdot})) - 2\,\ell n \frac{n_j}{n}$$

since the transformed points $A_r W^{-\frac{1}{2}}\underset{\sim}{x}_{ij}$ $i = 1,\ldots,k;\ j = 1,\ldots,n_i$ have pooled covariance matrix $\sum_{i=1}^{k} \frac{n_i}{n} A_r W^{-\frac{1}{2}} W_i W^{-\frac{1}{2}} A_r' = I_r$. The variables $\underset{\sim}{a}_1' W^{-\frac{1}{2}}\underset{\sim}{x},\ldots,\underset{\sim}{a}_r' W^{-\frac{1}{2}}\underset{\sim}{x}$, however, are <u>not</u> uncorrelated, not even given that $x \in \Pi_i$, $i = 1,\ldots,p$.

However, similar to (4.6) we do have a weighted loss of distance

$$(6.8) \qquad L_r(q_1,\ldots,q_k) = \sum_{i=1}^{k} q_i(-\hat{S}_i(\underset{\sim}{x}) + \hat{S}_i(A_r W^{-\frac{1}{2}}\underset{\sim}{x}))$$

where $q_1,\ldots,q_k$ are non-negative weights that add to one and A_r is a $r \times p$ matrix with orthonormal rows $\underset{\sim}{a}_1',\ldots,\underset{\sim}{a}_r'$. We use L_r rather than $\mathcal{L}_r$ to emphasize we are dealing with the training set which is a finite population. Now L_r may be interpreted as follows: Before x is chosen in the finite population, we do not know to which subpopulation Π_i it belongs. That is, we do not know which of the differences

$-\hat{S}_i(\underset{\sim}{x}) + \hat{S}_i(A_r W^{-\frac{1}{2}}\underset{\sim}{x})$ are most important. It them seems logical to weight these differences by their probabilities of occurrence n_i/n and then to consider the average loss of distance $\sum_{i=1}^{k} \frac{n_i}{n}(-\hat{S}_i(\underset{\sim}{x}) + \hat{S}_i(A_r W^{-\frac{1}{2}}\underset{\sim}{x})) =$ $= L_r(\frac{n_1}{n}, \ldots, \frac{n_k}{n})$. There is some gain in generality if we initially consider $L_r(q_1, \ldots, q_k)$ given by (6.8) above.

We now find the expected value of the average loss of distance $L_r(q_1, \ldots, q_k)$ with respect to the probability measure that assigns probability $1/n$ to each point of the learning set.

Lemma 6.1.

Let

$$D = \sum_{i=1}^{k} (\frac{n_i}{n} + q_i)(\bar{\underset{\sim}{x}}_{i.} - \bar{\underset{\sim}{x}})(\bar{\underset{\sim}{x}}_{i.} - \bar{\underset{\sim}{x}})' \quad . \tag{6.9}$$

Then for $1 \le r \le p$,

$$E\, L_r(q_1, \ldots, q_k) = p - r + \mathrm{tr}\, W^{-1} D - \sum_{i=1}^{r} \underset{\sim}{a}_i' W^{-\frac{1}{2}} D W^{-\frac{1}{2}} \underset{\sim}{a}_i \, .$$

Proof:

$$E\left\{ \sum_{j=1}^{k} q_j(-\hat{S}_j(\underset{\sim}{x}) + \hat{S}_j(A_r W^{-\frac{1}{2}}\underset{\sim}{x}) \right\} = \tag{6.10}$$

$$= \sum_{i=1}^{k} \frac{n_i}{n} \sum_{j=1}^{k} q_j E\{-\hat{S}_j(\underset{\sim}{x}) + \hat{S}_j(A_r W^{-\frac{1}{2}}\underset{\sim}{x}) \mid x \in \Pi_i\}$$

$$= \sum_{i=1}^{k} \frac{n_i}{n} \sum_{j=1}^{k} q_j\, E\{(\underset{\sim}{x} - \bar{\underset{\sim}{x}}_{j.})' W^{-1} (\underset{\sim}{x} - \bar{\underset{\sim}{x}}_{j.})$$

$$-(A_r W^{-\frac{1}{2}}(\underset{\sim}{x} - \bar{\underset{\sim}{x}}_{j.}))'(A_r W^{-\frac{1}{2}}(\underset{\sim}{x} - \bar{\underset{\sim}{x}}_{j.})) \mid \underset{\sim}{x} \in \Pi_i\}$$

$$= \sum_{i=1}^{k} \frac{n_i}{n} \sum_{j=1}^{k} q_j \{\mathrm{tr} W^{-1}[W_i + (\bar{\underset{\sim}{x}}_{i.} - \bar{\underset{\sim}{x}}_{j.})(\bar{\underset{\sim}{x}}_{i.} - \bar{\underset{\sim}{x}}_{j.})']$$

$$-\mathrm{tr}\, A_r\, W^{-\frac{1}{2}}[W_i + (\bar{\underset{\sim}{x}}_{i.} - \bar{\underset{\sim}{x}}_{j.})(\bar{\underset{\sim}{x}}_{i.} - \bar{\underset{\sim}{x}}_{j.})']W^{-\frac{1}{2}} A_r' \}$$

$$= p - r + \mathrm{tr}\, W^{-1} \sum_{i=1}^{k} \sum_{j=1}^{k} \frac{n_i}{n} q_j (\bar{\underset{\sim}{x}}_{i.} - \bar{\underset{\sim}{x}}_{j.})(\bar{\underset{\sim}{x}}_{i.} - \bar{\underset{\sim}{x}}_{j.})'$$

$$-\mathrm{tr}\, A_r\, W^{-\frac{1}{2}}[\sum_{i=1}^{k} \sum_{j=1}^{k} \frac{n_i}{n} q_j (\bar{\underset{\sim}{x}}_{i.} - \bar{\underset{\sim}{x}}_{j.})(\bar{\underset{\sim}{x}}_{i.} - \bar{\underset{\sim}{x}}_{j.})']W^{-\frac{1}{2}} A_r'$$

It is easy to see that

$$\sum_{i=1}^{k}\sum_{j=1}^{k} \frac{n_i}{n} q_j(\underset{\sim}{\bar{x}}_{i.} - \underset{\sim}{\bar{x}}_{j.})(\underset{\sim}{\bar{x}}_{i.} - \underset{\sim}{\bar{x}}_{j.})' = D$$

and the result follows. ∎

Due to the conclusion of Lemma 6.1, the results of Sections 4 and 5 can be applied to the finite population. In particular, we obtain

Theorem 6.2.

For fixed $r (1 \leq r \leq p)$, and L_r given by (6.8), the variables $\underset{\sim}{a}_1' W^{-\frac{1}{2}} \underset{\sim}{x} \ldots \underset{\sim}{a}_r' W^{-\frac{1}{2}} \underset{\sim}{x}$ minimize $E\, L_r(q_1, \ldots, q_k)$ when $\underset{\sim}{a}_1, \ldots, \underset{\sim}{a}_r$ constitute an orthonormal set of eigenvectors of the matrix $W^{-\frac{1}{2}} D W^{-\frac{1}{2}}$ corresponding to the r largest eigenvalues and then

$$E(L_r) = p - r + \delta_{r+1} + \ldots + \delta_p$$

where $\delta_1 \geq \ldots \geq \delta_p$ are the eigenvalues of the matrix $W^{-\frac{1}{2}} D W^{-\frac{1}{2}}$.

Theorem 6.3.

For each $i = 1, \ldots, p$, the linear combination $\underset{\sim}{a}_i' W^{-\frac{1}{2}} \underset{\sim}{x}$ that minimizes $E\, L_1(q_1, \ldots, q_k)$, subject to (6.6), is obtained, when and only when, $\underset{\sim}{a}_i$ is a normalized eigenvector of the matrix $W^{-\frac{1}{2}} D W^{-\frac{1}{2}}$ corresponding to the ith largest eigenvalue and perpendicular to $\underset{\sim}{a}_1, \ldots, \underset{\sim}{a}_{i-1}$.

The following corollary gives a training set optimality property of Fisher's between-within method.

Corollary 6.4.

When $q_i = n_i/n$, $i = 1, \ldots, k$ the sequential application of the $L_1(n_1/n, \ldots, n_k/n)$ criterion yields, in p stages, the p canonical variables produced by the between-within method of Fisher.

See Figure 6.1 for a geometric explaination of this approximation. Note how our criterion takes account of the scatter about each of the sample means and weights distances according to the relative occurrence of the samples, when $q_i = n_i/n$. Also notice, that with this choice of q_i's, the matrix D takes the form $2\Sigma(\frac{n_i}{n})(\bar{x}_{i.} - \bar{x})(\bar{x}_{i.} - \bar{x})' = 2B$ where B

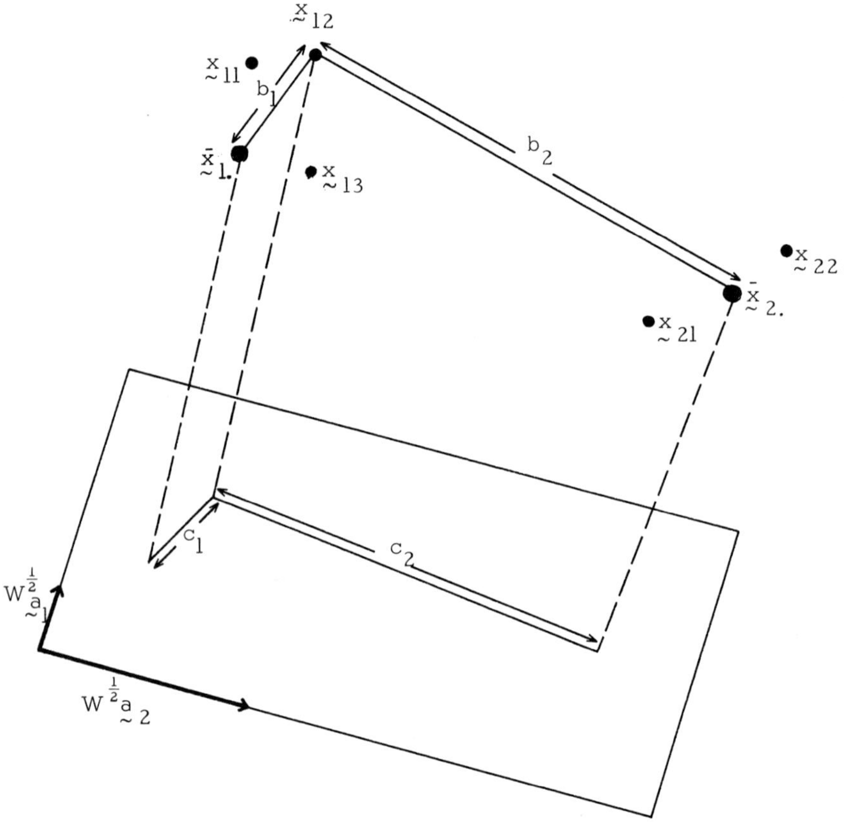

Loss of squared distance: $-\hat{S}_i(\underset{\sim}{x}_{12}) + S_i(A_2 W^{-\frac{1}{2}}\underset{\sim}{x}_{12}) = b_i^2 - c_i^2$, $i=1,2$,

where c_i and b_i and measured according to $\langle\ ,\ \rangle_{W^{-1}}$.

x_{12} contributes $\sum_{i=1}^{2} n_i(b_i^2 - c_i^2)/n^2$ to the expected weighted loss.

Figure 6.1. The weighted loss of distance criterion with $q_i = n_i/n$.

is Fisher's between-groups cross products matrix. The sample sizes n_i, however, can be taken completely arbitrary. There appears to be a lack of information in the literature as how to relate the prior with the sample sizes n_i. Thus, for instance, Lachenbruch (1975), p. 66 gives a particular form of the B matrix making no mention of the prior, and Kshirsagar and Arseven (1975), although they mention the prior, do not relate it in any way to the sample sizes n_i. No mention of this topic appears in Anderson (1958) or Rao (1973).

We may compare the matrix D above with the population matrix $D = 2\,\Sigma\, p_i(\underset{\sim}{\mu}_i - \bar{\underset{\sim}{\mu}})(\mu_i - \bar{\mu})'$ corresponding to the choice of weights $q_i = p_i$, $i = 1,\ldots,k$ in Theorem 4.3. This choice produces canonical variables which sequentially minimize the weighted loss of distance (4.4). It then seems logical to select n_i/n so that it is, if possible, close to p_i. Moreover if the p_i and the n_i/n are not consonant, it would then seem more desirable to try to approximate the population matrix $D = 2\Sigma\, p_i(\underset{\sim}{\mu}_i - \bar{\underset{\sim}{\mu}})(\underset{\sim}{\mu}_i - \bar{\underset{\sim}{\mu}})'$ by the matrix $2\sum_{i=1}^{k} p_i(\bar{\underset{\sim}{x}}_{i.} - \bar{\underset{\sim}{x}})(\bar{\underset{\sim}{x}}_{i.} - \bar{\underset{\sim}{x}})'$ and then use the eigenvectors of the matrix $W^{-\frac{1}{2}} \sum_{i=1}^{k} p_i(\bar{\underset{\sim}{x}}_{i.} - \bar{\underset{\sim}{x}})(\bar{\underset{\sim}{x}}_{i.} - \bar{\underset{\sim}{x}})' W^{-\frac{1}{2}}$ to find a reasonable hyperplane for use in the classification of new observations rather than those of the matrix $W^{-\frac{1}{2}} \sum_{i=1}^{k} \frac{n_i}{n}(\bar{\underset{\sim}{x}}_{i.} - \bar{\underset{\sim}{x}})(\bar{\underset{\sim}{x}}_{i.} - \bar{\underset{\sim}{x}})' W^{-\frac{1}{2}}$ as in Theorem 6.2. This is due to the fact that while it is true that Theorem 6.2 holds true for the finite population consisting of the learning set, this finite population does not mimic any more the original population. It may even be reasonable to replace $\bar{\underset{\sim}{x}}$ by $\Sigma\, p_i \bar{\underset{\sim}{x}}_{i.}$. If the p_i's are not known, but estimates are available, the estimates could be used.

We conclude this section by recording the following property.

<u>Lemma 6.5.</u>

Whatever the choice of the weights q_i, the p canonical variables derived in Theorem 6.3 produce a total decomposition of each score $\hat{S}_j(\underset{\sim}{x})$ as

$$-\hat{S}_j(\underset{\sim}{x}) = \sum_{i=1}^{p} (\underset{\sim}{a}_i' W^{-\frac{1}{2}}(\underset{\sim}{x} - \bar{\underset{\sim}{x}}_{j.})) - 2\,\ell n(n_j/n).$$

7. The Fisher Between-Within Method as Optimal Projection.

In this section, we obtain one further interpretation and optimality property of the p canonical variables given by Fisher's between-within method. The method produces p canonical variables $\underset{\sim}{Z}_i = \ell_i'\underset{\sim}{X}$ where

$$(7.1) \qquad \underset{\sim}{\ell}_i = \Sigma^{-\frac{1}{2}} \underset{\sim}{a}_i, \qquad i = 1, \ldots, p$$

and $\underset{\sim}{a}_1, \ldots, \underset{\sim}{a}_p$ constitute a set of orthonormal eigenvectors of the matrix $\Sigma^{-\frac{1}{2}}(\sum_{i=1}^{k} p_i(\underset{\sim}{\mu}_i - \bar{\underset{\sim}{\mu}})(\underset{\sim}{\mu}_i - \underset{\sim}{\mu})')\Sigma^{-\frac{1}{2}}$ corresponding, respectively, to the eigenvalues $\lambda_1 \geq \ldots \geq \lambda_p$. The expression $\sum_{i=1}^{k} p_i(\underset{\sim}{\mu}_i - \bar{\underset{\sim}{\mu}})' \Sigma^{-1}(\underset{\sim}{\mu}_i - \bar{\underset{\sim}{\mu}})$ is the weighted sum of squares of the Mahalanobis distance of each subpopulation mean $\underset{\sim}{\mu}_i$, to the 'total' population mean $\bar{\underset{\sim}{\mu}}$, and is a measure of the spread of the population. What we seek is an orthonormal system of axis $\underset{\sim}{m}_1, \ldots, \underset{\sim}{m}_p$ such that most of the spread lies along $\underset{\sim}{m}_1$, while the second axis $\underset{\sim}{m}_2$ satisfies this property among all axes perpendicular to $\underset{\sim}{m}_1$ and so on.

Let $M = (\underset{\sim}{m}_1, \ldots, \underset{\sim}{m}_p)$. We take as the appropriate inner product the one defined in terms of Σ^{-1}. With this inner product, denoted by $\langle \cdot , \cdot \rangle$ the axis will be orthonormal if

$$(7.2) \qquad \langle \underset{\sim}{m}_i, \underset{\sim}{m}_j \rangle = \underset{\sim}{m}_i' \Sigma^{-1} \underset{\sim}{m}_j = \delta_{ij} \qquad i, j = 1, \ldots, k$$

where δ_{ij} stands for Kronecker's delta. This is equivalent to

$$(7.3) \qquad M' \Sigma^{-1} M = I .$$

Now any vector in p-space can be written in terms of the $\underset{\sim}{m}_j$'s. In particular, for $\underset{\sim}{\mu}_i$, we have

$$(7.4) \qquad \underset{\sim}{\mu}_i = \langle \underset{\sim}{\mu}_i, \underset{\sim}{m}_1 \rangle \underset{\sim}{m}_1 + \ldots + \langle \underset{\sim}{\mu}_i, \underset{\sim}{m}_p \rangle \underset{\sim}{m}_p.$$

Writing $\underset{\sim}{\mu}_i$ as in (7.4), its norm squared may be expressed as

$$(7.5) \qquad \|\underset{\sim}{\mu}_i\|^2_{\Sigma^{-1}} = \langle \underset{\sim}{\mu}_i, \underset{\sim}{\mu}_i \rangle = \underset{\sim}{\mu}_i' \Sigma^{-1} \underset{\sim}{\mu}_i = \sum_{j=1}^{p} \langle \underset{\sim}{\mu}_i, \underset{\sim}{m}_j \rangle^2 = \sum_{j=1}^{p} (\underset{\sim}{m}_j' \Sigma^{-1} \underset{\sim}{\mu}_i)^2.$$

From (7.5) we can decompose the p-dimensional distance $\sum_{i=1}^{k} p_i(\underset{\sim}{\mu}_i - \bar{\underset{\sim}{\mu}})'\Sigma^{-1}(\underset{\sim}{\mu}_i - \bar{\underset{\sim}{\mu}})$ as a sum of projections, defined by Σ^{-1}, rather than by the usual definition of perpendicularity.

Theorem 7.1.

Let the vectors $\underset{\sim}{\mu}_i - \bar{\underset{\sim}{\mu}}$ be projected onto an r dimensional hyperplane through the origin $(1 \le r \le \min(k-1, p))$ according to the inner product $\langle \cdot , \cdot \rangle$. Then if $\underset{\sim}{\nu}_i - \bar{\underset{\sim}{\nu}}$ denotes the projection of $\underset{\sim}{\mu}_i - \bar{\underset{\sim}{\mu}}$,

$$\sum_{i=1}^{k} p_i \, \| \underset{\sim}{\nu}_i - \bar{\underset{\sim}{\nu}} \|^2_{\Sigma^{-1}}$$

is a maximum when the hyperplane is the one spanned by the vectors $\Sigma^{\frac{1}{2}} \underset{\sim}{a}_i, \ldots, \Sigma^{\frac{1}{2}} \underset{\sim}{a}_r$ where $\underset{\sim}{a}_1, \ldots, \underset{\sim}{a}_r$ are as in (7.1).

Proof:

Let $\underset{\sim}{m}_1, \ldots, \underset{\sim}{m}_r$ be a set of orthonormal eigenvectors which span the hyperplane, then

$$\sum_{i=1}^{k} p_i \| \underset{\sim}{\nu}_i - \bar{\underset{\sim}{\nu}} \|^2_{\Sigma^{-1}} = \sum_{i=1}^{k} p_i \sum_{j=1}^{r} [\underset{\sim}{m}_j' \Sigma^{-1}(\underset{\sim}{\mu}_i - \bar{\underset{\sim}{\mu}}]^2$$

$$(7.6) \qquad = \sum_{j=1}^{r} \underset{\sim}{m}_j' \Sigma^{-1} [\sum_{i=1}^{k} \underset{\sim}{p}_i (\underset{\sim}{\mu}_i - \underset{\sim}{\mu})(\underset{\sim}{\mu}_i - \underset{\sim}{\mu})'] \Sigma^{-1} \underset{\sim}{m}_j$$

$$= \sum_{j=1}^{r} \underset{\sim}{m}_j' \Sigma^{-1} B \Sigma^{-1} \underset{\sim}{m}_j$$

where B is as in (4.7). Under the transformation

$$(7.7) \qquad \Sigma^{-\frac{1}{2}} \underset{\sim}{m} = \underset{\sim}{a}$$

expression (7.6) becomes

$$(7.8) \qquad \sum_{j=1}^{r} \underset{\sim}{a}_j' \Sigma^{-\frac{1}{2}} B \Sigma^{-\frac{1}{2}} \underset{\sim}{a}_j$$

and the condition in (7.3) becomes $A_r' A_r = I_r$ where $A = (\underset{\sim}{a}_1, \ldots, \underset{\sim}{a}_r)$. For $j = 1, \ldots, r$, the quantity $\underset{\sim}{a}_j' \Sigma^{-\frac{1}{2}} B \Sigma^{-\frac{1}{2}} \underset{\sim}{a}_j$ represents how much of the weighted distance $\sum_{i=1}^{k} p_i(\underset{\sim}{\mu}_i - \bar{\underset{\sim}{\mu}})' \Sigma^{-1}(\underset{\sim}{\mu}_i - \bar{\underset{\sim}{\mu}})$ lies along the axis $\underset{\sim}{m}_j$.

Our purpose is to maximize (7.8) subject to $\underset{\sim}{a}_j' \underset{\sim}{a}_i = \delta_{ij}$ where δ_{ij} stands for Kronecker's delta. By an application of the Courant-Fisher minimax theorem, the maximum is achieved when $\underset{\sim}{a}_1, \ldots, \underset{\sim}{a}_r$ constitute a set of orthonormal eigenvectors of the matrix $\Sigma^{-\frac{1}{2}} B \Sigma^{-\frac{1}{2}}$ corresponding, respectively to the r largest eigenvalues $\lambda_1 \geq \ldots \geq \lambda_r$ and then

$$\sum_{i=1}^{k} p_i \|\underset{\sim}{\nu}_i - \bar{\underset{\sim}{\nu}}\|^2_{\Sigma^{-1}} = \lambda_1 + \ldots + \lambda_r . \tag{7.9}$$

Theorem 7.1 has several interesting implications. From the proof, we notice that if $\underset{\sim}{a}_1, \ldots, \underset{\sim}{a}_p$ constitute a set of orthonormal eigenvectors of the matrix $\Sigma^{-\frac{1}{2}} B \Sigma^{-\frac{1}{2}}$ corresponding, respectively to the eigenvalues $\lambda_1 \geq \ldots \geq \lambda_p$ then from (7.1) and (7.7)

$$\underset{\sim}{\ell}_i = \Sigma^{-\frac{1}{2}} \underset{\sim}{a}_i = \Sigma^{-1} \underset{\sim}{m}_i \qquad i = 1, \ldots, p \tag{7.10}$$

so that Fisher's canonical variables $\underset{\sim}{Z}_i = \underset{\sim}{\ell}_i \underset{\sim}{X} = m_i' \Sigma^{-1} \underset{\sim}{X}$ $i = 1, \ldots, p$ may be interpreted as the coordinates of $\underset{\sim}{X}$ in the priviledged system of axis $\underset{\sim}{m}_1, \ldots, \underset{\sim}{m}_p$.

With respect to the inner product $\langle \cdot , \cdot \rangle$, the vectors $\underset{\sim}{m}_1, \ldots, \underset{\sim}{m}_r$ $(1 \leq r \leq p)$ determine an optimal hyperplane. Of all hyperplanes passing through the origin and with dimension r, the one spanned by $\underset{\sim}{m}_1, \ldots, \underset{\sim}{m}_r$ minimizes the weighted sum of squares of the distances of the vectors $\underset{\sim}{\mu}_1 - \bar{\underset{\sim}{\mu}}, \ldots, \underset{\sim}{\mu}_k - \bar{\underset{\sim}{\mu}}$ to the plane. A distance is weighted by the corresponding p_i. Equivalently, one may maximize the spread of the projected means from the projected grand centroid when distance is measured according to $(\cdot)' \Sigma^{-1} (\cdot)$. This is illustrated in Figure 7.1.

In the standardized principal components space, we have from the spectral decomposition of $\Sigma = P \Lambda P'$ with P orthogonal and Λ diagonal, that $\underset{\sim}{\mu}_i$ becomes $\underset{\sim}{\eta}_i = \Lambda^{-\frac{1}{2}} P' \underset{\sim}{\mu}_i$ and the weighted spread of the η's is

$$\sum_{i=1}^{k} p_i (\underset{\sim}{\eta}_i - \underset{\sim}{\eta})'(\underset{\sim}{\eta}_i - \underset{\sim}{\eta}) .$$

Theorem 4.6 and the discussion that follows now hold for the usual inner product and projection. The statistician is tilting the r-dimensional hyperplane in p space and watching the shadows of the $\underset{\sim}{\eta}_i$ until they achieve maximum weighted separation.

The sample analogs of these results are immediate.

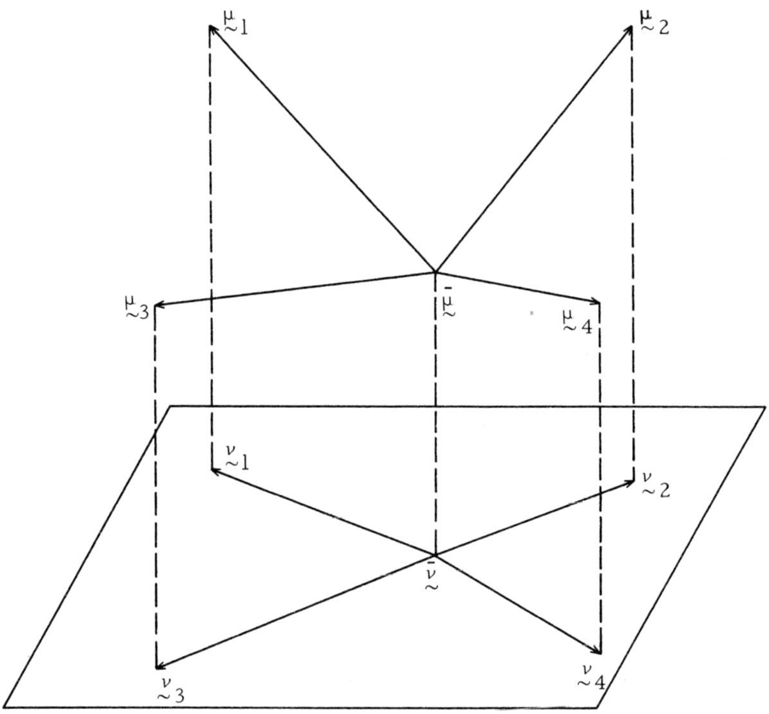

$$\max \quad \sum (p_i + q_i) \| \underset{\sim}{\nu}_i - \bar{\underset{\sim}{\nu}} \|^2_{\Sigma^{-1}} \quad \text{equivalent to}$$

$$\min \quad \sum (p_i + q_i) \| \underset{\sim}{\mu}_i - \bar{\underset{\sim}{\mu}} - (\underset{\sim}{\nu}_i - \bar{\underset{\sim}{\nu}}) \|^2_{\Sigma^{-1}} \ .$$

Figure 7.1. Mean separating property of Fisher's between-within method.

References

Anderson, T. W. (1958), An Introduction to Multivariate Statistical Analysis, John Wiley and Sons, Inc., New York.

Bellman, R. (1970), Introduction to Matrix Analysis, McGraw-Hill, New York.

Cooley, W. W. and Lohnes, P. R. (1971), Multivariate Data Analysis. John Wiley and Sons, Inc., New York.

Fisher, R. A. (1938), "The statistical utilization of multiple measurements", Ann. Eugen. **8**, 376-386.

Guseman, L. F., Jr., Peters, B. C., Jr., and Walker, H. F. (1975), "On Minimizing the Probability of Misclassification for Linear Feature Selection", Ann. Math. Statist. **47**, 661-668.

Kshirsagar, A. M. and Arseven, E. (1975), "A Note on the Equivalency of Two Discriminant Procedures", The American Statistician, **29** 38-39.

Lachenbruch, P. A. (1975), Discriminant Analysis, Hafner Press, New York.

Rao, C. R. (1973), Linear Statistical Inference and its Applications, John Wiley and Sons, Inc., New York.

Welch, B. L. (1939), "Note on Discriminant Functions", Biometrika, **31**, 218-220.

This research was supported by the Air Force Office of Scientific Research under Grant No. AF-AFOSR-72-2363D.

Department of Statistics
University of Wisconsin-Madison
Madison, Wisconsin 53706

A Simple Histogram Method for Nonparametric Classification

Pi Yeong Chi
J. Van Ryzin

Summary.

Suppose that there are ℓ populations which are mixed in proportions $p_1, p_2, \ldots, p_\ell$ respectively. The classification problem arises when an investigator wishes to classify an individual into one of these populations on the basis of the m-dimensional measurement on the individual. When the densities of the distributions and the mixing proportions are not specified completely, the non-error rate of any rule cannot be evaluated; nor can the optimal rule be formulated. This paper gives a Bayes risk consistent nonparametric procedure and a consistent estimate for the non-error rate when a training sample of size n is available and n becomes large. The procedure is based upon the idea of a histogram density estimator given by one of the authors [6, 11] but bypasses the direct density estimation calculations. The simplicity of the procedure is highly desirable in practical application. A Monte Carlo simulation is given here to compare the proposed procedures with the optimal rule in a variety of examples.

1. Introduction.

Assume we have populations $\pi_1, \pi_2, \ldots, \pi_\ell$ which are mixed in proportions $p_1, p_2, \ldots, p_\ell$, respectively ($p_i > 0$, and $\sum_{i=1}^{\ell} p_i = 1$).

The observation on the individual taken from the mixed population to be classified may be regarded as a random vector $Z = \binom{X}{I}$ in which I indexes the individual's population and X is a random variable with values in m-dimensional Euclidean space R^m. To classify an individual is to guess its population (the value of I) given that $X = x$, where X is the observed measurements of the individual.

Let the distribution function of Z be F_Z and $F_Z(x,i) = P\{X \leq x, I \leq i\}$. Let $P\{I = i\} = p_i$ and $F_i(x) = P\{X \leq x | I = i\}$, $i = 1,2, \ldots, \ell$, then the unconditional distribution function of X is $F(x) = \sum_{i=1}^{\ell} p_i F_i(x)$. Let F_i be absolutely continuous with respect to a σ-finite measure μ, and its density (Radon-Nikodym derivative) be f_i, $i = 1,2, \ldots, \ell$, respectively.

A classification procedure, or rule, D may be defined as a partition $D = (D_1, D_2, \ldots, D_\ell)$ of R^m. The rule D assigns an individual to population π_j if and only if the observed $X \in D_j$.

The probability of correct classification or non-error rate of a classification rule D is defined as

$$r(D) = \sum_{j=1}^{\ell} \int_{D_j} p_j f_j(x)\, d\mu(x) \quad . \tag{1.1}$$

A rule D^* achieves the <u>optimal non-error rate</u> if $r(D^*) = \sup_{D \in \mathcal{D}} r(D) = r^*$, where $\mathcal{D}$ is the collection of all possible classification rules.

If p_i and f_i's are not specified completely, the optimal rule and the non-error rate for any rule cannot be determined. But suppose that n individuals $Z_i = \binom{X_i}{I_i}$, $i = 1,2,\ldots,n$, can be sampled randomly from the mixed population and can be classified correctly. That is, not only the X_i's but also I_i's have been observed (by appropriate expenditure of time or money). A rule based on the sample $Z_1, Z_2, \ldots, Z_n$, called a "training sample", is then formulated with the hope that it will approximately maximize the non-error rate.

One approach to this problem is to use the sample from the i-th population to estimate the unknown density f_i respectively. The mixing proportions p_i can be estimated by $\hat{p}_i = N_i/n$, for $1 \leq i \leq \ell$, where N_i is the number of sampled individuals from population i and each N_i is a binomial random variable with expectation np_i. The resulting estimates are used as if they were the true values. Let $\hat{f}_i$

denote a parametric or nonparametric density estimator of f_i, $i = 1, 2, \ldots, \ell$ based on the training sample. Let $\hat{D}$ denote the rule such that

(1.2)
$$\hat{D}_j = \{x \in R^m : j \text{ is the smallest integer such that } \hat{p}_j\hat{f}_j(x) = \max_{1 \le j \le \ell} \hat{p}_j \hat{f}_j(x)\} .$$

If D is any rule in $\mathcal{D}$ and we define

(1.3)
$$\hat{r}(D) = \sum_{j=1}^{\ell} \int_{D_j} \hat{p}_j\hat{f}_j(x)\, d\mu(x),$$

then it can be shown that

(1.4)
$$\hat{r}(\hat{D}) = \sup_{D \in \mathcal{D}} \hat{r}(D) .$$

Let $\xrightarrow{a.s}$ and $\xrightarrow{p}$ denote convergence almost surely (with probability one) and in probability, respectively as a sample size $n \to \infty$.

Definition: (Van Ryzin [10] and Glick [5]): Let the rule $\bar{D}$ denote any rule based on observed sample values, then the rule $\bar{D}$ is Bayes risk consistent, BRC (or Bayes risk strongly consistent, BRSC) if $r(\bar{D})$ converges to r^* in probability (or with probability one) as sample size n approaches to infinity.

Theorem 1.1. (Glick [5, Sec. 5]): If the parametric or nonparametric density estimators $\hat{p}_i\hat{f}_i(x) \xrightarrow{p \text{ (or a.s.)}} p_i f_i(x)$ pointwise for almost all $x \in R^m$, and $\int_{R^m} \sum_{i=1}^{\ell} \hat{p}_i\hat{f}_i d\mu(x) \xrightarrow{p \text{ (or a.s.)}} 1$, $1 \le i \le \ell$, then

i) $\sup_{D \in \mathcal{D}} |\hat{r}(D) - r(D)| \xrightarrow{p \text{(or a.s.)}} 0$,

ii) $r(\hat{D}) \xrightarrow{p \text{(or a.s.)}} r^*$,

and

iii) $\hat{r}(\hat{D}) \xrightarrow{p \text{ (or a.s.)}} r^*$, as $n \to \infty$.

Furthermore, if $\int_{R^m} \sum_{i=1}^{\ell} \hat{p}_i\hat{f}_i \le 1$ a.s. (or $\le c$, constant), the convergence in i) is also in quadratic mean.

Remarks: The convergence ii) shown the $\hat{D}$ in BRC (or BRSC). The convergence iii) provides us with a sample based estimator of the non-error rate, i.e., $\hat{r}(\hat{D})$

The computations involved and the process of formulating a procedure in this method can be quite complicated if the investigator wants to partition R^m into D_i's, $1 \leq i \leq \ell$. We shall now give a computationally simple procedure in the one and two dimensional cases.

2. A Classification Rule Based on a Histogram Method in the One-Dimensional Case (m = 1).

Let $Y_1, Y_2, \ldots, Y_n$ $(Y_1 \leq Y_2 \leq \ldots \leq Y_n)$ be order statistics of the X_i's, $i = 1, 2, \ldots, n$. Let $Y_0 = -\infty$ and $Y_{n+1} = \infty$.

Let $\{k_n\}$ be a sequence of positive integers and define $J = \{0, 1, k_n+1, 2k_n+1, \ldots\}$. Define $A_n(x) = \max\{\alpha \mid \alpha \in J,\ Y_\alpha < x\}$ and $B_n(x) = \min\{A_n(x) + k_n, n\}$ if $x \leq Y_n$, with $B_n(x) = n+1$ if $x > Y_n$. In the remainder of this paper, we shall simply write $A_n(X) = A$ and $B_n(x) = B$ when convenient.

Let the sequences of integers $\{k_n\}$ always be taken to satisfy

$$\lim_{n\to\infty} \frac{\log n}{k_n} = 0 \quad \text{and} \quad \lim_{n\to\infty} \frac{k_n}{n} = 0\,. \tag{2.1}$$

Van Ryzin [11] defines the histogram density estimator,

$$\hat{f}(x) = \frac{F_n(Y_B) - F_n(Y_A)}{Y_B - Y_A} \tag{2.2}$$

where F_n represents the empiric unconditional distribution function of X.

The estimate in (2.2) for the above choice of A and B becomes

$$\hat{f}(x) = \begin{cases} \dfrac{k_n}{n(Y_{jk_n+1} - Y_{(j-1)k_n+1})} & \text{if } x \in U_j = (Y_{(j-1)k_n+1}, Y_{jk_n+1}] \\ & j = 1, \ldots, r_n \\ \dfrac{n - r_n k_n - 1}{n(Y_n - Y_{r_n k_n+1})} & \text{if } x \in U_{r_n+1} = (Y_{r_n k_n+1}, Y_n] \\ 0 & \text{if } x \leq Y, \text{ or } x > Y_n, \end{cases} \tag{2.3}$$

where r_n is that unique integer such that $n - k_n \leq r_n k_n + 1 < n$. The estimator in (2.3) is the same as that in the unsymmetric case of Van Ryzin [11, Case 2] with minor alterations at the end points. It is clear all the results derived by Van Ryzin [11] are valid here.

Note that the estimator $\hat{f}(x)$ defined by (2.3) is <u>constant</u> on the disjoint intervals

$$U_0 = (-\infty, Y_1],\quad U_1 = (Y_1, Y_{k_n+1}],\quad U_2 = (Y_{k_n+1}, Y_{2k_n+1}], \ldots,$$

(2.4) $$U_{r_n} = (Y_{r_n-1, h_n+1}, Y_{r_n k_n+1}],\quad U_{r_n+1} = (Y_{r_n k_n+1}, Y_n],$$

$$U_{r_n+2} = (Y_n, \infty)$$

and hence forms a <u>histogram.</u> The following results will be useful in what follows.

<u>Lemma 2.1.</u> (Van Ryzin [11, Lemma 1]): Let the sequence $\{k_n\}$ satisfy (2.1), and $F(x+\varepsilon) > F(x) > F(x-\varepsilon)$ for every $\varepsilon > 0$, then $Y_A, Y_B \xrightarrow{a.s.} x$, and $Y_B - Y_A \xrightarrow{a.s} 0$ as $n \to \infty$.

<u>Lemma 2.2.</u> (Van Ryzin [11, Corollary 2]): If the sequence $\{k_n\}$ satisfies (2.1) and $x \in C(f)$, then $\hat{f}(x) \xrightarrow{a.s.} f(x)$, where $C(f)$ is the set of all continuity points of f.

<u>Lemma 2.3.</u> (Kiefer and Wolfowitz [8, Theorem 2]): Let $Z_1, Z_2, \ldots, Z_n$ be independent $(n+1)$-dimensional random vectors with common distribution function F_Z. Let F_Z^n be the empiric distribution and be defined for any $z = (z_1, z_2, \ldots, z_{m+1})$. Define $D_n = \sup_z |F_Z^n(z) - F_Z(z)|$ and $G_n(r) = P(\sqrt{n}\, D_n \leq r)$, then for every m and F_Z, there exists a distribution function G such that the sequence of G_n converges at every continuity point of the latter as $n \to \infty$.

Since the observations $Z_i = \binom{X_i}{I_i}$, $i = 1, 2, \ldots, n$, of the training sample are independently, identically distributed with distribution function F_Z, let us denote $F_Z(z_1, z_2) = P\{X_i \leq z_1, I_i \leq z_2\}$ and the empiric function $F_Z^n(z_1, z_2) = \frac{1}{n}\sum_{i=1}^{n} I_{[X_i \leq z_1,\, I_i \leq z_2]}$, $I_{[S]}$ the indicator function of the set S Next, observe that if $x \in U_\nu$,

$$\hat{p}_j \hat{f}_j(x) = \frac{N_j}{n} \frac{K_{j\nu}}{N_j(Y_B - Y_A)} = \frac{K_{j\nu}}{n(Y_B - Y_A)} \tag{2.5}$$

where the U_ν, $\nu = 0, \ldots, r_n+2$ are defined in (2.4), $N_j = \sum_{i=1}^{n} I_{[I_i = j]}$ = the number of observations in the training sample belonging to population j, and $K_{j\nu} = \sum_{i=1}^{n} I_{[X_i \in U_\nu, I_i = j]}$ = the number of observations in the training sample belonging to population j and interval U_ν.

We now show $\hat{p}_j \hat{f}_j(x) \xrightarrow{p} p_j f_j(x)$ for all $x \in C(f) \cap C(f_j)$ if the sequence $\{k_n\}$ is properly chosen.

Theorem 2.1: If $x \in C(f) \cap C(f_j)$ and $\sqrt{n}/k_n = o(1)$, then $\hat{p}_j \hat{f}_j(x) \xrightarrow{p} p_j f_j(x)$ as $n \to \infty$, $j = 1, 2, \ldots, \ell$.

Proof:

Since $f(x) = 0$ implies $p_j f_j(x) = 0$ for all $1 \le j \le \ell$, by Lemma 2.2, $\hat{p}_j \hat{f}_j(x) \le \hat{f}(x) \xrightarrow{a.s.} f(x) = 0$. Hence in the remainder of the proof we assume $f(x) > 0$.

Observe that for $j = 1, \ldots, \ell$, if $x \in U_\nu$,

$$\hat{p}_j \hat{f}_j(x) = \frac{K_{j\nu}}{n(Y_B - Y_A)} = \frac{F_Z^n(Y_B, j) - F_Z^n(Y_A, j) - F_Z^n(Y_B, j-1) + F_Z^n(Y_A, j-1)}{Y_B - Y_A} = \tag{2.6}$$

$$= C_n^j + D_n^j$$

where

$$C_n^j = \left\{ \frac{F_Z^n(Y_B, j) - F_Z(Y_B, j)}{Y_B - Y_A} - \frac{F_Z^n(Y_A, j) - F_Z(Y_A, j)}{Y_B - Y_A} - \frac{F_Z^n(Y_B, j-1) - F_Z(Y_B, j-1)}{Y_B - Y_A} + \frac{F_Z^n(Y_A, j-1) - F_Z(Y_A, j-1)}{Y_B - Y_A} \right\}$$

and

$$D_n^j = \frac{[F(Y_B, j) - F(Y_A, j)] - [F(Y_B, j-1) - F(Y_A, j-1)]}{Y_B - Y_A} .$$

We complete the proof by showing $C_n^j \xrightarrow{p} 0$ and $D_n^j \xrightarrow{p} p_j f_j(x)$ as $n \to \infty$ for $j = 1, \ldots, \ell$.

Observe that

$$C_n^j \le \frac{4 \sup_z |F_Z^n(z) - F_Z(z)|}{Y_B - Y_A}$$

$$= \sqrt{n} \sup_z |F_Z^n(z) - F_Z(z)| \cdot \frac{4\sqrt{n}}{k_n} \cdot \frac{k_n}{n(Y_B - Y_A)},$$

where $z = (z_1, z_2)$.

Since $x \in X(f)$, we have by Lemma 2.2, that $\frac{k_n}{n(Y_B - Y_A)} \xrightarrow{a.s.} f(x)$ as $n \to \infty$. By Lemma 2.3, $\sqrt{n} \sup_z |F_Z^n(z) - F_Z(z)| \xrightarrow{L} G$, for some random variable G, where $\xrightarrow{L}$ denotes convergence in distribution. Hence since $\frac{\sqrt{n}}{k_n} \to 0$ as $n \to \infty$, we have $C_n^j \xrightarrow{p} 0$ as $n \to \infty$

(2.7) $$C_n^j \xrightarrow{p} 0 \quad \text{as} \quad n \to \infty$$

from the above bound on C_n^j. We now show $D_n^j \xrightarrow{p} p_j f_j(x)$ as $n \to \infty$, $j = 1, \ldots, \ell$. Note that

$$D_n^j = \frac{F(Y_B|I=j) \cdot P(I=j) - F(Y_A|I=j) \cdot P(I=j)}{Y_B - Y_A} \tag{2.8}$$

$$= p_j \cdot W_n^j(x),$$

where

$$W_n^j(x) = \frac{F_j(Y_B) - F_j(Y_A)}{Y_B - Y_A}.$$

Since $x \in C(f_j)$, let $I_\delta = (x - \delta, x + \delta)$, $\delta > 0$, be an interval such that $|f_j(x) - f_j(y)| < \varepsilon$ if $y \in I_\delta$, Then

$$P\{|W_n^j(x) - f_j(x)| \ge \varepsilon \text{ for some } n \ge N\}$$

$$\le P\{|W_n^j(x) - f_j(x)| \ge \varepsilon,\ Y_A, Y_B \in I_\delta \text{ for some } n \ge N\}$$
$$+ P\{Y_A \notin I_\delta \text{ for some } n \ge N\} + P\{Y_B \notin I_\delta \text{ for some } n \ge N\}.$$

The last two terms of this upper bound approach zero as $N \to \infty$ by Lemma 2.1 while the first term equals zero for by the mean value theorem since $W_n(x) = f_j(Y^*)$ for some $Y^* \in I_\delta$. Hence by (2.5), (2.6) and (2.7) we have $\hat{p}_j \hat{f}_j(x) \xrightarrow{p} p_j f_j(x)$ as $n \to \infty$, $j = 1, \ldots, \ell$, which completes the proof.

Motivated by the results of Theorems 1.1 and 2.1 we are led to the following easy to implement rule for classifying a new observation X.

Rule 2.1. To classify a new observation X based on our training sample $(X_1,I_1),\ldots,(X_n,I_n)$, partition the real line into the disjoint intervals $U_0,U_1,\ldots,U_{r_n},U_{r_n+1},U_{r_n+2}$ given by (2.4), classify X in U_0 or U_{r_n+2} arbitrarily and classify X in U_ν, $\nu = 1,\ldots,r_n+1$ as population π_j if j is the smallest integer $1,\ldots,\ell$ such that $K_{j\nu} = \max_{1\le i\le \ell} K_{i\nu} = M_\nu$, that is, $\hat{D}_j$ = union of such U_ν's.

Note that the rule is computationally very easy since Rule 2.1 assigns all points x in the interval U_ν, $\nu = 0,\ldots,r_n+2$ to the same population. Thus the algorithm merely stores these intervals and the population π_j to which each such interval belongs for classification and merely checks in which interval each new X is contained to determine its classification. Also note that U_ν is assigned to the first population π_j achieving a "majority vote" among the training sample.

Theorem 2.2: If k_n is taken such that $\sqrt{n}/k_n = o(1)$, and we assume that f_j is continuous almost everywhere for $i = 1,2,\ldots,\ell$, then $\hat{D} = (\hat{D}_1,\ldots,\hat{D}_\ell)$ defined by Rule 2.1 satisfies as $n\to\infty$,

i) $\sup_{D\in\mathscr{D}} |\hat{r}(D)-r(D)| \xrightarrow{p} 0$,

ii) $r(\hat{D}) \xrightarrow{p} r^*$,

and

iii) $\hat{r}(\hat{D}) \xrightarrow{p} r^*$.

Proof:

Since for $x \in U_\nu$,

$$K_{j\nu} = \max_{1\le i\le \ell} K_{i\nu}$$

$$\Longleftrightarrow \frac{K_{j\nu}}{(Y_B-Y_A)} = \max_{1\le i\le \ell} \frac{K_{i\nu}}{n(Y_B-Y_A)}$$

$$\Longleftrightarrow \hat{p}_j\hat{f}_j(x) = \max_{1\le i\le \ell} \hat{p}_i\hat{f}_i(x),$$

we see that Rule 2.1 is a rule of the form defined in (1.2). But by the law of large numbers and Theorem 2.1, we have $\hat{p}_j\hat{f}_j(x) \xrightarrow{p} p_j f_j(x)$, almost everywhere, as $n \to \infty$, and the proof is completed by Theorem 1.1 provided $\int_{R'} \sum_{j=1}^{\ell} \hat{p}_j\hat{f}_j(x)\, d\mu(x) \xrightarrow{p} 1$ as $n \to \infty$. But this is obvious since $\int_{R'} \sum_{j=1}^{\ell} \hat{p}_j\hat{f}_j(x)\, d\mu(x) = \frac{n-1}{n}$.

The estimated non-error rate $\hat{r}(\hat{D})$ can now be easily computed:

$$\hat{r}(\hat{D}) = \sum_{\nu=1}^{r_n+1} \int_{U_\nu} \max_{1 \leq j \leq \ell} \frac{K_{j\nu}}{n(Y_{B_n(x)} - Y_{A_n(x)})}\, d\mu(x) = \sum_{\nu=1}^{r_n+1} M_\nu / n, \tag{2.9}$$

where $\int_{U_\nu}$ is the integral over the interval U_ν and where $K_{j\nu}$ is defined in (2.4) and $M_\nu = \max_{1 \leq i \leq \ell} K_{i\nu}$.

Duda and Hart [3, Sec. 3.9] indicated that the sample for estimating the non-error rate should be different from the training sample in the conventional method because the estimated non-error rate will definitely be optimistic for small size training samples. However, (2.9) is a byproduct of Rule 2.1 and gives the desired features of computational ease and convergence to the optimal non-error rate when the training sample size is large. Later, using generated data, we will see that the non-error rate estimated by (2.9) is close to the non-error rate even if the training sample size is only fifty.

3. A Classification Rule Based on a Histogram Method in the Two-Dimensional Case (m=2).

Let $Z_i = \binom{X_i}{I_i}$, $i = 1, 2, \ldots, n$ be a training sample of size n taken from the mixed population, where $X_i = \binom{X_i^{(1)}}{X_i^{(2)}}$, and I_i is an integer-valued index which indicates from which population X_i comes.

We first define a bivariate histogram estimate $\hat{f}(x_1, x_2)$ of $f(x_1, x_2)$ given by Kim and Van Ryzin [6].

Let $Y_1^{(1)} \leq Y_2^{(1)} \leq \ldots \leq Y_n^{(1)}$ be the order statistics of $X_i^{(1)}$, $i = 1,2,\ldots,n$, and $Y_0^{(1)} = -\infty$ and $Y_{n+j}^{(1)} = +\infty$, $j = 1,2,\ldots$.

Let k_n be a sequence of positive integers. We shall simply write $k_n = k$ when convenient. Let us define

$$J_1 = \{0,1,[\sqrt{nk}]+1, 2[\sqrt{nk}]+1,\ldots\},$$

$$A_n(x_1) = \max\{\alpha_1 | \alpha_1 \varepsilon J_1,\ Y_\alpha^{(1)} < x_1,\ \alpha_1 \leq n\},$$

where $[\sqrt{nk}]$ means the largest integer $\leq \sqrt{nk}$.

Let $Y^{(2)} \leq Y_2^{(2)} \leq \ldots \leq Y_{[\sqrt{nk}]}^{(2)}$ be the order statistics of the second component of the X_i's such that X_i's are in the cylinder set $(Y_{A_n(x_1)}^{(1)}, Y_{A_n(x_1)+h}^{(1)}] \times (-\infty,\infty)$, where $h = h_n = [nk_n]$. Define

$$Y_0^{(2)} = -\infty,\quad Y_{[\sqrt{nk}]+j}^{(2)} = +\infty,\quad j = 1,2,\ldots .$$

$$J_2 = \{0,1,k+1,2k+1,\ldots\},\quad \text{and}$$

$$B_n(x_2,A_n(x_1)) = \max\{\alpha_2 | \alpha_2 \epsilon J_2,\ Y_{\alpha_2}^{(2)} < x_2,\ \alpha_2 \leq [\sqrt{nk}]\}.$$

Now, define $r_{1,n}$ as the largest integer such that $n - [nk] \leq r_{1,n}[\sqrt{nk}] + 1 < n$. Then, let $Y_{i,1}^{(2)} \leq Y_{i,2}^{(2)} \leq \ldots \leq Y_{i,[\sqrt{nk}]}^{(2)}$ be the order statistics of the second component of the $X_j = (X_j^{(1)}, X_j^{(2)})$'s which are in the cylinder set $(Y_{(i-1)[\sqrt{nk}]+1}, Y_{i[\sqrt{nk}]+1}] \times (-\infty,\infty)$, $i = 1,\ldots, r_{1,n}$.

In the remainder of this section, we shall simply write $A_n(x_1) = A$, $B_n(x_1) = B$, $h = h_n$, and $k = k_n$. We shall also omit the superscripts and simply write $Y_A^{(1)} = Y_A$, $Y_B^{(2)} = Y_B$ when convenient. As a consequence of the above definition we have for all n ,

$$P\{0 \leq A \leq n,\ Y_A < x_1 \leq Y_{A+h},$$
$$0 \leq B \leq [\sqrt{nk}],\ Y_B \leq x_2 \leq Y_{B+k}\} = 1 .$$

Define

$$\Delta_k F_n(x_1,x_2) = F_n(Y_{A+h},Y_{B+k}) - F_n(Y_A,Y_{B+k}) - F_n(Y_{A+h},Y_B) + (Y_A,Y_{B+k}),$$

where F_n is the empiric function, and

$$f(x_1,x_2) = \frac{\Delta_k F_n(x_1,x_2)}{(Y_{A+h} - Y_A)(Y_{B+k}-Y_B)} . \tag{3.1}$$

The density estimator $\hat{f}$ for f defined in (3.1) is slightly different from that of [6] but it is easy to show that all results from Kim and Van Ryzin [6] listed below are valid here. We add the condition that k_n satisfy:

$$\lim_n (\log n/k_n) = 0 . \tag{3.2}$$

<u>Lemma 3.1</u> [6, Lemma 1]: If $\{k_n\}$ satisfy (2.1) and (3.2) and $f(x_1,x_2) > 0$ at (x_1,x_2), then as $n \to \infty$, $Y_A \xrightarrow{a.s.} x_1$, and $Y_{A+h}-Y_A \xrightarrow{a.s.} 0$.

<u>Lemma 3.2</u> [6, Lemma 4]: Let $\{k_n\}$ satisfy (2.1) and (3.2). If $f(x_1,x_2) > 0$, then $Y_B \xrightarrow{a.s.} x_2$, $Y_{B+k} \xrightarrow{a.s.} x_2$, and $Y_{B+k}-Y_B$ $Y_{B+k} - Y_B \xrightarrow{a.s.} 0.$

<u>Lemma 3.3</u> [6, Main Theorem]: Let $\{k_n\}$ satisfy (2.1) and (3.2). Then $\hat{f}(x_1,x_2) \xrightarrow{a.s.} f(x_1,x_2)$ at $(x_1,x_2) \in C(f)$ as $n \to \infty$, where $C(f)$ is the set of continuity points of f in R^2.

Let $S_{A,B} = \{(x_1,x_2) \in R^2 : (x_1,x_2) \in (Y_{A+h},Y_A] \times (Y_{B+k},Y_B]\}$, and

$$N_j(S_{A,B}) = \sum_{i=1}^{n} \chi_{[(X_i^{(1)},X_i^{(2)}) \in S_{A,B},\ I_i = j]} .$$

Note that $N_j(S_{A,B})$ is merely the number of training samples in the block $S_{A,B}$ which belong to population j .

Define

$$\hat{p}_j \hat{f}_j(x_1,x_2) = \frac{N_j(S_{A,B})}{n(Y_{A+h}-Y_A)(Y_{B+k}-Y_B)} . \tag{3.3}$$

The following theorem shows that (3.3) consistently estimates $p_j f_j(x_1,x_2)$.

Theorem 3.1: If $(x_1,x_2) \in C(f) \cap C(f_j)$, and k_n is taken such that $\sqrt{n}/k_n = o(1)$, then we have

$$\hat{p}_j \hat{f}_j(x_1,x_2) \xrightarrow{p} p_j f_j(x_1,x_2) \text{ as } n \to \infty, \ j = 1,2,\ldots,\ell .$$

Proof :

If $f(x_1,x_2) = 0$, $p_j f_j(x_1,x_2) = 0$ for all $1 \le j \le \ell$. By Lemma 3.3, $\hat{f}_j(x_1,x_2) \le \hat{f}(x_1,x_2) \xrightarrow{a.s.} f(x_1,x_2) = 0$. In the remainder of the proof, we assume that $f(x_1,x_2) > 0$.

Let us denote $F_Z(z) = F_Z(z_1,z_2,z_3) = P(X_i^{(1)} \le z_1, X_i^{(2)} \le z_2, I_i \le z_3)$ and the empiric function $F_Z^n(z) = \frac{1}{n} \sum_{i=1}^n \chi_{[X_i^{(1)} \le z_1, X_i^{(2)} \le z_2, X_i \le z_3]}$.

Then, as $n \to \infty$,

$$\begin{aligned}
\hat{p}_j \hat{f}_j(x_1,x_2) &= \frac{N_j(S_{A,B})}{n(Y_{A+h}-Y_A)(Y_{B+k}-Y_B)} \\
&= \{[F_Z^n(Y_{A+h},Y_{B+k},j) - F_Z^n(Y_{A+h},Y_{B+k},j-1)] \\
&\quad - [F_Z^n(Y_A,Y_{B+k},j) - F_Z^n(Y_A,Y_{B+k},j-1)] \\
&\quad - [F_Z^n(Y_{A+h},Y_B,j) - F_Z^n(Y_{A+h},Y_B,j-1)] \\
&\quad + [F_Z^n(Y_A,Y_B,j) - F_Z^n(Y_A,Y_B,j-1)]\}/[(Y_{A+h}-Y_A)(Y_{B+k}-Y_b \\
&= C_n^j + D_n^j
\end{aligned}$$

where

$$C_n^j = \frac{1}{(Y_{A+h}-Y_A)(Y_{B+k}-Y_B)} \{[F_Z^n(Y_{A+h}, Y_{B+k}, j) - F_Z(Y_{A+h}, Y_{B+k}, j)]$$

$$- [F_Z^n(Y_{A+h}, Y_{B+k}, j-1) - F_Z(Y_{A+h}, Y_{B+k}, j-1)]$$

$$- [F_Z^n(Y_A, Y_{B+k}, j) - F_Z(Y_A, Y_{B+k}, j-1)]$$

$$+ [F_Z^n(Y_A, Y_{B+k}, j-1) - F_Z(Y_A, Y_{B+k}, j-1)]$$

$$+ [F_Z^n(Y_A, Y_B, j) - F_Z(Y_A, Y_B, j)]$$

$$- [F_Z^n(Y_A, Y_B, j-1) - F_Z(Y_A, Y_B, j-1)]\}$$

$$- [F_Z^n(Y_{A+h}, Y_B, j) - F_Z(Y_{A+h}, Y_B, j)]$$

$$+ [F_Z^n(Y_{A+h}, Y_B, j-1) - F_Z(Y_{A+h}, Y_B, j-1)] ,$$

and

$$D_n^j = \frac{1}{(Y_{A+h}-Y_A)(Y_{B+k}-Y_B)} \{F_Z(Y_{A+h}, Y_{B+k}, j) - F_Z(Y_{A+h}, Y_{B+k}, j-1)$$

$$-F_Z(Y_A, Y_{B+k}, j) + F_Z(Y_A, Y_{B+k}, j-1)$$

$$-F_Z(Y_{A+h}, Y_B, j) + F_Z(Y_{A+h}, Y_B, j-1)$$

$$+ F_Z(Y_A, Y_B, j) - F_Z(Y_A, Y_B, j-1)\}.$$

The proof now follows from Lemma 2.3 and 3.3. That is,

$$C_n^j \leq 8\sqrt{n} \sup_z |F_Z^n(z) - F_Z(z)| \cdot \frac{\sqrt{n}}{k_n} \frac{k_n}{n(Y_{A+h}-Y_A)(Y_{B+k}-Y_B)}$$

where as $n \to \infty$,

$$\hat{f}(x_1, x_2) = \frac{k_n}{n(Y_{A+h}-Y_A)(Y_{B+k}-Y_B)} \xrightarrow{a.s.} f(x_1, x_2)$$

by Lemma 3.3 and

$$\sqrt{n}\sup_z |F_Z(z)-F_Z^n(z)| \xrightarrow{L} \text{some random variable},$$

by Lemma 2.3, where $\xrightarrow{L}$ denotes converges in law. Hence, since $\sqrt{n}/k_n \to 0$, we have $C_n^j \xrightarrow{p} 0$, as $n \to \infty$. Also as in Theorem 2.1, we have

$$D_n^j = p_j \cdot W_n^j(x_1, x_2)$$

where

$$W_n^j(x_1,x_2) = \frac{F_j(Y_{A+h}, Y_{B+k}) - F_j(Y_A, Y_{B+k}) - F_j(Y_B, Y_{B+k}) + F_j(Y_A, Y_B)}{(Y_{A+h}-Y_A)(Y_{B+k}-Y_B)}.$$

But,

$$|W_n^j(x_1,x_2) - f_j(x_1,x_2)| =$$

$$\frac{1}{(Y_{A+h}-Y_A)(Y_{B+k}-Y_B)} \int_{Y_B}^{Y_{B+k}} \int_{Y_A}^{Y_{A+h}} |f_j(s,t) - f_j(x_1,x_2)|\, ds dt$$

$$< \sup_{(s,t)} \{|f(s,t)-f(x_1,x_2)| : Y_A < s,\ x_1 \le Y_{A+h}, Y_B < t,\ x_2 \le Y_{B+k}\} \xrightarrow{a.s.} 0$$

as $n \to \infty$, since as $n \to \infty$, $Y_{A+h} - Y_A \xrightarrow{a.s.} 0$ and $Y_{B+k} - Y_B \xrightarrow{a.s.} 0$ by Lemmas 3.1 and 3.2, respectively, and the fact that f is continuous at (x_1, x_2). Therefore $D_n^j \xrightarrow{a.s.} p_j f_j$, $j = 1, \ldots, \ell$ completing the proof.

Based on Theorem 3.1 we have the simple classification rule based on the density estimates $\hat{f}_j(x_1, x_2)$ <u>which does not need to compute these estimates but merely results in a convenient partitioning of 2-space into rectangles or blocks</u>, each of which is assigned to one of the ℓ populations. This rule is:

<u>Rule 3.1</u> <u>Classify rectangle or block</u> $S_{A,B}$ <u>to</u> π_j <u>if</u> j <u>is the smallest integer such that</u> $N_j(S_{A,B}) = \max_{1 \le i \le \ell} N_i(S_{A,B})$.

Note that rule 3.1 merely is a "majority vote rule" for the block $S_{A,B}$.

Theorem 3.2: Let k_n be taken as an increasing function of n such that $\sqrt{n}/k_n = o(1)$. Assume that the $f_j(x_1,x_2)$'s are continuous almost everywhere for $j = 1,2,\ldots,\ell$. Then, Rule 3.1 satisfies

i) $\sup_{D \in \mathcal{D}} |\hat{r}(D) - r(D)| \xrightarrow{p} 0$

ii) $r(\hat{D}) \xrightarrow{p} r^*$, as $n \to \infty$,

and

iii) $\hat{r}(\hat{D}) \xrightarrow{p} r^*$.

Proof:

Note that by using the same approach as in Theorem 2.2 one can show

$$\iint_{R^2} \hat{f}(x_1,x_2) \leq 1$$

and as $n \to \infty$, both

$$\iint_{R^2} \hat{f}(x_1,x_2) \xrightarrow{p} 1,$$

and

$$\hat{p}_j \hat{f}_j(x_1,x_2) \xrightarrow{p} p_j f_j(x_1,x_2)$$

by Theorem 3.1. Hence Theorem 1.1 applies.

By using the same type of derivation as in (2.9), we have the following simple consistent estimator of the non-error rate.

(3.4) $$\hat{r}(\hat{D}) = \sum_{\text{all blocks } S_{A,B}} \sum N_j^*(S_{A,B})$$

where $N_j^*(S_{A,B}) = \max_{1 \leq i \leq \ell} N_i(S_{A,B})$

= the number of observations from the jth population which dominates the other populations in the rectangular region or block $S_{A,B}$.

4. A Monte Carlo Study of the Classification Rules.

We have shown that, by properly choosing the sequence $\{k_n\}$, Rules 2.1 and 3.1 are Bayes risk consistent rules, and their estimated probabilities of correct classification (non-error rate) converge to the probability of correct classification of the optimal rule as the training sample size increases. We shall denote such a rule as a histogram rule (HR). Besides the large sample optimality, it is important to also know how such a rule performs when the training sample size is small or moderate. Samples from twelve mixed populations are generated in this section and some Monte Carlo comparisons are made. The non-error rates and the estimated non-error rates of these rules are calculated for sample sizes 50, 100, 200, and 400 for the distributions and the proportions of the mixtures as given below. For each sample size, 500 sets of independent training samples were generated by computer to obtain the Monte Carlo results given.

4.1 One-Dimensional Case (m = 1)

It has been shown that k_n can be taken as $c\, n^{\frac{1}{2}+\delta}$, where c is a finite constant, and $0 < \delta < \frac{1}{2}$ in Rule 2.1, by Theorem 2.2. When the sample size is small the choice of c and δ may be important to the performance of the rule. However, it will be seen later that the choice of these constants is not very sensitive for the examples used in the generated data.

Throughout our simulation, we work with the two-population classification problem only.

There are twelve mixed populations shown in Table 1. Samples generated from each population are of sizes 50, 100, 200, and 400. There are 500 samples for each sample size and each mixture. Three values of δ are selected: .05, .15, and .25. These δ's give enough coverage of k_n for practical use in small samples and satisfy the conditions of Theorem 2.2.

Table 1. Twelve Mixed Populations

Mixture Number	Population 1 Distribution	Population 1 Proportion	Population 2 Distribution	Population 2 Proportion
1	N(0,1)	.5	N(1,1)	.5
2	N(0,1)	.25	N(1,1)	.75
3	N(0,1)	.5	N(2,1)	.5
4	N(0,1)	.25	N(2,1)	.75
5	N(0,1)	.5	N(.5,1)	.5
6	N(0,1)	.25	N(.5,1)	.75
7	N(0,1)	.5	N(0,4)	.5
8	N(0,1)	.25	N(0,4)	.75
9	N(0,1)	.5	N(2,64)	.5
10	N(0,1)	.25	N(2,64)	.75
11	N(0,1)	.5	N(1,4)	.5
12	$\frac{1}{2}$N(0,1)+ $\frac{1}{2}$N(10,1)	.5	$\frac{1}{2}$N(5,1)+ $\frac{1}{2}$N(15,1)	.5

The non-error rates (NER's) of each sample for rule 2.1 were calculated for each of the three δ's with $c = 1$, Tables 2 through 5 give the mean estimated non-error rates (NER's) for the 500 samples for each of the three choices of δ for the sample sizes $n = 50$, 100, 200 and 400 respectively. Also, the standard error of the repeated estimated mean NER's is given in these tables below the mean entry. The range and median of the means is also given. In column 2 of these tables, the optimal non-error rate, r^*, of the Bayes rule as given by Welch [12] is calculated. The non-error rate r^* is the best obtainable rate. In column 3 of each table the estimated mean non-error rate and its standard deviation are given for the "plug-in" rule. This is the rule one would use if we plug-in the usual estimators (sample means and standard deviations) in the optimal Bayes rule assuming normal populations. Thus, column 3 represents the "best parametric rule" one can obtain assuming each of the two populations is normal. In each table, the best histogram rule is underlined.

In examining Tables 2 through 5, certain conclusions seem appropriate. First, the choice of δ does not seem too critical since all ranges are small except for mixture 12. Second, the mean non-error rates improve (as expected by Theorem 2.1) as the sample size increases. Third, the standard errors of each of the means for the histogram rules are small and are less than .012, .012, .011 and .010 for sample sizes n = 50, 100, 200 and 400 respectively. Finally, and most importantly, the loss in NER over the optimal rule for the best histogram rules is favorable in all cases with that of the "plug-in rule", which is the best one can do in a parametric setting. On the other hand, if one uses the parametric approach blindly by assuming both population 1 and 2 in the mixture are normally distributed and estimate f_i by plugging-in its sample means and variances, the results can be disastrous. This is shown by the special mixture 12. The mean error rates for the plug-in rules are .56, .53, .52 and .51 for sample sizes 50, 100, 200 and 400 respectively, while the optimal rule for non-error rate of .99. When the histogram rule is used for the same set of data, the best mean non-error rates are .79, .84, .87 and .90 for sample sizes 50, 100, 200 and 400 respectively.

The results of Tables 2 through 5 for NER are summarized for the median case among the three δ's in Table 6. Table 6 also gives the similar results on the median case among the three δ's for the mean estimated non-error rate (ENER) for each simulated sample of size 500 for each mixture. The results in general for the ENER were quite similar to those for NER and hence only the summary results of Table 6 are given for ENER.

Table 2. The Mean Non-Error Rate of 500 Samples (n = 50)

Mixture Number		Optimal Rule r^*	Plug-in Rule	Histogram Rule $\delta = .05$	.15	.25	Range	Median
1	Mean	.6915	.6657	.6187	.6239	.6413	.0276	.6239
	S.d.		.0013	.0023	.0021	.0022		.0021
2	Mean	.7775	.7128	.6805	.6997	.7145	.0340	.6997
	S.d.		.0170	.0116	.0105	.0090		.0105
3	Mean	.8413	.8342	.7826	.7817	.7748	.0078	.7817
	S.d.		.0044	.0088	.0094	.0096		.0094
4	Mean	.8730	.8637	.7964	.7857	.7426	.0538	.7857
	S.d.		.0077	.0107	.0108	.0100		.0108
5	Mean	.5987	.5601	.5326	.5386	.5476	.0150	.5386
	S.d.		.0077	.0077	.0083	.0089		.0083
6	Mean	.7510	.6885	.6607	.6907	.7087	.0480	.6907
	S.d.		.0156	.0118	.0111	.0107		.0111
7	Mean	.6613	.5931	.5319	.5250	.5230	.0089	.5250
	S.d.		.0115	.0106	.0110	.0111		.0110
8	Mean	.7500	.7395	.6540	.6641	.6865	.0325	.6641
	S.d.		.0115	.0113	.0112	.0011		.0112
9	Mean	.8818	.8742	.7924	.7712	.7311	.0362	.7712
	S.d.		.0042	.0104	.0113	.0109		.0113
10	Mean	.8587	.8360	.7595	.7413	.7269	.0326	.7413
	S.d.		.0062	.0107	.0106	.0105		.0106
11	Mean	.6950	.6610	.5971	.6019	.6084	.0077	.6019
	S.d.		.0084	.0106	.0110	.0108		.0110
12	Mean	.9907	.5552	.7861	.7830	.6461	.1400	.7830
	S.d.			.0117	.0139	.0112		.0139

Table 3. The Mean Non-Error Rate of 500 Samples (n = 100)

Mixture Number		Optimal Rule r^*	Plug-in Rule	Histogram Rule (HR) $\delta=.05$	.15	.25	Range	Median
1	Mean	.6915	.6743	.6352	.6408	.6425	.0073	.6408
	S.d.		.0025	.0016	.0017	.0014		.0017
2	Mean	.7775	.7476	.7208	.7234	.7263	.0055	.7234
	S.d.		.0136	.0991	.0077	.0070		.0077
3	Mean	.8413	.8377	.8055	.8072	.7818	.0254	.8055
	S.d.		.0136	.0072	.0070	.0068		.0072
4	Mean	.8730	.8691	.8345	.8351	.7839	.0512	.8345
	S.d.		.0031	.0074	.0082	.0094		.0074
5	Mean	.5987	.5634	.5339	.5362	.5494	.0155	.5362
	S.d.		.0067	.0074	.0077	.0051		.0077
6	Mean	.7510	.7134	.6918	.7096	.7109	.0191	.7096
	S.d.		.0129	.0096	.0081	.0096		.0081
7	Mean	.6613	.6149	.5507	.5363	.5318	.0189	.5363
	S.d.		.0105	.0102	.0107	.0111		.0107
8	Mean	.7500	.7441	.6757	.6980	.7083	.0326	.6980
	S.d.		.0098	.0097	.0097	.0096		.0097
9	Mean	.8818	.8778	.8423	.8219	.7326	.1097	.8219
	S.d.		.0028	.0081	.0093	.0089		.0093
10	Mean	.8587	.8418	.7806	.7743	.7984	.0241	.7806
	S.d.		.0042	.0094	.0097	.0097		.0094
11	Mean	.5950	.6670	.6141	.6088	.6176	.0088	.6141
	S.d.		.0068	.0092	.0100	.0105		.0092
12	Mean	.9907	.5342	.8405	.8098	.6743	.1662	.8098
	S.d.		.0094	.0109	.0121	.0116		.0121

Table 4. The Mean Non-Error Rate of 500 Samples (n = 200)

Mixture Number		Optimal Rule, r^*	Plug-in Rule	Histogram Rule (HR) δ=.05	.15	.25	Range	Median
1	Mean	.6915	.6798	.6571	.6665	.6603	.0094	.6603
	S.d.		.0011	.0011	.0009	.0008		.0008
2	Mean	.7775	.7641	.7490	.7413	.7399	.0091	.7413
	S.d.		.0090	.0070	.0052	.0048		.0052
3	Mean	.8413	.8398	.8209	.8235	.8145	.0090	.8209
	S.d.		.0022	.0055	.0024	.0058		.0055
4	Mean	.8730	.8711	.8492	.8532	.8285	.0247	.8492
	S.d.		.0022	.0059	.0013	.0071		.0059
5	Mean	.5987	.5693	.5387	.5481	.5489	.0102	.5481
	S.d.		.0059	.0072	.0074	.0081		.0074
6	Mean	.7510	.7269	.7116	.7259	.7385	.0269	.7259
	S.d.		.0089	.0087	.0075	.0071		.0075
7	Mean	.6613	.6407	.5826	.5737	.5512	.0314	.5737
	S.d.		.0078	.0090	.0101	.0110		.0101
8	Mean	.7500	.7488	.6989	.7179	.7286	.0297	.7179
	S.d.		.0115	.0079	.0079	.0081		.0079
9	Mean	.8818	.8794	.8579	.8514	.7819	.0760	.8514
	S.d.		.0021	.0057	.0060	.0097		.0060
10	Mean	.8587	.8458	.7915	.7965	.7889	.0076	.7915
	S.d.		.0044	.0084	.0096	.0104		.0084
11	Mean	.6950	.6736	.6278	.6296	.6334	.0056	.6296
	S.d.		.0047	.0082	.0090	.0091		.0090
12	Mean	.9907	.5198	.8694	.8080	.8374	.0614	.8374
	S.d.		.0071	.0092	.0102	.0091		.0091

Table 5. The Mean Non-Error Rate of 500 Samples (n = 400)

Mixture Number		Optimal Rule Rule r^*	Plug-in Rule	Histogram Rule (HR) $\delta=.05$	.15	.25	Range	Median
1	Mean	.6915	.6817	.6682	.6717	.6707	.0035	.6707
	S.d.		.0003	.0008	.0007	.0011		.0011
2	Mean	.7775	.7684	.7660	.7491	.7447	.0213	.7491
	S.d.		.0034	.0047	.0053	.0035		.0053
3	Mean	.8413	.8407	.8279	.8287	.8346	.0067	.8287
	S.d.		.0016	.0045	.0046	.0041		.0046
4	Mean	.8730	.8720	.8609	.8604	.8535	.0074	.8604
	S.d.		.0015	.0046	.0044	.0044		.0044
5	Mean	.5987	.5706	.5427	.5522	.5486	.0095	.5486
	S.d.		.0050	.0063	.0070	.0073		.0073
6	Mean	.7510	.7300	.7180	.7329	.7391	.0211	.7329
	S.d.		.0044	.0073	.0061	.0070		.0061
7	Mean	.6613	.6521	.5940	.5983	.5983	.0043	.5983
	S.d.		.0056	.0077	.0089	.0101		.0089
8	Mean	.7500	.7499	.7113	.7310	.7386	.0273	.7310
	S.d.		.0064	.0068	.0063	.0062		.0063
9	Mean	.8818	.8801	.8710	.8704	.8383	.0327	.8704
	S.d.		.0018	.0039	.0041	.0053		.0041
10	Mean	.8587	.8480	.7774	.8103	.7774	.0329	.7774
	S.d.		.0034	.0065	.0076	.0065		.0065
11	Mean	.6950	.6755	.6345	.6400	.6441	.0096	.6400
	S.d.		.0047	.0072	.0078	.0084		.0078
12	Mean	.9907	.5122	.8944	.8967	.7545	.1422	.8944
	S.d.		.0053	.0084	.0088	.0090		.0084

Table 6. The Medians of the Mean Non-Error Rates and the Mean Estimated Non-Error Rates of the HR Among Three δ's.

Mixture Number	Optimal Rule, r^*	n = 50			n = 100			n = 200			n = 400		
		PR	NER	ENER	PR	NER	ENER	PR	NER	ENER	PR	NER	ENER
1	.6915	.6657	.6239	.7124	.6743	.6408	.7017	.6798	.6603	.7059	.6817	.6707	.7065
2	.7775	.7128	.6997	.7528	.7476	.7234	.7528	.7641	.7413	.7503	.7684	.7491	.7531
3	.8413	.8342	.7817	.8221	.8377	.8055	.8365	.8398	.8209	.8259	.8407	.8287	.8386
4	.8730	.8637	.7857	.8010	.8691	.8345	.8160	.8711	.8492	.8265	.8720	.8604	.8287
5	.5987	.5601	.5386	.6364	.5634	.5362	.6546	.5693	.5481	.6461	.5706	.5486	.6436
6	.7510	.6885	.6907	.7570	.7134	.7096	.7551	.7269	.7259	.7519	.7300	.7329	.7525
7	.6613	.5931	.5250	.6070	.6149	.5263	.6027	.6407	.5737	.5880	.6521	.5983	.5825
8	.7500	.7395	.6641	.7246	.7441	.6980	.7455	.7488	.7179	.7502	.7499	.7310	.7510
9	.8818	.8742	.7712	.7708	.8778	.8219	.7969	.8794	.8514	.8206	.8801	.8704	.8401
10	.8587	.8360	.7413	.7856	.8418	.7806	.7985	.8458	.7915	.8020	.8408	.7774	.8067
11	.6950	.6610	.6019	.6731	.6670	.6141	.6690	.6736	.6296	.6592	.6755	.6400	.6528
12	.9907	.5552	.7830	.8359	.5342	.8098	.8459	.5198	.8374	.8805	.5122	.8944	.9396

PR = Plug-in Rule

4.2 Two-Dimensional Case (m = 2)

Suppose we have a two-population classification problem with two-dimensional measurements. Let the two populations which form the mixture be bivariate normal distributions. Let us write these distributions and the proportion with which they occur $N(\mu_{i1},\mu_{i2},\sigma_{i1},\sigma_{i2},\rho_i)$ and p_i, for i = 1 and 2, respectively. The μ's are population means, the σ's are population standard deviations and the ρ is the population correlation coefficients.

Two hundred samples of sizes 50, 100, 200, and 400 were generated through the use of the method in [2] by Bhattacharyya, Johnson, and Neave.

Due to the high cost of computation of the NER and ENER only one mixed population is generated. This mixture consists of two populations with distributions N(0,0,1,1,.5) and N(2,2,2,2,.5) with proportions .333 and .667, respectively.

By considering the sample size and the restrictions on choosing k_n in Theorem 3.2, k_n is chosen as $n^{.55}/2$ for Rule 3.1.

Table 7 shows the histogram rule, Rule 3.1, works well even with a relatively small sample size (n = 50).

Table 7. The Mean and Standard Error of the Mean of the Non-Error Rates and the Estimated Non-Error Rates of 200 Samples.†

Sample Size	Optimal Rule, r*		Plug-in Rule	NER	ENER
50	.8243	Mean	.8023	.7596	.7585
		(s.e.)	(.0012)	(.0025)	(.0027)
100	.8243	Mean	.8138	.7765	.7880
		(s.e.)	(.0053)	(.0017)	(.0022)
200	.8243	Mean	.8195	.7869	.8028
		(s.e.)	(.0003)	(.0012)	(.0015)
400	.8243.	Mean	.8220	.7962	.8082
		(s.e.)	(.0001)	(.0008)	(.0012)

†These samples are from a mixed population which consists of two populations with distribution N(0,0,1,1,.5) and N(2,2,2,2,.5) with proportions .333 and .667, respectively.

5. Conclusion

It has been shown that the non-error rates of the proposed histogram rules converge in probability to the optimal Bayes rule as the training sample increases. The estimated non-error rates can be calculated easily as byproducts of these rules. Since the differences of the non-error rates and the estimated non-error rates of these rules converge to 0 , this allows us to formulate the rules and at the same time to estimate the performance of these rules when a training sample is given.

From the simulated data, it is shown that the histogram rules work quite well for all cases considered and it is not very sensitive to different values of δ used in the rules. Tables 6 and 7 show that the estimated non-error rates are not too biased in the direction of overestimating the Bayes risk for small training sample sizes. This is because the choice of k_n in the rules here is restricted to $k_n > \sqrt{n}$.

If someone is concerned about the choice of k_n for the histogram method and can obtain an additional set of random samples besides the training sample, he can pick k_n by trying different values of k_n such that k_n maximizes the number of correctly classified individuals in the new sample using the rule formulated by the training sample.

The density estimators given by Anderson [1], Gessaman [4], and Loftsgarrden and Quesenberry [9] are quite similar to those of Van Ryzin [11] and Kim and Van Ryzin [6]. Hence the idea used in the histogram rule here may be applied to these estimators. However, computationally they may not have the ease of the rules presented here. Another interesting problem is to find the approximate sample size for these rules so that the difference between the non-error rate of the rule and the non-error rate of the optimal rule is within a specified constant, with a specified probability. Hence further investigations are needed.

References

1. Anderson, T. W., Some Nonparametric Multivariate Procedures Based on Statistically Equivalent Blocks, Multivariate Analysis, ed. P. R. Krishnaiah, Academic Press, New York, 1966.

2. Bhattacharyya, G. K., Johnson, R. A., and Neave, H. R., A Comparative Power Study of the Bivariate Rank Sum Test and T^2, Technometrics, 13, (1971), 191-198.

3. Duda, R. O. and Hart, P. E., Pattern Classification and Scene Analysis, John Wiley & Sons, Inc., New York, 1973.

4. Gessaman, M. P., A consistent Nonparametric Multivariate Density Estimator Based on Statistically Equivalent Blocks, Ann. Math. Statist., 41, (1970), 1344-1346.

5. Glick, N., Sample-Based Classification Procedures Derived from Density Estimators, J. Amer. Statist. Assoc., 67, (1972), 116-122.

6. Kim, B. K., and Van Ryzin, J., A Bivariate Histogram Estimator, Technical Report No. 444, Dept. of Statistics, University of Wisconsin, Madison, 1976.

7. Kim, B. K. and Van Ryzin, J., Uniform Consistency of a Histogram Density Estimator and Modal Estimation, Comm. Statist., 4, (1975), 303-315.

8. Kiefer, J. and Wolfowitz, J., On the Deviations of the Empiric Distribution Function of Vector Chance Variables, Trans. Amer. Math. Soc., 87, (1958), 173-186.

9. Loftsgarrden, D. O. and Quesenberry, C. P. A Non-Parametric Estimate of a Multivariate Density Function, Ann. Math. Statist., 36, (1965), 1049-1051.

10. Van Ryzin, J., Non-Parametric Bayesian Decision Procedures for (Pattern) Classification with Stochastic Learning, Fourth Prague Conference on Information Theory, Statistical Decision Functions and Random Processes, Sept. 1965, Publishing House of the Czechoslovak Academy of Sciences, Prague, (1967), 479-494.

11. Van Ryzin, J., A Histogram Method of Density Estimation, Comm. Statist., 2, (1973), 493-506.

12. Welch, B. L., Note on Discriminant Functions, Biometrika, 2, (1939), 218-220.

This research was supported in part by DHEW, PHS, National Institute of Health under Grant 5 R01 CA 18332-02.

National Institute of Environmental Health Sciences
Research Triangle Park, North Carolina

Department of Statistics
University of Wisconsin-Madison
1210 West Dayton Street
Madison, Wisconsin 53706

Optimal Smoothing of Density Estimates

Grace Wahba

1. Introduction.

1.1. Density estimation in classification.

The Neyman-Pearson lemma tells us that, if we want to classify an object as coming from one of two possible populations with associated densities f_1 and f_2, then we should base the classification on the likelihood ratio f_1/f_2. Frequently this leads to simple and effective algorithms. For example if each of the two densities can be assumed to be multivariate normal then the likelihood ratio is constant along hyperplanes or hyperquadratics in Euclidean p-space. Then the optimal classification rule consists of determining which side of a certain hyperplane or hyperquadratic of constant likelihood the measurement vector to be classified lies. If the means and covariances are not known a priori, the problem of "learning" the optimal classification rule from the data reduces to "learning" an appropriate hyperplane or hyperquadratic.

This procedure is in common use, and gives satisfactory results for a wide class of non-normal densities. However, if the underlying densities have, for example, C or S shaped ridges or multiple modes, then one would like to estimate the likelihood f_1/f_2. (See Chi and Van Ryzin [5] who present a simple consistent non-parametric procedure for estimating f_1/f_2 in several dimensions.)

In this paper we approach the problem of estimating the likelihood ratio from the point of view of estimating the individual densities. For the purpose of classification there is no obvious advantage to estimating the densities as an intermediate step, other than that the state of the art of density estimation seems to be further along than that of likelihood ratio estimation. In fact we would be pleased if the methods described here could be applied directly to the estimation of the likelihood ratio.

1.2. The major types of density estimation.

There is a very extensive literature on one dimensional and multi-dimensional density estimation, we mention only a few of the many papers. The early papers, on kernel methods, are Whittle [47], Rosenblatt [27], Parzen [24]. Orthogonal series estimates have been discussed by Watson [44], Kronmal and Tarter [20], and recently by Brunk [4] and Crain [8]. Boneva, Kendall and Stefanov [3] introduced spline methods, see also Lii and Rosenblatt [21], Wahba [35,39]. The k-nearest neighbor methods were introduced by Loftsgarten and Quesenberry [22], see also Van Ryzin [31], Wahba [32,39]. Penalized maximum likelihood methods were treated by Good and Gaskins [15], and de Montricher, Tapia and Thompson, [10] showed their relationship to spline methods. Recently, Walter and Blum [42] have provided a framework which unifies some of these methods, see also Parzen [25]. A survey and some Monte Carlo studies appear in Wegman [45,46]. For an extensive bibliography and survey, see Cover and Wagner [7]. The above list has no pretentions to completeness.

1.3. The smoothing parameter in kernel, orthogonal series, k-nearest neighbor, histospline, and penalized maximum likelihood methods.

The (univariate) kernel estimate f is of the form

$$\hat{f}(x) = \frac{1}{nh} \sum_{i=1}^{n} K\left(\frac{X_i - x}{h}\right)$$

where $X_1, X_2, \ldots, X_n$ are independent, identically distributed observations from the density f, and $K(\cdot)$ is a "hill" function integrating to one and satisfying some regularity conditions. See Parzen [24]. The experimenter must choose the parameter h, which will control the visual smoothness of the resulting density. Mathematically, h controls the tradeoff between the squared bias and the variance. The optimal h from the point of view of mean square error depends on both the sample size and on the unknown density. Woodroofe [48] has given a theoretical approach to choosing h from the data, but it appears that the method is not practical. In practice, if f is a univariate density, h can frequently be chosen visually in a satisfactory manner.

The orthogonal series estimate for densities supported on [0,1] is given by

$$\hat{f}(x) = \sum_{\nu=1}^{r} \hat{f}_\nu \, \phi_\nu(x)$$

where $r << n$, the $\hat{f}_\nu$ are the sample Fourier coefficients

$$\hat{f}_\nu = \frac{1}{n} \sum_{i=1}^{n} \phi_\nu(X_i)$$

and the $\{\phi_\nu\}$ are a set of $\mathcal{L}_2[0,1]$ orthonormal functions. See Kronmal and Tarter [20]. The parameter r, which must be chosen, controls the tradeoff between the square bias and the variance, small r means small variance but possibly large bias, large r means small bias at the expense of large variance. Again the optimal r depends on the sample size and the unknown density. Similarly every non-parametric density estimate (or estimate of the likelihood ratio) has a parameter (sometimes hidden!) which must be chosen by the experimenter, and which controls the tradeoff between the square bias and the variance. In the k-nearest neighbor methods, the parameter is k, and in the Boneva-Kendall-Stefanov histospline (or the histogram, for that matter) the parameter is the "bin size". In penalty function methods the parameter is a multiplier on the penalty.

Probably one major reason why multidimensional density estimates are not commonly used for classification in practice despite the fact that there is so much theoretical interest in the subject, is that the resulting classification methods cannot be made to "work" unless the smoothing parameter is chosen reasonably well, and this is hard to do visually in more than one dimension.

In this paper we present a "new" class of density estimates.†

---

†The example we treat in detail in Section 4 has previously been proposed by Cogburn and Davis [6] in connection with spectral density estimation.

In truth, a new class of density estimates is not really needed at the present state of the art of density estimation. There are plenty of perfectly good old ones around. Furthermore, if the density to be estimated is known to be smooth, then there is an upper bound to convergence rates for mean square error and most of the methods mentioned above are known to essentially attain it. So no one is likely to come up with a startlingly better nonparametric method. To be specific, let $\mathcal{C}_{m,p}$ be the class of densities defined by

$$\mathcal{C}_{m,p} = \{f : f \text{ a density}, \ f, f', \ldots, f^{(m-1)} \text{ abs. cont.}, \\ f^{(m)} \in \mathcal{L}_p, \ \|f^{(m)}\|_p \le M\},$$

where $\|\cdot\|_p$ is the norm in $\mathcal{L}_p$ and M is a fixed constant. Let f_n be any density estimate based on n independent observations from $f \in \mathcal{C}_{m,p}$. Then the rate of convergence of $\sup_{f \in \mathcal{C}_{m,p}} E(f_n(x) - f(x))^2$ cannot be better than $n^{-(2m-2/p)/(2m+1-2/p)-\varepsilon}$ for arbitrarily small ε. See Farrell [11], Wahba [34]. Furthermore, if the smoothing parameter is chosen correctly, then kernel methods, orthogonal series methods, histospline methods and certain examples of k-nearest neighbor methods are all known to achieve the rate $n^{-(2m-2/p)/(2m+1-2/p)}$. See Wahba [32], [35], [39]). Parzen [24] gave the rate $n^{-4/5}$ in 1963 for the kernel method for the case $m = 2$, $p = \infty$.

Why, then, do we test the reader's patience with yet another class of estimates? The method presented here appears to be as good as some of the others floating around, for medium sample sizes (on the basis of convergence rate calculations and some very preliminary Monte Carlo results). The difference between this class of methods and the others we know of is that it comes with a viable algorithm for estimating the optimal (integrated mean square error) smoothing parameter from the data.[†]

Denote by $f_{n,\lambda}$ the density estimate to be proposed in this paper, where n is the sample size, and λ is the smoothing parameter to

[†] I. J. Good has a procedure for selecting the parameter that goes with a penalized maximum likelihood method [15], and Tarter and Kronmal [49] have done the same for orthogonal series methods.

be chosen. Let $X_1, X_2, \ldots, X_n$ be n independent observations from a density f. We provide an estimate $\hat{\lambda} = \hat{\lambda}(X_1, X_2, \ldots, X_n)$ for λ^* where $\lambda^* = \lambda^*(n;f)$ is defined as the minimizer of the "integrated" mean square error,

$$E \frac{1}{n} \sum_{i=1}^{n} (f_{n,\lambda}(\tfrac{i}{n}) - f(\tfrac{i}{n}))^2 .$$

Some (relatively weak) theorems are presented on the properties of this estimator. In Monte Carlo studies the estimate works better than the Theorems indicate. Let $x_1, x_2, \ldots, x_n$ be realizations of $X_1, X_2, \ldots, X_n$, and $f_{n,\lambda}(t; z_1, z_2, \ldots, z_n)$ indicate the dependence of $f_{n,\lambda}$ on $z_1, z_2, \ldots, z_n$, where $z_i = X_i$ or x_i. We have observed in related studies [9][14] that $\hat{\lambda}(x_1, x_2, \ldots, x_n)$ approximates the minimizer of

$$\frac{1}{n} \sum_{i=1}^{n} (f_{n,\lambda}(\tfrac{i}{n}; x_1, x_2, \ldots, x_n) - f(\tfrac{i}{n}))^2 \tag{1.1}$$

<u>better</u> than it estimates the minimizer of

$$E \frac{1}{n} \sum_{i=1}^{m} (f_{n,\lambda}(\tfrac{i}{n}; X_1, X_2, \ldots, X_n) - f(\tfrac{i}{n}))^2 \, !$$

In the experiments reported on here we found that $\hat{\lambda}(x_1, \ldots, x_n)$ came almost as close to the minimizer of (1.1) as it is possible to get, we typically observed that the relative inefficiency of $\hat{\lambda}$, defined as

$$\frac{\frac{1}{n} \sum_{i=1}^{n} (f_{n,\hat{\lambda}(x_1,\ldots,x_n)}(\tfrac{i}{n}; x_1, x_2, \ldots, x_n) - f(\tfrac{i}{n}))^2}{\inf_{\lambda} \frac{1}{n} \sum_{i=1}^{n} (f_{n,\lambda}(\tfrac{i}{n}; x_1, x_2, \ldots, x_n) - f(\tfrac{i}{n}))^2} \tag{1.2}$$

was between 1.0 and 1.1! (In these cases $n = 170$ and f is a mixture of Beta densities.)

The estimate $\hat{\lambda}$ of λ is based on the method of generalized cross validation (GCV), and this method has general applicability in other contexts, including curve smoothing (Wahba and Wold [40][41], surface smoothing (Wahba [36]), ridge regression (Golub, Heath, and Wahba [14]), and the approximate solution of linear operator equations when the data are noisy (Wahba [37]). These problems are all related to the problems of estimating the mean of a multivariate normal vector, which has received much attention by Stein and co-workers, see Hudson [18] for a bibliography. Feinberg and Holland consider a similar problem for estimating cell probabilities [12].

In Section 2 of this paper we review the general problem of recovering a smooth curve from noisy data, without any reference whatever to density estimation. A smoothing parameter must be chosen, and we present the GCV method for choosing it from the data, and report some of the known theoretical properties of the method. In Section 3 we show how density estimation can be cast in the context of Section 2, that of recovering a smooth curve from noisy data, and in certain special cases we are able to obtain the corresponding theoretical properties of the estimate for λ in the context of density estimation. In Section 4 we present a few typical examples from a very modest Monte Carlo study. In Section 5 we outline future work remaining to be done on the method.

2. Recovering a Smooth Curve from Noisy Data.

In this section we discuss the problem of recovering a smooth curve from noisy data, without reference to density estimation. In Section 3 we show how the density estimation problem can be put in the context of this section and the results of this section used to smooth density estimates optimally.

2.1. The models.

Consider the model

$$y(t) = f(t) + \varepsilon(t), \qquad t \in [0,1] \tag{2.1}$$

where f and ε are independent zero mean Gaussian stochastic processes with

$$E\, f(s)\, f(t) = b\, Q(s,t), \qquad s,t \in [0,1], \tag{2.2}$$

b is an unknown positive constant and Q is a known strictly positive definite continuous covariance, and

$$\begin{aligned} E\, \varepsilon(s)\varepsilon(t) &= \sigma^2, & s &= t \\ &= 0, & s &\neq t. \end{aligned} \tag{2.3}$$

Suppose y of (2.1) is observed for $t = t_1, t_2, \ldots, t_n$ and it is desired to recover f. Then it follows from elementary principles that

(2.4a)
$$\begin{aligned} E\{f(t) \mid y(t_i) &= y_i,\ i = 1,2,\ldots,n\} \\ &= (Q_{t_1}(t), Q_{t_2}(t), \ldots, Q_{t_n}(t))(Q_n + n\lambda I)^{-1} y \end{aligned}$$

where

$$Q_{t_i}(t) = Q(t_i, t) ,$$

Q_n is the $n \times n$ matrix with ijth entry

(2.4b)
$$\begin{aligned} [Q_n]_{ij} &= Q(t_i, t_j), \\ \lambda &= \sigma^2/nb , \end{aligned}$$

and

$$y = (y_1, y_2, \ldots, y_n)' .$$

Next, suppose that instead of being a stochastic process, f is an element of $\mathscr{H}_Q$, where $\mathscr{H}_Q$ is the reproducing kernel Hilbert space (rkhs) with reproducing kernel (rk) Q. Define

(2.5)
$$f_{n,\lambda}(t) = (Q_{t_1}(t), \ldots, Q_{t_n}(t))(Q_n + n\lambda I)^{-1} y .$$

Then (see Kimeldorf and Wahba [19]) $f_{n,\lambda}$ is the solution to the minimization problem: Find $f \in \mathscr{H}_Q$ to minimize

(2.6)
$$\frac{1}{n} \sum_{j=1}^{n} (f(t_j) - y_j)^2 + \lambda \|f\|_Q^2 , \qquad (y_j = y(t_j))$$

where $\|\cdot\|_Q$ is the norm in $\mathscr{H}_Q$. Here λ may be thought of as controlling the tradeoff between smoothness as measured by $\|f\|_Q^2$ and infidelity to the data, as measured by

$$\frac{1}{n} \sum_{j=1}^{n} (f(t_j) - y_j)^2 .$$

Thus, the "f is a stochastic process" point of view" and "f is an element of $\mathscr{H}_Q$" point of view, lead to the same form of algorithm for recovering f from noisy data, only the meaning given to λ is different. (We remark that it is well known that sample functions f from the stochastic model are, with probability 1, not in $\mathscr{H}_Q$).

Assume that Q is given and σ^2 and b are unknown. If f is a stochastic process then one must estimate $\lambda = \sigma^2/nb$ from the data in order to estimate f. On the other hand, if f is known to be in $\mathcal{H}_Q$, then we would like to estimate λ which minimizes some measure of the error. We adopt as our measure the average square error $T(\lambda)$ given by

$$T(\lambda) = \frac{1}{n}\sum_{j=1}^{n}(f_{n,\lambda}(t_j) - f(t_j))^2 . \tag{2.7}$$

We have obtained elsewhere (Wahba and Wold [40, 41], Wahba [37], Craven and Wahba [9], and Golub, Heath and Wahba [14]) an estimate $\hat{\lambda} = \hat{\lambda}(y_1, y_2, \ldots, y_n)$ called the generalized cross validation (GCV) estimate of λ which has the remarkably nice property that if f is the stochastic process of (2.2) then $\hat{\lambda}$ estimates σ^2/nb, and if $f \in \mathcal{H}_Q$ then $\hat{\lambda}$ estimates the minimizer of $T(\lambda)$!

2.2. The Generalized Cross Validation (GCV) estimate $\hat{\lambda}$ for the smoothing parameter λ, and some of its properties.

In this subsection we give the GCV estimate $\hat{\lambda}$ for λ, and report some of its properties. We defer until the end of this section a discussion of where the estimate came from.

To define the estimate, let $A(\lambda)$ be the $n \times n$ matrix defined by

$$A(\lambda) = Q_n(Q_n + n\lambda I)^{-1} \tag{2.8}$$

and define the function $V(\lambda)$ as

$$V(\lambda) = \frac{\frac{1}{n}\|(I - A(\lambda))y\|_n^2}{[\frac{1}{n}\mathrm{Tr}\,(I - A(\lambda))]^2} \tag{2.9}$$

where $\|\cdot\|_n$ is the Euclidean n-space norm, and Q_n has been defined in (2.4b).

Definition of $\hat{\lambda}$: The GCV estimate $\hat{\lambda}$ of λ is defined as the value of λ which minimizes $V(\lambda)$ defined in (2.8) and (2.9).

The first property of $\hat{\lambda}$ is easy to verify, and we state the results as a theorem.

<u>Theorem 1.</u> Let f be the stochastic process given by (2.2). Let E_N denote expectation with respect to the noise random variables

$\varepsilon(t_1), \varepsilon(t_2), \ldots, \varepsilon(t_n))$ and E_S expectation with respect to the "signal" random variables $\{f(t_1), f(t_2), \ldots, f(t_n))\}$. Then $E_S E_N V(\lambda)$ and $E_S E_N T(\lambda)$ are both minimized for $\lambda = \sigma^2/nb$.

Proof: Let $\lambda_{\nu n}$, $\nu = 1, 2, \ldots, n$ be the eigenvalues of Q_n. Then

$$(2.10) \qquad E_N V(\lambda) = \frac{\frac{1}{n}\|(I - A)f\|^2 + \sigma^2 \operatorname{Tr}(I - A)^2}{(\frac{1}{n}\operatorname{Tr}(I - A))^2}$$

$$(2.11) \qquad E_S E_N V(\lambda) = \frac{\frac{b}{n}\operatorname{Tr}(I - A)\,Q_n\,(I - A) + \frac{\sigma^2}{n}\operatorname{Tr}(I - A)^2}{(\frac{1}{n}\operatorname{Tr}(I - A))^2}$$

$$= b\left(\sum_{\nu=1}^{n} \frac{(\lambda_{\nu n}/n) + (\sigma^2/nb)}{((\lambda_{\nu n}/n) + \lambda)^2}\right) \Big/ \left(\sum_{\nu=1}^{n} \frac{1}{((\lambda_{\nu n}/n) + \lambda)}\right)^2.$$

The minimum is achieved when $U_1 U_2' = U_2 U_1'$ where U_1 and U_2 are the numerator and denominator, respectively, in (2.11), or

$$(2.12) \qquad \left(\sum \frac{(\lambda_{\nu n}/n) + (\sigma^2/nb)}{((\lambda_{\nu n}/n) + \lambda)^2}\right)\left(\sum \frac{1}{((\lambda_{\nu n}/n) + \lambda)}\right)\left(\sum \frac{1}{((\lambda_{\nu n}/n) + \lambda)^2}\right) = \left(\sum \frac{1}{((\lambda_{\nu n}/n) + \lambda)}\right)^2 \left(\sum \frac{(\lambda_{\nu n}/n) + (\sigma^2/nb)}{((\lambda_{\nu n}/n) + \lambda)^3}\right)$$

and it is easily seen by setting $\lambda = \sigma^2/nb$ in (2.12) that $\lambda = \sigma^2/nb$ is a solution, and it can be verified that it is a minimizer. We can write

$$T(\lambda) = \frac{1}{n}\|A(\lambda)y - f\|_n^2$$

and

$$(2.13) \qquad E_N T(\lambda) = \frac{1}{n}\|(I - A)f\|_n^2 + \frac{\sigma^2}{n}\operatorname{Tr} A^2$$

$$E_S E_N T(\lambda) = b\{\frac{1}{n} \operatorname{Tr} (I - A) Q_n (I - A) + \frac{\sigma^2}{nb} \operatorname{Tr} A^2\}$$

(2.14)

$$= b(\sum' \frac{(\lambda_{\nu n}/n)\lambda^2}{((\lambda_{\nu n}/n)+\lambda)^2} + \frac{\sigma^2}{nb} \sum \frac{(\lambda_{\nu n}/n)^2}{((\lambda_{\nu n}/n)+\lambda)^2}$$

which is also minimized for $\lambda = \sigma^2/nb$. ∎

We remark that with the stochastic model (2.1-2.3) the maximum likelihood estimate for λ is given by the minimizer of

$$M(\lambda) = \frac{y'(I - A(\lambda))y}{[\operatorname{Det}(I - A(\lambda))]^{1/n}} \tag{2.15}$$

and it can be verified that Theorem 1 is also true for this estimate. That is, the minimizer of $E_S E_N M(\lambda)$ is $\lambda = \sigma^2/nb$. This is the approach taken by Anderssen and Bloomfield in their pioneering papers [1][2].

What happens if $f \in \mathcal{H}_Q$? We have the property that the minimizer of $E_N V(\lambda)$ tends to the minimizer of $E_N T(\lambda)$ as $n \to \infty$ <u>for any</u> $f \in \mathcal{H}_Q$ under very general conditions on the mesh $\{t_i\}$, and on Q.

<u>Theorem 2.</u>

Let $t_i = t_{in}$, $i = 1,2,\ldots,n$, $n = 1,2,\ldots$, satisfy $\int_0^{t_{in}} w(u)du = \frac{i}{n}$, where $w(u)$ is a strictly positive continuous function with $\int_0^1 w(u)du = 1$, and suppose that the probability measure associated with the covariance Q is equivalent (see Root [28], Hajek [17], Wahba [33] for definitions and examples) to that corresponding to some continuous time stochastic process $\{X(t), t \in [0,1]\}$, satisfying an mth order linear differential equation

$$(L_m X)(t) \sum_{j=0}^{m} a_j(t) X^{(j)}(t) = \frac{dW(t)}{dt}, \qquad t \in [0,1], \tag{2.16}$$

for some m, where W is the Wiener process, $a_m(t) > 0$, and the a_j's satisfy some regularity conditions. Then, for any fixed $f \in \mathcal{H}_Q$, the minimizer, λ^*, say of $E_N V(\lambda)$, and the minimizer, $\tilde{\lambda}$, say of $E_N T(\lambda)$ satisfy

$$\lambda^* = \tilde{\lambda}\,(1 + o(1)) \tag{2.17}$$

where $o(1) \to 0$ as $n \to \infty$.

Proof:

A proof for the case $t_{in} = i/n$ and $f \in \mathcal{H}_Q$ satisfies further "very smooth" conditions roughly equivalent to $f^{(\nu)}$ abs. cont., $\nu = 0,1,\ldots,$ $2m-1$, $L_m^* L_m f \in \mathcal{L}_2$, may be found in Wahba [37]. The case of general t_{in} and general $f \in \mathcal{H}_Q$ with $L_m = D^m$ will appear in Craven and Wahba [9] ($D = \frac{d}{dt}$). A sketch of a proof in the special case $\mathcal{H}_Q$ a space of periodic functions, $t_{in} = i/n$, and $L_m = D^m$ appears in Wahba and Wold [41] and in more detail in Section 4 . ∎

We remark that the equivalence condition on Q can be reformulated, roughly, that $\mathcal{H}_Q$ is topologically equivalent to a Hilbert space of functions $\{f : f^{(\nu)}$ abs. cont., $\nu = 0,1,\ldots.m-1,\ L_m f \in \mathcal{L}_2[0,1]\}$. Since topological equivalence means "same convergent sequences" it is not important which Q from a particular equivalence class is used in practice, when n is large. For computational reasons, then, one chooses the simplest Q which in most applications will turn out to be one corresponding to $L_m = D^m$. Then $f_{n,\lambda}$ will be a polynomial spline of degree $2m-1$ (See [19]). The practical minded statistican or numerical analyst who has small or medium sized data sets would probably choose the lowest order method which is nontrivial - i.e., $m = 2$. We recommend $m = 2$ for small to medium sized data sets no matter how many derivatives one can assume f has.

It can be shown (See Section 3) that if f satisfies the "very smooth" conditions given above plus some boundary conditions, then λ^* and $\tilde{\lambda}$ go to zero at the rate $n^{-2m/(4m+1)}$. However, the maximum likelihood estimate can be shown to go to zero at a faster rate than $n^{-2m/(4m+1)}$, and for this reason we use the GCV estimate instead.

We pause to indicate where the mysterious function $V(\lambda)$ with the magical properties came from. The fundamental idea began with cross-validation (also known as predictive sample reuse) as discussed in Geisser [13], Stone [29]. Suppose a particular λ is a good choice for the smoothing parameter. Let $f_{n,\lambda}^{(k)}$ be the solution of the minimization problem of (2.6) with the kth data point left out. Then $f^{(k)}_{n,\lambda}(t_k)$ should be a good predictor of the missing data point $y(t_k)$, and we measure this by

$$(2.18)\qquad V_0(\lambda) = \sum_{k=1}^{n} (f^{(k)}_{n,\lambda}(t_k) - y(t_k))^2 w_k(\lambda)$$

where the weights $w_k(\lambda)$ will be discussed shortly. If Q is periodic and stationary, that is, $Q(s,t) = q((s-t)\bmod 1)$ for some q; and $t_j = j/n$ then Q_n is a circulant matrix and using this fact it can be shown that $V_0(\lambda) \equiv V(\lambda)$ with the weights $w_k(\lambda) \equiv 1$. (See Wahba and Wold [41]). In general, if the $w_k(\lambda)$ are chosen so that

$$w_k(\lambda) = \left[\frac{(1 - a_{kk}(\lambda))}{1 - \frac{1}{n}\sum_{j=1}^{n} a_{jj}(\lambda)}\right]^2$$

where the $a_{kk}(\lambda)$ are the kk^{th} entries of $A(\lambda)$, then it can be shown that $V_0(\lambda) = V(\lambda)$. See Craven and Wahba [9]. The weights $w_k(\lambda)$ were chosen because they are just those weights required to give the result of Theorem 2. We challenge the reader to find $V(\lambda)$ as the result of applying some reasonable optimality principle, we don't know what this principle is, but think it exists!

3. Density Estimation as a Problem in Recovering a Smooth Curve from Noisy Data.

3.1. The density estimate.

Let Q be a r.k. satisfying the hypotheses of Theorem 2. Let the Mercer-Hilbert-Schmidt expansion (see Riesz-Nagy [26]) of Q be

$$(3.1)\qquad Q(s,t) = \sum_{\nu=0}^{\infty} \lambda_\nu \phi_\nu(s)\phi_\nu(t)\ .$$

(That is, $\{\lambda_\nu\}$ and $\{\phi_\nu\}$ are the eigenvalues and eigenfunctions of the Hilbert-Schmidt operator with Hilbert-Schmidt kernel Q). We will always assume here that $\phi_0(t) \equiv 1$, and this assures that $\int_0^1 \phi_\nu(s)ds \equiv 0$, $\nu = 1,2,\ldots,$. If Q is associated with L_m, of Theorem 2, then Q behaves like a Green's function for a $2m^{th}$ order linear differential operator, and then it is known [23] that the eigenvalues $\{\lambda_\nu\}$ of Q go to zero at the rate ν^{-2m}. The $\{\phi_\nu\}$ are infinitely differentiable [26].

Let $X_1, X_2, \ldots, X_n$ be n independent observations from a density f supported on $[0,1]$ and let the sample (generalized) Fourier coefficients $\hat{f}_\nu$ of f be defined by

$$\hat{f}_\nu = \frac{1}{n} \sum_{j=1}^{n} \phi_\nu(X_j), \qquad \nu = 1, 2, \ldots, n,$$

$$\hat{f}_0 = f_0 = 1.$$

It is easy to see that $\hat{f}_\nu$ is an unbiased estimate of $f_\nu = \int_0^1 \phi_\nu(t) f(t)\,dt$, with variance going to zero at the rate $1/n$. We define $f_n(t)$, the sample inverse Fourier transform of f by

$$f_n(t) = \sum_{\nu=0}^{n} \hat{f}_\nu \, \phi_\nu(t) \tag{3.2}$$

$$= \frac{1}{n} \sum_{j=0}^{n} K_n(t, X_j), \tag{3.3}$$

where

$$K_n(s,t) = \sum_{\nu=0}^{n} \phi_\nu(s)\, \phi_\nu(t). \tag{3.4}$$

If $f \in \mathscr{H}_Q$, it can be shown that $\|f\|_Q^2 = \sum_{\nu=0}^{\infty} \frac{f_\nu^2}{\lambda_\nu} < \infty$, and then $f_n(t)$ is very nearly unbiased for $f(t)$ if n is large since

$$E\, f_n(t) = \int_0^1 K_n(t,u) f(u)\,du = \sum_{\nu=0}^{n} f_\nu \phi_\nu(t) = f(t) + o(1), \tag{3.5}$$

where

$$|o(1)| = \left| \sum_{\nu=n+1}^{\infty} f_\nu \phi_\nu(t) \right| \le \sum_{\nu=n+1}^{\infty} \frac{f_\nu^2}{\lambda_\nu} \sum_{\nu=n+1}^{\infty} \lambda_\nu \phi_\nu^2(t) \le \sum_{\nu=n+1}^{\infty} \frac{f_\nu^2}{\lambda_\nu} \cdot Q(t,t). \tag{3.6}$$

The estimate $f_n(t)$ is not consistent for $f(t)$, however. Letting Z be a random variable with density f, we have

$$E f_n(s) f_n(t) = \frac{1}{n} E\, K_n(s,Z)\, K_n(t,Z) + \left(1 - \frac{1}{n}\right) E K_n(s,Z)\, E K_n(t,Z), \tag{3.7}$$

$$\operatorname{cov} f_n(s)\, f_n(t) = \tag{3.8}$$

$$\frac{1}{n}\left[\int_0^1 K_n(s,u) K_n(t,u) f(u)\,du - \int_0^1 K_n(s,u) f(u)\,du \int_0^1 K_n(t,u) f(u)\,du \right].$$

Now since the eigenfunctions ϕ_ν of Q are analytic, and f is "nice" we may write

$$(3.9) \quad \int_0^1 K_n(\tfrac{i}{n},u)\, K_n(\tfrac{k}{n},u)\, f(u)du$$

$$\simeq \frac{1}{n}\sum_{\ell=1}^{n} K_n(\tfrac{i}{n},\tfrac{\ell}{n})\, K_n(\tfrac{k}{n},\tfrac{\ell}{n})\, f(\tfrac{\ell}{n})$$

$$= \frac{1}{n}\sum_{\ell=0}^{n} f(\tfrac{\ell}{n}) \sum_{\mu=0}^{n} \phi_\mu(\tfrac{i}{n})\, \phi_\mu(\tfrac{\ell}{n}) \sum_{\nu=0}^{n} \phi_\nu(\tfrac{k}{n})\, \phi_\nu(\tfrac{\ell}{n}) .$$

Now, since

$$(3.10) \quad \frac{1}{n+1}\sum_{j=0}^{n} \phi_\mu(\tfrac{j}{n})\, \phi_\nu(\tfrac{j}{n}) \simeq \int_0^1 \phi_\mu(s)\, \phi_\nu(s)ds \begin{array}{ll} = 1 & \mu = \nu \\ = 0, & \mu \neq \nu , \end{array}$$

$$\mu, \nu = 0,1,\ldots,n ,$$

we also must have

$$(3.11) \quad \frac{1}{n+1}\sum_{\mu=0}^{n} \phi_\mu(\tfrac{j}{n})\, \phi_\mu(\tfrac{\ell}{n}) \begin{array}{lll} \simeq 1 , & & j = \ell \\ \simeq 0 , & & j \neq \ell \end{array}$$

Thus, we observe from the right hand side of (3.9)

$$\frac{1}{n}\int_0^1 K_n(s,u)\, K_n(t,u)\, f(u)du \begin{array}{lll} \simeq f(t) , & & s = t \\ \simeq 0 , & & |s-t| = \frac{j}{n}, \; j = 1,2,\ldots,n \end{array}$$

which results in

$$(3.12) \quad \begin{array}{rcll} E\, f_n(t) & \simeq & f(t) & \\ \operatorname{cov} f_n(\tfrac{j}{n})\, f_n(\tfrac{k}{n}) & \simeq & f(\tfrac{j}{n}) , & j = k \\ & \simeq & 0 , & j \neq k . \end{array}$$

Thus we have the approximate model for $f_n(t)$:

$$(3.13) \quad f_n(t) \simeq f(t) + \varepsilon(t)$$

where

$$(3.14) \quad \begin{array}{rcll} E\, \varepsilon(\tfrac{i}{n}) & \simeq & 0 & \\ E\, \varepsilon(\tfrac{i}{n})\, \varepsilon(\tfrac{j}{n}) & \simeq & f(\tfrac{i}{n}) , & i = j \\ & \simeq & 0 , & i \neq j . \end{array}$$

We will take as "data" the "observations"

$$(3.15)\qquad y(\tfrac{i}{n}) \equiv y_i = f_n(\tfrac{i}{n}), \qquad i = 1,2,\ldots,n$$

where $f_n(t)$ is defined by (3.2), and proceed to look at the problem of obtaining an estimate of f as that of recovering f given the noisy "ordinate" data y_i.

The model given by (3.12) - (3.15) is not exactly the same as that considered in Section 2 since here $E\,\varepsilon^2(\frac{i}{n})$ depends on i in general whereas in (2.3) $E\,\varepsilon^2(\frac{i}{n}) = \sigma^2$. However, at least in the stationary periodic case (see immediately following (3.18)) and below, we will be able to obtain the density estimate version of Theorems 1 and 2.

Following (2.4) and (2.5) we begin by considering as a density estimate

$$(3.16)\qquad f_{n,\lambda}(t) = (Q_{t_i}(t),\ldots,Q_{t_n}(t))(Q_n + n\lambda I)^{-1} y$$

where $t_i = i/n$, and

$$y = (f_n(\tfrac{1}{n}), f_n(\tfrac{2}{n}), \ldots, f_n(\tfrac{n}{n})),$$

and λ is going to be chosen by the method of generalized cross-validation. We want to obtain an approximate form for (3.16) which is easier to compute, and then modify the result to use the information that $f_0 = \int_0^1 f(u)du \equiv 1$. By observing that since

$$Q(s,t) = \sum_{\nu=0}^{\infty} \lambda_\nu\, \phi_\nu(s)\, \phi_\nu(t)$$

then

$$(3.17)\qquad Q_n \approx \Gamma D \Gamma'$$

where here Γ is the $n\times n$ matrix with νjth entry $\frac{1}{\sqrt{n}}\phi_{\nu-1}(\frac{j}{n})$, $\nu, j = 0,1,2,\ldots,n-1$, and D is the diagonal matrix with $\nu\nu$th entry $n\lambda_{\nu-1}$, $\nu = 1,2,\ldots,n$. Substituting (3.17) into (3.16) gives the approximate expression for (3.16) evaluated at $t = \frac{k}{n}$, $k = 1,2,\ldots,n$,

$$(3.18)\qquad f_{n,\lambda}(\tfrac{k}{n}) \simeq \sum_{\nu=0}^{n-1} \frac{\hat{f}_\nu\, \phi_\nu(\frac{k}{n})}{1+\lambda/\lambda_\nu}$$

where

$$\hat{f}_\nu = \frac{1}{n}\sum_{i=0}^{n-1} \phi_\nu\left(\frac{i}{n}\right)y_i . \tag{3.19}$$

The expression we actually want to use is

$$f_{n,\lambda}(t) = 1 + \sum_{\nu=1}^{n} \frac{\hat{f}_\nu}{(1+\lambda/\lambda_\nu)} \phi_\nu(t) . \tag{3.20}$$

This is an approximation to

$$E\{f(t) \mid y(t_i) = y_i , \ i = 1,2,\ldots,n, \ \int_0^1 f(u)du = 1\}$$

for the stochastic model of Section 2. For the model $f \in \mathcal{H}_Q$, this is an approximation to the solution of a constrained version of the minimization problem of (2.6), namely, find $f \in \mathcal{H}_Q$ to minimize

$$\frac{1}{n}\sum_{j=1}^{n} (f(t_j) - y_j)^2 + \lambda \|f\|_Q^2$$

subject to

$$\int_0^1 f(u)du = 1 \tag{3.22}$$

(See [19] for details).

Our density estimate will be $f_{n,\lambda}(t)$ given by (3.20), where the crucial smoothing parameter λ will be chosen according to the method of generalized cross-validation for densities, to be described in the next section. The estimate is a form of an orthogonal series estimate; instead of truncating the series we "taper" it. The function $f_{n,\lambda}$ may be thought of as the result of putting f_n through a "low-pass" filter, where the shape of the filter is determined by Q and the "half power point", ν_0, of the filter is determined by $\lambda_{\nu_0} = \hat{\lambda}$. On the other hand, letting

$$K_{n,\lambda}(s,t) = 1 + \sum_{j=1}^{n} \frac{\phi_\nu(s)\,\phi_\nu(t)}{1 + \lambda/\lambda_\nu} , \tag{3.23}$$

we have

$$f_{n,\hat{\lambda}} \simeq \frac{1}{n}\sum_{j=1}^{n} K_{n,\hat{\lambda}}(t,X_i) \tag{3.24}$$

which is a window estimate in the case $K_{n,\lambda}(s,t) = K_{n,\lambda}(s-t)$, and $\hat{\lambda}$ controls the width of the window. (An example will appear later.) In general, $f_{n,\lambda}$ is a δ-function estimate of the type considered recently by Walter and Blum [42].

3.2. Generalized cross-validation for estimating λ in the density estimation case.

Returning to the expression for $V(\lambda)$ in Section 2 for the curve smoothing problem we can write (2.9) as

$$V(\lambda) = \frac{\sum_{\nu} \frac{\lambda^2 |\tilde{f}_\nu|^2}{((\lambda_{\nu n}/n)+\lambda)^2}}{\left[\frac{1}{n}\sum_{\nu} \frac{\lambda}{((\lambda_{\nu n}/n)+\lambda)}\right]^2} \tag{3.25}$$

where the $\{\lambda_{\nu n}\}$ are the eigenvalues of Q_n as in (2.10) and (2.11), and

$$\begin{pmatrix} \tilde{f}_1 \\ \tilde{f}_2 \\ \vdots \\ \tilde{f}_n \end{pmatrix} = \tilde{\Gamma}^T y$$

where $\tilde{\Gamma}$ is the orthogonalizing matrix for Q_n, $\tilde{\Gamma}^T Q_n \tilde{\Gamma} = \mathrm{diag}\{\lambda_{\nu n}\}$. In the density estimation context the expression we want to use to approximate (3.25) is

$$V(\lambda) \equiv \frac{\sum_{\nu=1}^{n} \frac{\lambda^2 |\hat{f}_\nu|^2}{(\lambda_\nu + \lambda)^2}}{\left[\frac{1}{n}\sum_{\nu=1}^{n} \frac{\lambda}{(\lambda_\nu+\lambda)}\right]^2} \tag{3.26}$$

where

$$\hat{f}_\nu = \frac{1}{n}\sum_{i=1}^{n} \phi_\nu(X_i)$$

is an estimate for f_ν, as before. (It can be shown that the $\{\lambda_{\nu n}\}$ "behave like" $n\lambda_\nu$, $\nu = 1, 2, \ldots, n$.)

Since the "noise model" of (3.12)-(3.15) is not exactly the same as that of Section 2 it does not necessarily follow that the minimizer of $V(\lambda)$ has the properties described in Section 2. It turns out however, that if $\mathcal{H}_Q$ is a space of periodic functions in $[0,1]$, then we can get the same properties for the minimizer of $V(\lambda)$ in the density estimation case as in Section 2. For the remainder of this section we let

$$Q(s,t) = \sum_{\nu=-\infty}^{\infty} \lambda_\nu \phi_\nu(s) \phi_\nu^*(t) \tag{3.27}$$

where $\phi_\nu(s) = e^{2\pi i \nu s}$, $\nu = 0, \pm 1, 2, \ldots$, and

$$\lambda_0 = 1$$

$$\lambda_\nu = \lambda_{-\nu} = |P_m(2\pi i \nu)|^{-2}, \qquad \nu = 1, 2, \ldots$$

where $P_m(z)$ is an mth degree polynomial with all its zeros inside the unit circle. Q satisfies the hypotheses of Theorem 2, with

$$L_m X = P_m(D)X$$

where $D = \frac{d}{dt}$ and X satisfies the periodic boundary conditions $X^{(\nu)}(1) \equiv X^{(\nu)}(0)$, $\nu = 0, 1, \ldots, m-1$. Furthermore, it can be verified that

$$\begin{aligned} \|f\|_Q^2 &= \left[\int_0^1 f(u)du\right]^2 + \sum_{\substack{\nu=-\infty \\ \nu \neq 0}}^{\infty} |P_m(2\pi i \nu)|^2 |f_\nu|^2 \\ &= \left[\int_0^1 f(u)du\right]^2 + \int_0^1 [(L_m f)(u)]^2 du , \end{aligned} \tag{3.28}$$

where

$$f_\nu = \int_0^1 e^{-2\pi i \nu t} f(t)dt, \qquad \nu = \pm 1, 2, \ldots . \tag{3.29}$$

Also, the solution to the constrained minimization problem of (3.22) in this space is just the solution to: Find $f \in \mathcal{H}_Q$ to minimize

$$\frac{1}{n} \sum_{i=1}^{n} (f(t_i) - y_i)^2 + \lambda \int_0^1 [(L_m f)(u)]^2 du \tag{3.30}$$

subject to

$$\int_0^1 f(u)du = 1 .$$

Here

$$\hat{f}_\nu = \frac{1}{n} \sum_{j=1}^{n} e^{-2\pi i \nu X_j} \tag{3.31}$$

and it is appropriate to take $V(\lambda)$ as

$$(3.32)\qquad V(\lambda) = \frac{\sum_{\nu=1}^{n/2} \frac{\lambda^2 |\hat{f}_\nu|^2}{(\lambda_\nu + \lambda)^2}}{\left[\frac{1}{n} \sum_{\nu=1}^{n/2} \frac{\lambda}{(\lambda_\nu + \lambda)} \right]^2},$$

where we are supposing n even for simplicity. Then

$$(3.33)\qquad E|\hat{f}_\nu|^2 = \frac{1}{n^2} E \sum_{j=1}^{n} \sum_{k=1}^{n} e^{2\pi i\nu(X_j - X_k)} = (1 - \frac{1}{n})|f_\nu|^2 + \frac{1}{n}$$

and

$$(3.34)\qquad EV(\lambda) = \frac{(1 - \frac{1}{n}) \sum_{\nu=1}^{n/2} \frac{\lambda^2 |f_\nu|^2}{(\lambda_\nu + \lambda)^2} + \frac{1}{n} \sum_{\nu=1}^{n/2} \frac{\lambda^2}{(\lambda_\nu + \lambda)^2}}{\left[\frac{1}{n} \sum_{\nu=1}^{n/2} \frac{\lambda}{(\lambda_\nu + \lambda)} \right]^2}$$

Letting

$$y = (f_n(\tfrac{1}{n}), f_n(\tfrac{2}{n}), \ldots, f_n(\tfrac{n}{n}))'$$

$$f = (f(\tfrac{1}{n}), f(\tfrac{2}{n}), \ldots, f(\tfrac{n}{n}))'$$

$$\varepsilon = (\varepsilon(\tfrac{1}{n}), \varepsilon(\tfrac{2}{n}), \ldots, \varepsilon(\tfrac{n}{n}))'$$

where $f_n(\cdot)$ and $\varepsilon(\cdot)$ are defined by (3.2) and (3.13), we have

$$T(\lambda) \stackrel{\text{def}}{=} \frac{1}{n} \sum_{j=1}^{n} (f_{n,\lambda}(\tfrac{i}{n}) - f(\tfrac{i}{n}))^2 \simeq \frac{1}{n} \|A(\lambda)y - f\|_n^2$$

$$(3.35)\qquad E\,T(\lambda) \simeq \frac{1}{n} \|(I-A)f\|_n^2 + E\,\varepsilon' A^2 \varepsilon$$

$$\simeq \frac{1}{n} \|(I-A)f\|_n^2 + \frac{1}{n} \sum^{n} f(\tfrac{i}{n}))\, \tilde{a}_{ii}(\lambda),$$

where $A(\lambda) = Q_n(Q_n + n\lambda I)^{-1}$ as in (2.8), and $\tilde{a}_{ii}(\lambda)$ is the ii^{th} entry of $A^2(\lambda)$. Since Q_n is circulant A is circulant, A^2 is circulant and $\tilde{a}_{ii}(\lambda) \equiv \frac{1}{n} \operatorname{Tr} A^2(\lambda)$. Since $\frac{1}{n} \sum_{i=1}^{n} f(\frac{i}{n}) \approx 1$, we have

$$(3.36)\qquad E\,T(\lambda) \simeq \frac{1}{n} \|(I-A)f\|_n^2 + \frac{1}{n} \operatorname{Tr} A^2(\lambda).$$

Inspection of (3.34) and (3.36) with the aid of (3.17) reveals that (3.34) and (3.36) are essentially the same as (2.10) and (2.13), the expressions for $E_N V(\lambda)$ and $E_N T(\lambda)$, with $\sigma^2 = 1$.

Let us put a "phoney prior" on $f(s)$ induced by

$$(3.37) \qquad f(s) = 1 + \sum_{\substack{\nu=-\infty \\ \nu \neq 0}}^{\infty} f_\nu \phi_\nu(s)$$

where $f_\nu = f^*_{-\nu}$ and otherwise the f_ν are independent zero mean complex normal random variables (See Goodman [14]) with $E|f_\nu|^2 = b\lambda_\nu$. We have in effect introduced the prior convariance $bQ(s,t)$ on f and then conditioned on $\int_0^1 f(u)du = 1$. The prior is "phoney" of course, because the "sample functions" are not necessarily positive even though they do integrate to 1. Proceeding boldly despite the phoniness of the prior, and letting E_S be expectation with respect to this prior, we have

$$(3.38) \qquad E_S\, E\, V(\lambda) \cong b \left(\sum_{\nu=1}^{n/2} \frac{\lambda_\nu + 1/nb}{(\lambda_\nu + \lambda)^2} \right) \Big/ \left(\sum_{\nu=1}^{n/2} \frac{1}{(\lambda_\nu + \lambda)} \right)^2$$

$$(3.39) \qquad E_S\, E\, T(\lambda) \cong b \left(\sum_{\nu=1}^{n/2} \frac{\lambda^2 \lambda_\nu}{(\lambda_\nu + \lambda)^2} + \frac{1}{nb} \sum_{\nu=1}^{n/2} \frac{\lambda_\nu^2}{(\lambda_\nu + \lambda)^2} \right.$$

and the right hand sides of these expressions are minimized for the same λ, namely $\lambda = 1/nb$.

We now return to a more defensible assumption, namely, f is a density in $\mathscr{V}_Q$.

<u>Theorem 2'.</u>

Let Q be given by (3.27). Let λ^* and $\tilde{\lambda}$ be the minimizers of $EV(\lambda)$ and $ET(\lambda)$ as given by the right hand sides of (3.34) and (3.36). Let f be any density in $\mathscr{V}_Q$. Then as $n \to \infty$

$$(3.40) \qquad \lambda^* = \tilde{\lambda}(1 + o(1)).$$

<u>Outline of Proof.</u>

Set $(1 - \frac{1}{n}) \simeq 1$, in the expression for $V(\lambda)$ in (3.34). It can be shown that if $n \to \infty$, $\lambda \to 0$ in such a way that $n\lambda^{1/2m} \to \infty$, then

$$(3.41) \quad 1 - \frac{1}{n} \sum \frac{\lambda}{\lambda_\nu + \lambda} \equiv \frac{1}{n} \sum \frac{\lambda_\nu}{\lambda_\nu + \lambda} \equiv \frac{1}{n} \sum \frac{1}{(1 + |P_m(2\pi\nu)|^2 \lambda)} \simeq \frac{1}{n\lambda^{1/2m}} \tilde{k}_m (1 + o(1)),$$

(3.42) $$1 - \frac{2}{n}\sum \frac{\lambda}{\lambda_\nu+\lambda} + \frac{1}{n}\sum \left(\frac{\lambda^2}{(\lambda_\nu+\lambda)^2}\right) = \frac{1}{n}\sum \left(\frac{\lambda_\nu^2}{(\lambda_\nu+\lambda)^2}\right) \equiv \frac{1}{n}\sum \frac{1}{(1+|P_m(2\pi\nu)|^2\lambda)^2}$$

$$= \frac{1}{n\lambda^{1/2m}} k_m(1 + o(1))$$

where

$$\tilde{k}_m = \frac{1}{2\pi}\int_0^\infty \frac{dx}{(1 + x^{2m})}, \quad k_m = \frac{1}{2\pi}\int_0^\infty \frac{dx}{(1+x^{2m})^2},$$

and $o(1) \to 0$ as $n \to \infty$. Setting

(3.43) $$\psi_f(\lambda) = \lambda^2 \sum \frac{f_\nu^2}{(\lambda_\nu + \lambda)^2}$$

and substituting (3.41) and (3.42) into (3.34) and (3.36) one obtains,

(3.44) $$E\,V(\lambda) \simeq \{\psi_f(\lambda) + (1 - \frac{2\tilde{k}_m}{n\lambda^{1/2m}} + \frac{k_m}{n\lambda^{1/2m}})\}/[1 - \frac{\tilde{k}_m}{n\lambda^{1/2m}}]^2$$

(3.45) $$E\,T(\lambda) \simeq \psi_f(\lambda) + \frac{k_m}{n\lambda^{1/2m}}$$

The minimum $\tilde{\lambda}$ of $E\,T(\lambda)$ given by the right hand side of (3.45) occurs for the (smallest) solution of

(3.46) $$\psi_f'(\lambda) = \frac{k_m}{2mn\lambda^{(2m+1)/2m}}$$

Differentiating $E\,V(\lambda)$ given by the right hand side of (3.44) gives

$$\frac{d}{d\lambda} V(\lambda) = \frac{1}{U_1}\{\psi_f'(\lambda) - \frac{k_m}{2mn\lambda^{(2m+1)/2m}}[1 - \frac{2\tilde{k}_m}{k_m}(1 - \frac{U_2}{U_1^{1/2}})]\}$$

where

$$U_1 = (1 - \frac{\tilde{k}_m}{n\lambda^{1/2m}})^2$$

$$U_2 = \psi_f(\lambda) + (1 - \frac{1}{n\lambda^{1/2m}}(2\tilde{k}_m - k_m)).$$

Provided $n\lambda \to \infty$ in such a way that $n\lambda^{1/2m} \to \infty$, we have

$$U_1 = 1 + o(1)$$

$$U_2 = 1 + o(1)$$

and

$$\frac{U_2}{U_1^{1/2}} = 1 + o(1) .$$

It can be shown that the minimizers of $EV(\lambda)$ and $ET(\lambda)$ have the property that $\lambda \to 0$, $n\lambda^{1/2m} \to \infty$, and so

$$\frac{d}{d\lambda} EV(\lambda) = [1 + o(1)]\{\psi_f'(\lambda) - \frac{k_m}{2mn\lambda^{1/2m}}[1 + o(1)]\},$$

so that for large n, $EV(\lambda)$ and $ET(\lambda)$ have asymptotically the same minimizer, that is, $\lambda^* = \tilde{\lambda}(1+o(1))$.

If $\sum \frac{f_\nu^2}{\lambda_\nu^2} < \infty$, which in this case entails that $f^{(\nu)}$ abs. cont., $\nu = 0,1,\ldots\ 2m-1$, and

$$\int ((L_m^* L_m f)(t))^2 dt \equiv C_f < \infty ,$$

then

$$\psi_f(\lambda) = (\lambda^2 \sum \frac{f_\nu^2}{\lambda_\nu^2})(1 + o(1)) = \lambda^2 C_f(1 + o(1))$$

and it can be shown (See [38]) that λ^* and $\tilde{\lambda}$ satisfy

$$\lambda^2 C_f = \frac{k_m}{2mn\lambda^{(2m+1)/2m}}(1 + o(1))$$

or

$$\lambda^* = \tilde{\lambda}(1 + o(1)) = (\frac{k_m}{2mnC_f})^{2m/4m+1}(1 + o(1))$$

and then

$$ET(\lambda^*) = \frac{k_m^{4m/(4m+1)}}{n^{4m/(4m+1)}}[\frac{1}{(2mC_f)^{4m/(4m+1)}} + (2mC_f)^{1/(4m+1)}](1 + o(1)). \tag{3.47}$$

We note that the above above assumptions on f say that $f \in C_{2m,2}$ ($C_{2m,2}$ was defined in the introduction) and so if f_{n,λ^*} is to be in the class of "good" estimates we should have, for mean square error at a point

$$E(f_{n,\lambda^*}(t) - f(t))^2 \simeq C\, n^{-(4m-1)/(4m)} .$$

The rate $n^{-4m/(4m+1)}$ of (3.47) reflects the fact that slightly better convergence rates obtain for integrated mean square error over a finite interval than mean square error at a point.

We remark on the importance of the periodic assumption. In principle f can be estimated by (3.16) with the minimizer of $V(\lambda)$ given by (3.25) used as the estimate of λ. The practical problem occurs in computing the eigenfunctions. If the ϕ_ν are taken as complex exponentials, and the true density is not periodic, then the resulting $f_{n,\lambda}$ will not approximate f at the endpoints. Of course $f^{(\nu)}(1) = f^{(\nu)}(0) = 0$, $\nu = 0,1,\ldots,m-1$, are a perfectly good and quite reasonable set of periodic boundary conditions. With respect to the theoretical properties of the minimizer of $V(\lambda)$, we used $\mathrm{Var}\,\hat{f}_\nu = \text{const.}$ and $A(\lambda)$ circulant to get Theorem 2', these properties do not hold in general when $\mathscr{V}_Q$ is not a periodic space. Without this simplification the expressions for $EV(\lambda)$ and $ET(\lambda)$ are slightly more messy. We believe that in general the minimizer of $V(\lambda)$ estimates the minimizer of $\tilde{T}(\lambda)$ for $\tilde{T}$ some other quadratic form in the errors, but do not have a demonstration.

4. Preliminary Experimental Results.

The method was tried experimentally on five densities satisfying periodic boundary conditions, by generating a set of pseudo-random numbers distributed according to each density. We present details of the results from three of the examples. (The other two were not substantially different). We let Q be the periodic covariance of (3.27) with $m = 2$ and $\lambda_\nu^{-1} = (2\pi\nu)^{2m}$, $\nu = \pm 1, 2, \ldots$. This corresponds to the assumption that f and f' are continuous, $f'' \in \mathscr{L}_2$, and $f^{(\nu)}(1) = f^{(\nu)}(0)$, $\nu = 0,1$. The density estimate we use is then an approximation to the solution of the problem: Find f such that f, f' continuous, $f'' \in \mathscr{L}_2$, $f^{(\nu)}(1) = f^{(\nu)}(0)$, $\nu = 0,1$, $\int_0^1 f(u)du = 1$, to minimize

$$(4.1) \qquad \frac{1}{n}\sum_{j=1}^{n}\left(f\left(\frac{j}{n}\right) - y_j\right)^2 + \lambda\int_0^1 (f''(u))^2 du ,$$

where

$$y_i = f_n\left(\frac{i}{n}\right)$$

$$(4.2)\qquad f_n(t) = 1 + \sum_{\substack{\nu=-n/2\\ \nu\neq 0}}^{n/2} \hat{f}_\nu e^{2\pi i\nu t}$$

$$\hat{f}_\nu = \frac{1}{n}\sum_{j=1}^{n} e^{-2\pi i\nu X_j}$$

The approximation to the solution of the minimization problem actually being computed is

$$(4.3)\qquad f_{n,\lambda}(t) = 1 + \sum_{\substack{\nu=-n/2\\ \nu\neq 0}}^{n/2} \frac{\hat{f}_\nu}{(1+(2\pi\nu)^4\lambda)} e^{2\pi i\nu t} .$$

The GCV estimate $\hat{\lambda}$ for λ is found by computing

$$(4.4)\qquad V(\lambda) = \frac{\frac{1}{n}\sum_{\substack{\nu=-n/2\\ \nu\neq 0}}^{n/2} \frac{|\hat{f}_\nu|^2}{(\lambda_\nu+\lambda)^2}}{\left[\frac{1}{n}\sum_{\substack{\nu=-n/2\\ \nu\neq 0}}^{n/2} \frac{1}{(\lambda_\nu+\lambda)}\right]^2}, \quad \lambda_\nu = (2\pi\nu)^{-4}$$

at increments $\lambda = 10^{j/3}$ for $j = -21$ to 3 and determining the global minimum $\hat{\lambda}$ by inspection (to the nearest $10^{1/3}$).

The densities tried were all mixtures of beta densities and the three cases presented are

Case I $\quad f \sim \frac{1}{3}\beta(10,5) + \frac{1}{3}\beta(7,7) + \frac{1}{3}\beta(5,10), \quad n = 174$

Case II $\quad f \sim \frac{6}{10}\beta(12,7) + \frac{4}{10}\beta(3,11), \quad n = 174$

Case IV $\quad f \sim \frac{1}{3}\beta(20,5) + \frac{1}{3}\beta(12,12) + \frac{1}{3}\beta(5,20) \quad n = 170$

Figure 1 presents plots of the true densities along with histograms of the Monte Carlo realizations of n independent observations from each density. The "bin size" of the histogram was chosen by eye to give a pleasing picture.

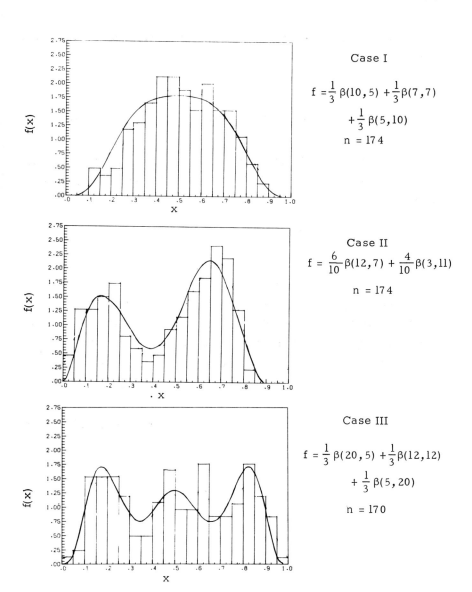

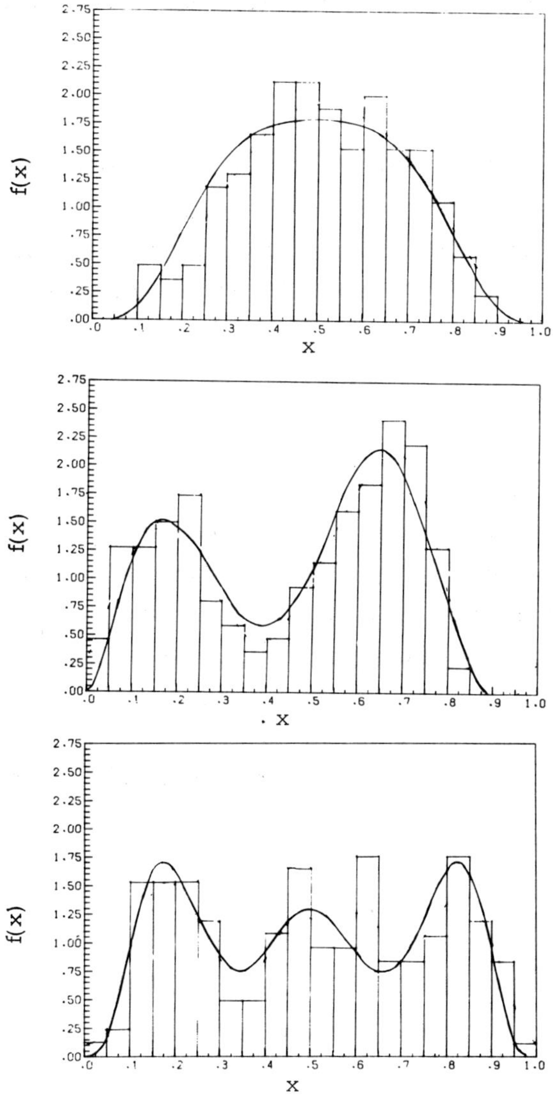

Case I

$$f = \frac{1}{3}\beta(10,5) + \frac{1}{3}\beta(7,7) + \frac{1}{3}\beta(5,10)$$

$n = 174$

Case II

$$f = \frac{6}{10}\beta(12,7) + \frac{4}{10}\beta(3,11)$$

$n = 174$

Case III

$$f = \frac{1}{3}\beta(20,5) + \frac{1}{3}\beta(12,12) + \frac{1}{3}\beta(5,20)$$

$n = 170$

True Density and Histogram of n Computer-Generated Observations

Figure 1

Figure 2 gives plots of $V(\lambda)$ of (4.4) and $T(\lambda)$ vs. $\log_{10}\lambda$. $T(\lambda)$ was computed by

$$T(\lambda) = \frac{1}{n}\sum_{i=1}^{n}\left(f\left(\frac{i}{n}\right) - f_{n,\lambda}\left(\frac{i}{n}\right)\right)^2 .$$

The reader can see that the minimizer of $V(\lambda)$ is close to the minimizer of $T(\lambda)$.

Table 1 gives the computed numbers on which Figure 2 was based in the neighborhood of the minima, along with the (approximate) relative inefficiency

$$\frac{T(\hat{\lambda})}{\inf_j T(\lambda = 10^{j/3})} .$$

The minima of V and T are marked "*".

Figure 3 gives plots of $f_n(t)$ and $f_{n,\hat{\lambda}}(t)$ (defined by (4.2) and (4.3) respectively). Note that $f_{n,\hat{\lambda}}$ is a smoothed version of f_n. Figure 4 compares $f_{n,\hat{\lambda}}$ and the original density f. The periodicity of $\hat{f}_{n,\lambda}$ is evident in Case II of Figure 4, where it can be seen that $\hat{f}'_{n,\lambda}(0) = f'_{n,\lambda}(1) \neq 0$.

Once the computer programs were debugged runs reported here cost less than \$30 to run at the weekend rate, and most of this expense went toward determining $T(\lambda)$.

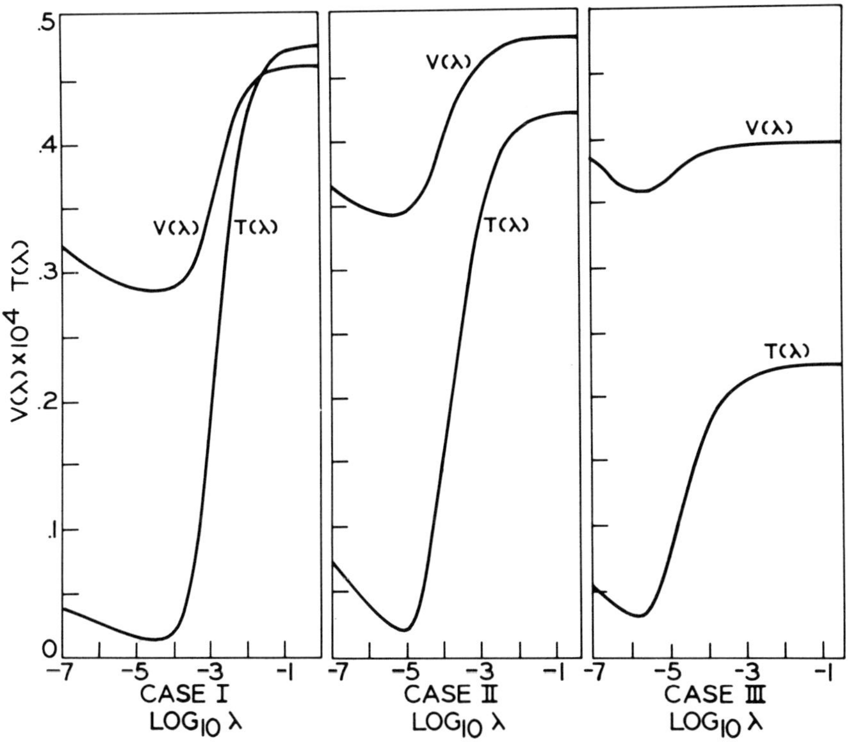

V(λ), THE CROSS VALIDATION FUNCTION AND
T(λ), THE INTEGRATED SQUARE ERROR, VS $\mathrm{LOG}_{10}\,\lambda$

FIG. 2

λ	$V(\lambda)$	$T(\lambda)$	Relative Inefficiency (defined by(1.2))
.464-05	.291-04	.203-01	
.100-04	.288-04	.171-01	
.215-04	.287-04	.153-01	
.464-04	.286-04*	.149-01*	
.100-03	.287-04	.179-01	$\frac{.149}{.149} = 1.0$
.215-03	.294-04	.337-01	
.464-03	.313-04	.815-01	
.200-02	.348-04	.171+00	
.215-02	.389-04	.278+00	
	Case I		
.464-06	.352-04	.505-01	
.100-05	.347-04	.398-01	
.215-05	.344-04	.301-01	
.464-05	.343-04*	.222-01	
.100-04	.345-04	.202-01*	$\frac{.222}{.202} = 1.1$
.215-04	.354-04	.355-01	
.464-04	.374-04	.819-01	
.100-03	.402-04	.154+00	
.215-03	.428-04	.227+00	
	Case II		
.100-06	.383-04	.546-01	
.215-06	.374-04	.453-01	
.464-06	.367-04	.371-01	
.100-05	.361-04	.317-01*	$\frac{.330}{.317} = 1.04$
.215-05	.359-04*	.330-01	
.464-05	.360-04	.458-01	
.100-04	.366-04	.728-01	
.215-04	.375-04	.108+00	
.464-04	.383-04	.143+00	
	Case III		

Table I

$V(\lambda)$ and $T(\lambda)$ vs. λ in the neighborhood of the minima, and the Relative Inefficiency of $\hat{\lambda}$.

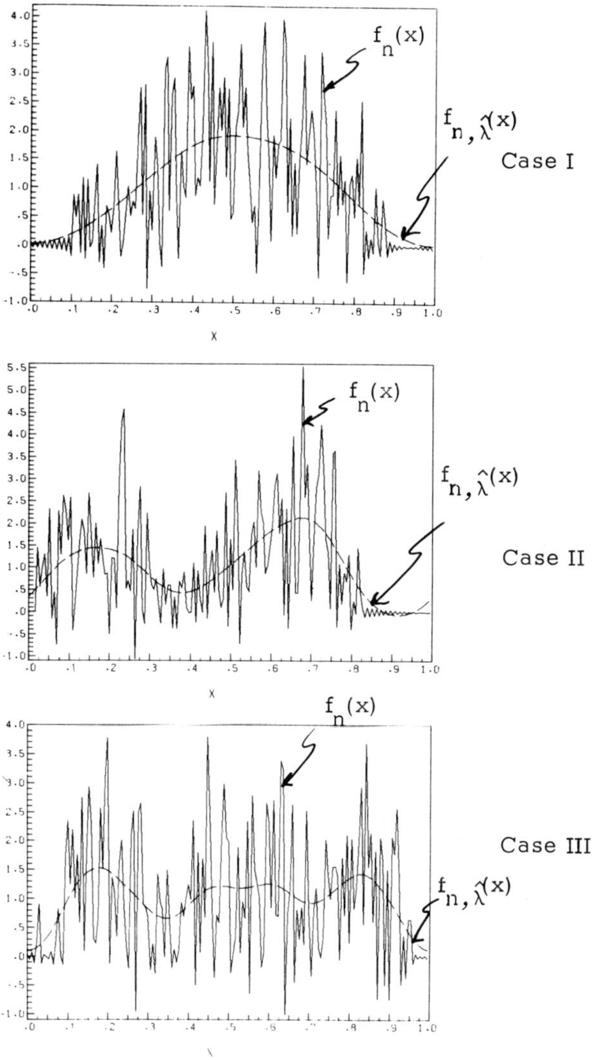

Sample Inverse Fourier Transform, $f_n(x)$, and the Density Estimate $f_{n,\hat{\lambda}}(x)$.

Figure 3

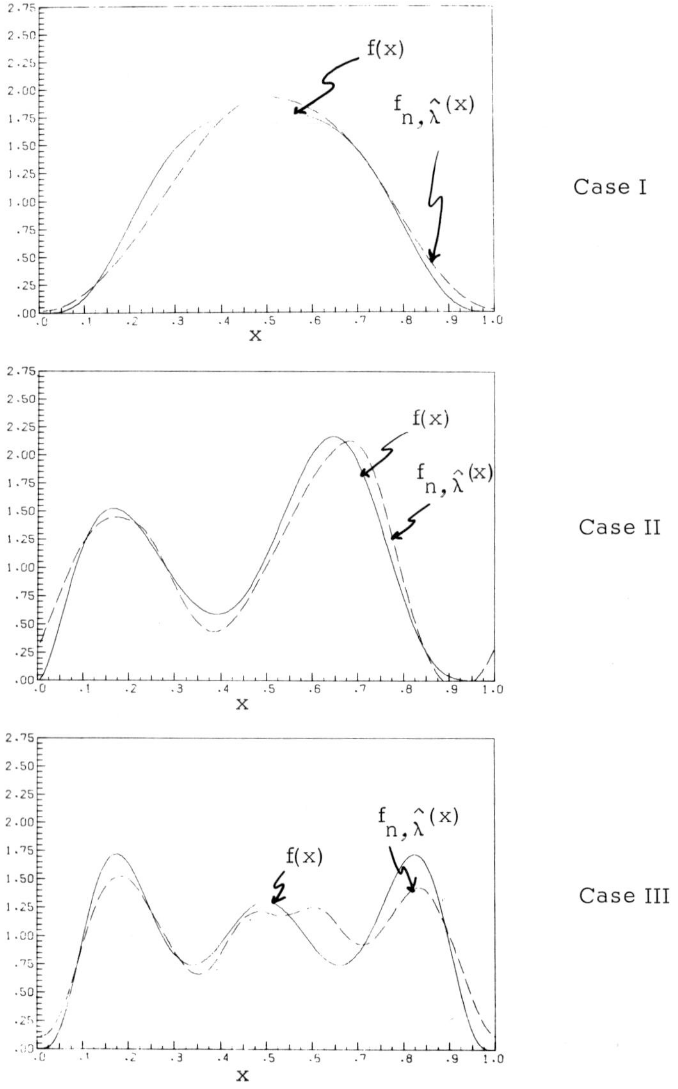

True and Estimated Densities

Figure 4

5. Further Work.

The first question that arises is: How does this method compare with other methods for estimating a density, under comparable assumptions on the unknown density. We believe that the method will give about the same results as a corresponding kernel or histospline method for large n, provided that the smoothing parameter in each method is chosen as well as possible (i.e. to minimize the average mean square error knowing the true density). A fair study of how two methods might compare in practice, however, requires that both methods have an objective procedure for choosing their respective smoothing parameters from the sample.

It remains to determine theoretically exactly what the minimizer of $V(\lambda)$ estimates in the non-periodic case.

It remains to develop cheap calculational procedures when $\mathcal{H}_Q$ is not a periodic space. Any function on $[0,1]$ with $f^{(\nu)}$ abs. cont., $\nu = 0,1,\ldots,m-1$, $f^{(m)} \in \mathcal{L}_2$, can be decomposed into a polynomial of degree m and an element of the periodic space $\mathcal{H}_Q$ in (3.27) with $\lambda_\nu = (2\pi\nu)^{-2m}$. We are attempting to use this to develop approximate computational procedures that can handle the non-periodic case easily.

When f is known to be periodic and furthermore, $f^{(\nu)}(0) = f^{(\nu)}(1) = 0$, $\nu = 0,1,\ldots,m-1$, then $f_{n,\lambda}$ can be taken as the solution to the minimization problem of (4.1) subject to these additional constraints. (The solution can be written down using results in e.g. Kimeldorf and Wahba [19]). $V(\lambda)$ would be unchanged, and Case II of Figure 4 would not have that annoying increase in $f_{n,\hat{\lambda}}$ near $x = 1.0$.

In principle the method extends immediately to f a (doubly periodic) density on the unit square, where f is in the tensor product space of 2 periodic rkhs on the unit interval. See Wahba [36]. It is apparent that all of the theoretical results will hold, since they depend only on the eigenvalues of the r.k. and the Fourier coefficients of f. We think the method may actually be computationally feasible, with very clever programming, in up to 3 or 4 dimensions, but of course at some point an overwhelmingly large number of observations would be required to estimate a multi-dimensional density by this method. One approach to simplifying the computation in the multi-dimensional approach would be to use a simple approximation to the window of (3.24) to compute $f_{n,\hat{\lambda}}$.

Acknowledgment.

The computer programs for the study in Section 4 were ably written by Mr. Dick Jones. General assistance was provided by Mr. Michael Akritas.

References

1. Anderssen, R. S. and Bloomfield, P. (1974), A time series approach to numerical differentiation, Technometrics 16 (1), 69-75.
2. Anderssen, B. and Bloomfield, P. (1974), Numerical differentiation procedures for non-exact data, Numer. Math. 22, 157-182.
3. Boneva, L., Kendall, D. and Stefanov, I. (1971), Spline transformations: Three new diagnostic aids for the statistical data analyst. J. Roy. Statist. Soc. 33, 1-70.
4. Brunk, H. D. (1976), Univariate density estimation by orthogonal series, TR#51, Dept. of Statistics, Oregon State University, Corvallis, Oregon.
5. Chi, P.Y. and Van Ryzin, John, A histogram method for nonparametric classification, this volume.
6. Cogburn, R. and Davis, H.T. (1974), Periodic splines and spectra estimation, Ann. Statist. 2, 1108-1126.
7. Cover, T. M. and Wagner, I. T. (1976), Topics in statistical pattern recognition, in Digital Pattern Recognition, Vol. 10 of Communications in Statistics, eds. Fu, Keidel and Wolter, Springer Verlag, 15-46.
8. Crain, B. R. (1976), Matrix density estimation, Commun. Statist. A5(1), 89-96.

9. Craven, P. and Wahba, G. (1976), Smoothing noisy data with spline functions: Estimating the correct degree of smoothing by the method of generalized cross-validation, in preparation.

10. de Montricher, G. F., Tapia, R. A., and Thompson, J. R. (1975), Nonparametric maximum likelihood estimation of probability densities by penalty function methods, Ann. Statist. 6, 1329-1348.

11. Farrell, R. H. (1972), On best obtainable asymptotic rates of convergence in estimation of a density function at a point, Ann. Math. Statist. 43, 170-180.

12. Fienberg, S. and Holland, P. (1972), On the choice of flattening constants for estimating multinomial probabilities, J. Multivariate Anal. 2, 1, 127-134.

13. Geisser, S. (1975), The predictive sample reuse method with applications, JASA, 70, 350, 320-328.

14. Golub, G., Heath, M. and Wahba, G. (1975), Cross-validation and optimum ridge regression, abstract in Abstracts of Papers to be presented at the SIAM-SIGUM 1975 Fall meeting, December 3, 4, 5, 1975, San Francisco.

15. Good, I. J. and Gaskins, R. A. (1971), Nonparametric roughness penalties for probability densities, Biometrika 58, 255-277.

16. Goodman, N. R. (1963), Statistical analysis based on a certain multivariate complex Gaussian distribution, Ann. Math. Statist. 34, 152-177.

17. Hájek, Jaroslav (1962), On linear statistical problems in stochastic processes, Czech. Math. J., 12 (87), 404-444.

18. Hudson, H. M. (1974), Empirical Bayes estimation, Technical Report #58, Stanford University Dept. of Statistics, Stanford, California.

19. Kimeldorf, George and Wahba, Grace (1971), Some results on Tchebycheffian spline functions, J. Math. Anal. Appl. 33, 82-95.

20. Kronmal, R. and Tarter, M. (1968), The estimation of probability densities and cumulatives by Fourier series mehtods, J. Amer. Statist. Assoc. 63, 925-952.

21. Lii, K.-S. and Rosenblatt, M. (1974), Asymptotic behavior of a spline estimate of a density function, manuscript, University of California, San Diego.

22. Loftsgarten, D. O. and Quesenberry, C. P. (1965), A non-parametric estimate of a multivariate density function, Ann. Math. Statist. 36 , 1049-1051.

23. Naimark, M. A. (1968), Linear differential operators, Part II, Ungar, New York.

24. Parzen, E. (1962), On the estimation of a probability density function and mode, Ann. Math. Statist. 33, 1065-1076.

25. Parzen, E. (1973), Relations between methods of non parametric probability density estimation, State University of N.Y., Buffalo, manuscript.

26. Riesz, F. and Sz.-Nagy, B. (1955), Functional Analysis, Unger, New York.

27. Rosenblatt, M. (1956), Remarks on some non-parametric estimates of a density function, Ann. Math. Statist. 27, 832-837.

28. Root, W. L. (1962), Singular Gaussian measures in detection theory. Time Series Analysis, Proceedings of a Symposium held at Brown University, ed. M. Rosenblatt, Wiley, New York, 292-314.

29. Stone, M. (1974), Cross-validatory choice and assessment of statistical prediction, JRSS, Series B, 36, 2, 111-147.

30. Van Ryzin, J. (1966), Bayes risk consistency of classification procedures using density estimation, Sankhya, Ser. A 28, 261-270.

31. Van Ryzin, J. (1973), A histogram method of density estimation, Commun. Statist, 12, 493-506.

32. Wahba, Grace (1971), A polynomial algorithm for density estimation, Am. Math. Statist. 42, 1870-1886.

33. Wahba, G. (1974), Regression design for some equivalence classes of kernels, Ann. Statist. 2, 5 , 925-934.

34. Wahba, Grace (1975), Optimal convergence properties of variable knot kernel, and orthogonal series methods for density estimation, Ann. Statist. 3, 15-29

35. Wahba, G. (1975), Interpolating spline methods for density estimation I. Equi-spaced knots, Ann. Stat. 3, 1, 30-48.

36. Wahba, G. (1975), A canonical form for the problem of estimating smooth surfaces, Univ. of Wisconsin-Madison, Department of Statistics, Technical Report #420.

37. Wahba, G. (1975), Practical approximate solutions to linear operator equations when the data are noisy, University of Wisconsin-Madison, Department of Statistics, Technical Report #430, to appear, SIAM J. Num. Anal.

38. Wahba, G. (1975), Smoothing noisy data by spline functions, Numer. Math. 24, 303-394.

39. Wahba, G. (1976), Histosplines with knots which are order statistics, J. Roy. Stat. Soc. Series B. 38, 2, 140-151.

40. Wahba, G and Wold, S. (1975), A completely automatic French curve: Fitting spline functions by cross-validation. Comm. Statist. 4. (1), 1-17.

41. Wahba, G. and Wold, S. (1975), Periodic splines for spectral density estimation: The use of cross-validation for determining the degree of smoothing, Comm. Statist. 4, 2, (125-141.

42. Walter, G. and Blum, J. (1976), Probability density estimation using delta-sequences, manuscript, University of Wisconsin-Milwaukee.

43. Watson, G.S. and Leadbetter, M. R. (1963), On the estimation of the probability density I, Ann. Math. Statist. 34, 480-491.

44. Watson, G. S. (1969), Density estimation by orthogonal series, Ann. Math. Statist. 40, 1496-1498.

45. Wegman, E. (1972), Nonparametric probability density estimation I, a survey of available methods, Technometrics.

46. Wegman, E. (1972), Nonparametric probability density estimation II. A comparison of density estimation methods, J. Statist. Comput. Simul. 1, 225-245.

47. Whittle, P. (1958). On the smoothing of probability density functions, J. R. Statist. Soc., B. 20, 334-343.

48. Woodroofe, M. (1970). On choosing a delta-sequence, Ann. Math. Statist., 41, 166-171.

49. Tarter, M. E., and Kronmal, R. A. (1976), An introduction to the implementation and theory of nonparametric density estimation, The American Statistician, 30, 3, 105-112.

Supported by the United States Air Force under Grant No. AFOSR 72-2363-C.

Department of Statistics
University of Wisconsin-Madison
Madison, Wisconsin 53706

Index

A
B 7
C 8
D 9
E 0
F 1
G 2
H 3
I 4
J 5